Annals of Mathematics Studies

Number 133

An Introduction to G-Functions

by

Bernard Dwork, Giovanni Gerotto,
and Francis J. Sullivan

PRINCETON UNIVERSITY PRESS

————

PRINCETON, NEW JERSEY

1994

The Annals of Mathematics Studies are edited by
Luis A. Caffarelli, John N. Mather, and Elias M. Stein

Princeton University Press books are printed on acid-free paper
and meet the guidelines for permanence and durability of the Committee on Production
Guidelines for Book Longevity of the Council on Library Resources

Printed in the United States of America

2 4 6 8 10 9 7 5 3 1

Library of Congress Cataloging-in-Publication Data

Dwork, Bernard M.
      An introduction to G-functions / by Bernard Dwork,
Giovanni Gerotto, Francis J. Sullivan.
         p.   cm. — (Annals of mathematics studies ; no. 133)
      On t.p. "G" is italicized.
      Includes bibliographical references and index.
      ISBN 0-691-03675-6 — ISBN 0-691-03681-0 (pbk.)
      1. H-functions. 2. p-adic analysis. I. Gerotto, Giovanni.
II. Sullivan, Francis J., 1943– . III. Title. IV. Series.
QA242.5.D96 1994
515'.55—dc20   94-2414

The publisher would like to acknowledge the authors of this volume for providing the
camera-ready copy from which this book was printed

Dedicated to the memory of

Yvette Amice (1936–1993)

CONTENTS

PREFACE

This book is based upon a course given with the collaboration of G. Gerotto and F. J. Sullivan during the academic year 1992/93 at the University of Padova.

The plan was to present completely the p-adic foundations of the basic properties of G-functions, assuming only a basic knowledge of algebra. In particular we give a complete proof of the main result: the analytic continuation of a G-series is again a G-series. Algebraic number theory enters in an essential way only in the use of Čebotarev's density theorem in the proof of Katz's Theorem (Theorem III.6.1). The language of algebraic number theory is used starting with Chapter VII, but only the most basic facts are used: normalization of valuations, product formula, relation between local and global degrees.

B. DWORK

INTRODUCTION

What is a G–function? Modifying the standard terminology, we say that

$$z = \sum_{j=0}^{\infty} A_j X^j \in K[[X]]$$

is a *G–series at the origin* (defined over an algebraic number field K), if

a) There exists a non-zero $L \in K(X)[d/dX]$ such that $Lz = 0$;

b) For each imbedding of K into $\mathbf{C}$, the series z has a non-zero radius of convergence;

c) There exists a sequence of positive integers $\{c_s\}$ such that

$$\sup_s \left(\frac{1}{s} \log c_s\right) < \infty$$

and such that $A_j c_s \in \mathcal{O}_K$, the ring of integers of K, for all $j \leq s$.

In this definition we may replace K by $\mathbf{Q}^{\mathrm{alg}}$, the algebraic closure of $\mathbf{Q}$ since each element of $\mathbf{Q}^{\mathrm{alg}}(X)$ is necessarily defined over an algebraic number field, and this property of L then forces the coefficients A_j to all lie in a finite extension of $\mathbf{Q}$. Condition (b) is needed only if the origin is a singular point of L.

If $\zeta \in \mathbf{Q}^{\mathrm{alg}}$ then we may use the term "*G–series at* ζ" without ambiguity. It is easy to check that the G–series at the origin form a ring (see Proposition VIII.2.1 (ii)).

Easy examples of G–series are the expansions at 1 of $\log X$ and of X^α for $\alpha \in \mathbf{Q}$. A celebrated theorem of Eisenstein (see Dwork–van der Poorten [1]) assures us that any element of $\mathbf{Q}^{\mathrm{alg}}[[X]]$ algebraic over $\mathbf{Q}(X)$ must be a G–series.

We say that an irreducible element of $\mathbf{Q}^{\mathrm{alg}}(X)[d/dX]$ is an *irreducible G–operator* if it has a solution which is a G–series at some point, say, the origin. It follows from the remarkable theorem of the Chudnovskys' (see Chapter VIII) and the work of André, Christol, Baldassarri and Chiarellotto that if L is an irreducible G–operator then for

$\zeta \in \mathbf{Q}^{\mathrm{alg}}$ each formal solution $y \in \mathbf{Q}^{\mathrm{alg}}[[X - \zeta]]$ of $Ly = 0$ is again a G–series. (By Theorem VIII.1.5, L satisfies the Galočkin condition, hence, by Theorem VII.2.1, L satisfies the Bombieri condition, so by Theorem VII.3.3 it satisfies the local Bombieri condition at ζ and the assertion then follows from Theorem VII.4.2).

We extend the notion of "G–series at ζ" to include the case in which $\zeta \in \mathbf{C}$ is transcendental. For such a ζ we say that $y \in \mathbf{Q}^{\mathrm{alg}}(\zeta)[[X - \zeta]]$ is a G–series if

    a) There exists a non-zero $L \in \mathbf{Q}^{\mathrm{alg}}(X)[d/dX]$ such that $Ly = 0$;

    b) No archimedean condition;

    c) This condition is the same as previously with $\mathcal{O}_K$ replaced by

$$\mathcal{O}_\zeta = \{\lambda \in \mathbf{Q}^{\mathrm{alg}}(\zeta) \,|\, |\lambda|_{v,\mathrm{gauss}} \leq 1 \text{ for all finite valuations } v \text{ of } \mathbf{Q}^{\mathrm{alg}}\} \,.$$

Here $|-|_{v,\mathrm{gauss}}$ denotes the v-adic Gauss norm relative to ζ (for the definition see Section I.4).

It follows from the Galočkin condition that if ζ is transcendental over the field $\mathbf{Q}$ and if $y \in \mathbf{Q}^{\mathrm{alg}}(\zeta)[[X - \zeta]]$ is a solution of $Ly = 0$ where L is an irreducible G–operator then y is a G–series at ζ in the above sense. We can show by an easy specialization argument that conversely if L is an irreducible element of $\mathbf{Q}^{\mathrm{alg}}(X)[d/dX]$ which has a solution which is a G–series at a transcendental point $\zeta \in \mathbf{C}$ then L is indeed an irreducible G–operator.

Finally we extend the notion of G–series (at each fixed $\zeta \in \mathbf{C}$) so as to make the set of such series a $\mathbf{C}$–vector space. With these definitions we may conclude that the analytic continuation of a G–series at $x = 0$ to $x = \zeta$ is again a G–series provided we exclude a finite number of singular points.

We may define a G–function to be a multivalued, locally analytic function on $\mathbf{C}\backslash S$ (with card $S < \infty$) which at some point is represented by a G–series. Equivalently, a G–function is the solution of an irreducible G–operator. We may eliminate the emphasis on irreducible operators. We say that $L \in \mathbf{Q}^{\mathrm{alg}}(X)[d/dX]$ is a G–operator if **all** solutions at some non singular point are G–series. Then all solutions at any non-singular point are G–series. If L is a G–operator then all its singularities are regular and the exponents are rational. Thus if the origin is a (possibly) singular point, there exists a basis $(y_1, \dots, y_n)$ for the solutions at $X = 0$ of the form

$$(y_1, \dots, y_n) = (z_1, \dots, z_n)X^A \qquad (*)$$

where $z_1, \dots, z_n$ lie in $K[[X]]$ and $A \in \mathcal{M}_n(\mathbf{Q}^{\mathrm{alg}})$ with rational eigenvalues. In fact if L is a G–operator then $z_1, \dots, z_n$ are G–series and the converse is also valid.

So much for the definition of G–functions. How can we determine whether an operator is a G–operator? Since the singularities must be regular we may exclude e^X, $J_\alpha(X)$ and hypergeometric functions $_kF_m$ with $k < m + 1$. Since the exponents must be rational we may exclude X^α and $_2F_1(a, b, c, X)$ if $\alpha \notin \mathbf{Q}$ or if a, b, or c is irrational. There is no general method for verifying that a given operator L is a G–operator. Three methods known to us are:

1. If L_1 and L_2 are G–operators then so is the composition $L_1 \circ L_2$.

2. A sufficient condition is that in the representation $(*)$ of the basis of solutions, the eigenvalues of A be rational and the series $z_1, \ldots, z_n$ satisfy the local Bombieri condition, i.e., it is sufficient that

$$\sum_v \log \frac{1}{R_v} < \infty \, ,$$

where R_v is the common radius of convergence of all the z_i for each finite valuation v.

Remark.   In all known cases of G–operators, $R_v = 1$ for almost all v. In general it is very difficult to compute R_v directly. For example, the second order differential equation satisfied by

$$_2F_1(a, b, c, X) = \sum_{s=0}^{\infty} \frac{(a)_s (b)_s}{(c)_s s!} X^s$$

is a G–operator if a, b, $c \in \mathbf{Q}$ because this solution and a second independent one are known so explicitly (Section IV.8) that we can check that $R_p = 1$ for almost all p and $R_p > 0$ for all p. We do not know how to verify directly that the same is valid for the power series solution at a non-singular point, say at $x = 2$.

3. If L comes from geometry then it is known that a strong Frobenius operator exists for almost all v and hence by a classical argument $R_v = 1$ for almost all v. This is the most important general example of G–operators and it is believed that no other examples exist. This is the situation in which L is associated to a differential module M having a filtration of submodules whose factor modules are isomorphic, as differential modules, to factor modules of the relative cohomology, in some dimension, of a family of varieties defined over $K(X)$. A very general example of such G–modules is associated with the construction given in Section 7 of Chapter II. Let f be an arbitrary element of $K(X)[Y_1, \ldots, Y_n]$, put

$$g = Y_0 f$$

and let R be the subring of $K(X)[Y_0, Y_1, \ldots, Y_n]$ generated by the monomials in g. Define operators E_i on R by

$$E_i = Y_i \partial / \partial Y_i ,$$

for $i = 0, 1, \ldots, n$, and set

$$g_i = E_i g , \qquad D_i = E_i + g_i .$$

Let

$$W = R / \sum_{i=0}^{n} D_i R .$$

Then W is a finite dimensional $K(X)$–vector space which is a differential module via the action of the operator $\sigma_X = \partial / \partial X + \partial g / \partial X$. This construction can be generalized, replacing each D_i by $D_{i,a} = D_i + a_i$ where $(a_0, \ldots, a_n) \in \mathbf{Q}^{n+1}$.

The object of these lectures is to provide an introduction to the theory of G–functions at the level of a first year graduate course. It is our hope that the reader will acquire all the tools needed to study applications to number theory, in particular to the theory of irrationality of special values of functions. We have been strongly influenced by Bombieri [1] and by André [1]. It is our hope that the present volume will make their work more accessible.

Our work is divided into three parts. The first two chapters are devoted to the elementary theory of p-adic fields and functions. The previous paragraph provides a motivation for the inclusion of the proof of the rationality of the zeta function. Chapters III to VI, the second part of this volume, are devoted to the theory of ordinary linear differential equations over finite fields and p-adic fields. The main results here are the notion of nilpotence, Katz's theorem on global nilpotence, and the effective (p-adic) bounds for ordinary disks and for disks with one regular singularity. The basic problem is that of determining p-adic radii of convergence and rate of growth of formal solutions at the boundary of the disk of convergence.

The third part of the book, Chapters VII and VIII, deals with global problems, i.e., how do radii of convergence vary with valuation and what profit we may draw from knowing the variation.

We do not treat the applications to number theory.

It is our belief that the p–adic theory of differential equations is still in its infancy. We mention two questions.

1. (Modified) Robba Conjecture

Let $G \in \mathcal{M}_n(\mathbf{Q}(X))$ and let p be a fixed prime. We will use only the p–adic valuation. Suppose that all the singularities of $D - G$ in $D(0, 1^-)$ lie in $D(0, R)$ and that for each α with $R < |\alpha| < 1$, the solution matrix at α of $D - G$ converges in $D(\alpha, |\alpha|^-)$. Then (conjecturally) there exists a solution matrix $V X^A$ where A is a constant matrix and V is analytic on the annulus.

2. Accessory parameter problem

Consider the general n^{th} order scalar differential operator, L, with $m + 1$ singularities, all regular, and a fixed set of exponents all in $\mathbf{Q} \cap \mathbf{Z}_p$. Taking three of the singularities to be 0, 1, ∞, the parameters consist of the $m - 2$ unspecified singularities and the $(n - 1)[n(m - 1) - 2]/2$ accessory parameters. It is known (Schmitt [1]) that the variety of nilpotence for the mod p–reduction is a complete intersection of dimension $n - 2$. We ask for the set of parameters for which the solutions of $Ly = 0$ at a generic point converge in a disk of radius unity.

LIST OF SYMBOLS

Throughout the text $\mathbf{C}$, $\mathbf{R}$, $\mathbf{Q}$, $\mathbf{F}_q$, $\mathbf{Z}$ and $\mathbf{N}$ denote the fields of complex, real and rational numbers, the finite field with q elements, the ring of rational integers and the set of natural numbers, respectively.

If K is a field then $K[X]$, $K[[X]]$, $K(X)$ and $K((X))$ are respectively the polynomials, formal power series, rational functions, and (finitely negative) formal Laurent series in X with coefficients from K.

The algebraic closure of the field K is denoted by K^{alg}.

We use $G\ell(n, A)$, $\mathcal{M}_{n,m}(A)$ and $\mathcal{M}_n(A)$ to denote respectively the general linear group of order n, the set of $n \times m$ matrices and the ring of $n \times n$ matrices over the commutative ring A. If $M \in \mathcal{M}_n(A)$ then ${}^t M$ (or M^t), $\det M$, $\operatorname{Tr} M$ and $\operatorname{adj} M$ are the transpose, determinant, trace and adjoint of M respectively.

The symbol I_n denotes the $n \times n$ identity matrix.

If Ω/K is a finite extension of fields then $\mathrm{N}_{\Omega/K}$ and $\mathrm{Tr}_{\Omega/K}$ are the norm and trace maps respectively. If Ω/K is a Galois extension, $\mathcal{G}(\Omega/K)$ denotes the Galois group of Ω/K.

If S is a set $\#S$ denotes the cardinality of S, and if σ is a permutation, $\operatorname{sgn}(\sigma)$ is the sign of σ. If A and B are sets, we denote by B^A the set of maps from A into B.

As usual the symbols $\oplus$ and $\wedge$ denote direct sums and wedge products respectively, and $\Lambda(M)$ denotes the exterior algebra of the module M. The submodule generated by the subset X of the module M will be indicated by $\langle X \rangle$.

In discussion of differential modules over fields containing $K(X)$ the symbol D is used for d/dX while δ indicates $X\, d/dX$.

If X is a metric space with distance dist, for any $a \in X$ and $\rho > 0$ we will denote the open and closed disks with center a and radius ρ by $D(a, \rho^-)$ and $D(a, \rho^+)$ respectively, that is we set

$$D(a, \rho^-) = \{x \in X \mid \operatorname{dist}(a, x) < \rho\}$$
$$D(a, \rho^+) = \{x \in X \mid \operatorname{dist}(a, x) \leq \rho\}\,.$$

We occasionally use the simpler notation $D(a, \rho)$ for $D(a, \rho^+)$.

An Introduction to G-Functions

CHAPTER I

VALUED FIELDS

1. Valuations.

A *field* is a set K with two operations $+$ and $\cdot$ such that $(K, +)$ and $(K^\times = K \backslash \{0\}, \cdot)$ are commutative groups related by the distributive law

$$a \cdot (b + c) = a \cdot b + a \cdot c \qquad \text{for all } a, b, c \in K.$$

A *valuation* on K is a map

$$K \longrightarrow \mathbf{R}_+ = \{z \in \mathbf{R} \mid z \geq 0\}$$
$$x \longmapsto |x|$$

satisfying all of the following:

(V1) $|x| = 0$ if and only if $x = 0$,

(V2) $|xy| = |x||y|$,

(V3)   if $|x| \leq 1$ then $|1 + x| \leq 1$.

(Sometimes this type of valuation is referred to as an "ultrametric" or "non-archimedean" valuation.)

The valuation is said to be *trivial* if $|x| = 1$ for all $x \neq 0$.

Every valuation $|-|$ defines a topology on K: a fundamental system of neighborhoods of 0 in this topology is given by the sets

$$V_\varepsilon = D(0, \varepsilon^+) = \{x \mid x \in K, \ |x| \leq \varepsilon\} \quad \text{for every } \varepsilon > 0 \ .$$

If $a \in K$, the fundamental neighborhoods of a are the sets $a + V_\varepsilon = D(a, \varepsilon^+)$. This topology is called the *induced topology*. As usual K becomes a metric space with distance $(a, b) \to |a - b|$.

The following properties of valuations are easily verified:

- $|-x| = |x|$ for every $x \in K$,
- if $\omega \in K$, with $\omega^m = 1$, then $|\omega| = 1$ and $|\omega x| = |x|$ for every $x \in K$,
- if $m = \underbrace{1 + 1 + \cdots + 1}_{m\times}$ then $|m| \leq 1$.

Note that this last property holds even when the field K has *characteristic* 0, that is, when K contains an isomorphic copy of the rational integers (otherwise K contains a copy of the field of integers modulo p for some positive rational prime p, in which case the field is said to be of *characteristic $p > 0$*).

Proposition 1.1.   *If x, $y \in K$ then*

$$|x + y| \leq \sup\{|x|, |y|\}.$$

If $|x| \neq |y|$ then equality holds.

Proof.   Suppose that $|x| \leq |y| \neq 0$. Then $|x/y| \leq 1$ so that by (V3) we have $\left|1 + \dfrac{x}{y}\right| \leq 1$. It now follows by (V2) that

$$|x + y| = \left|y\left(1 + \frac{x}{y}\right)\right| = |y|\left|1 + \frac{x}{y}\right| \leq |y|.$$

Suppose now that $|y| < |x|$. Then we must show that $|x + y| = |x|$. Since $|x + y| \leq \sup\{|x|, |y|\} = |x|$ let us assume that $|x + y| < |x|$. From the identity $x = (x + y) - y$ it follows that $|x| \leq \sup\{|x + y|, |y|\} < |x|$ so that we obtain the desired contradiction. **Q.E.D.**

If $|-|$ is a valuation, then it's easy to verify that $|-|^\gamma$ is also a valuation for every $\gamma \in \mathbf{R}_+$, and that the induced topologies are the same. We say that two valuations $|-|_1$ and $|-|_2$ are *equivalent* if they induce the same topology on K.

Lemma 1.2.   *If $|-|_1$ and $|-|_2$ are equivalent valuations on K, then there exists a $\gamma \in \mathbf{R}_+$ such that*

$$|x|_1 = |x|_2^\gamma$$

for every $x \in K$.

Proof.   If $|x|_1 < 1$ then $\lim\limits_{n \to \infty} x^n = 0$ in the topology induced by $|-|_1$. Since the induced topologies coincide, we must have $\lim\limits_{n \to \infty} x^n = 0$ in the topology induced by $|-|_2$, so that $|x|_2 < 1$. Let $x_0 \in K$ be such that $0 < |x_0|_1 < 1$ (we are tacitly assuming that $|-|_1$ is non-trivial, since that case is easily handled); then $0 < |x_0|_2 < 1$. Let $y \in K$ satisfy $0 < |y|_1 < 1$ and let

$$q = \frac{\log |y|_1}{\log |x_0|_1} \in \mathbf{R}_+.$$

Let $\dfrac{m}{n} \geq q$; then $\dfrac{\log |y|_1}{\log |x_0|_1} \leq \dfrac{m}{n}$ implies that $|y|_1^n \geq |x_0|_1^m$ so that $1 \geq \left| \dfrac{x_0^m}{y^n} \right|_1$; this implies that $\left| \dfrac{x_0^m}{y^n} \right|_2 \leq 1$ whence $|y|_2^n \geq |x_0|_2^m$ and finally $\dfrac{\log |y|_2}{\log |x_0|_2} \leq \dfrac{m}{n}$. In the same way one obtains that $\dfrac{\log |y|_1}{\log |x_0|_1} \geq \dfrac{m}{n}$ implies $\dfrac{\log |y|_2}{\log |x_0|_2} \geq \dfrac{m}{n}$. Therefore,

$$\frac{\log |y|_2}{\log |x_0|_2} = \frac{\log |y|_1}{\log |x_0|_1}$$

(since the two numbers represent the same Dedekind cut). If we put

$$\gamma = \frac{\log |x_0|_2}{\log |x_0|_1} \quad \text{(independent of } y\text{)},$$

we get $|y|_2 = |y|_1^\gamma$, for every $y \in K$ with $|y|_1 < 1$. If $|y|_1 > 1$ then $|1/y|_1 < 1$ so that $|1/y|_2 = |1/y|_1^\gamma$ and $|y|_2 = |y|_1^\gamma$. If $|y|_1 = 1$ then it is now evident that $|y|_2 = 1$. **Q.E.D.**

We define

$$G = G_K = \text{the } valuation\ group \text{ of } |-|$$
$$= \{z \in \mathbf{R} \mid z = |x| \text{ for some } x \in K,\ x \neq 0\},$$
$$\mathcal{O} = \mathcal{O}_K = \text{the } valuation\ ring \text{ of } |-| = \{x \in K \mid |x| \leq 1\},$$
$$\mathfrak{p} = \mathfrak{p}_K = \text{the } valuation\ ideal \text{ of } |-| = \{x \in K \mid |x| < 1\}.$$

Note that G is a (multiplicative) group by property (V2), $\mathcal{O}$ is a ring by Proposition 1.1, and $\mathfrak{p}$ is an ideal by (V2) and Proposition 1.1. The ideal $\mathfrak{p}$ is the unique maximal ideal of $\mathcal{O}$ because if $x \in \mathcal{O}$, $x \notin \mathfrak{p}$, then $|x| = 1$ and $|x^{-1}| = 1$ so that $x^{-1} \in \mathcal{O}$. The field $\mathcal{O}/\mathfrak{p}$ plays an important role in what follows, so we define

$$\overline{K} = \text{the } residue\ field \text{ of } K = \mathcal{O}/\mathfrak{p}.$$

Examples.

• Let $K = \mathbf{Q}$, the field of rational numbers, and let p be a positive prime. If $x \in K$ we can write $x = p^\nu \dfrac{m}{n}$ with $(m, p) = (n, p) = 1$ and $\nu \in \mathbf{Z}$. We define

$$|x|_p = p^{-\nu}.$$

To check that this defines a valuation on K it suffices to verify that if $\nu \geq 0$ then $|1 + x|_p \leq 1$. We will refer to this valuation as the *p-adic valuation* of $\mathbf{Q}$.

• Let $K = \mathbf{C}(X) = \{f(X)/g(X) \mid f, g \in \mathbf{C}[X]\}$ the field of rational functions in one variable with coefficients in the field $\mathbf{C}$ of complex numbers, and let

$$|f(X)| = 2^{\deg f}, \qquad |g(X)| = 2^{\deg g}$$

$$|f/g| = 2^{\deg f - \deg g}.$$

It is not difficult to verify that this does indeed give a well-defined map and that this map is a valuation. This valuation induces the trivial valuation on $\mathbf{C}$.

2.   Complete Valued Fields.

We say that the sequence $(a_i)_{i \in \mathbf{N}}$, with $a_i \in K$, is a *Cauchy sequence in K* if for every $\varepsilon > 0$, there exists an $m \in \mathbf{N}$ such that

$$|a_{i+h} - a_i| \leq \varepsilon$$

whenever $i \geq m$ and $h \in \mathbf{N}$.

We say that the field K is *complete* if every Cauchy sequence in K is convergent in K. The set $\mathcal{R}$ of all Cauchy sequences in K is a ring under componentwise addition and multiplication:

$$(a_i) + (b_i) = (a_i + b_i)$$
$$(a_i) \cdot (b_i) = (a_i \cdot b_i).$$

The set $\mathcal{N} = \{(a_i) \mid \lim\limits_{i \to \infty} a_i = 0\}$ is an ideal of $\mathcal{R}$. It is actually a maximal ideal of $\mathcal{R}$. Indeed, let (a_i) be a Cauchy sequence in K not in

$\mathcal{N}$. Since $a_i \neq 0$ for sufficiently large i, there exists a Cauchy sequence (c_i) in K such that

$$\begin{cases} (c_i) \equiv (a_i) & \text{mod } \mathcal{N} \\ c_i \neq 0 & \text{for every } i \text{ .} \end{cases}$$

It is now easy to prove that the sequence $(1/c_i)$ is a Cauchy sequence in K so that $\mathcal{N}$ is a maximal ideal. Moreover, the field $K' = \mathcal{R}/\mathcal{N}$ contains a subfield which is isomorphic to K, namely the field $\tilde{K}$ consisting of the classes of the constant sequences (a_i) with $a_i = a \in K$ for all i.

If $(a_i) \in \mathcal{R}$, $(a_i) \notin \mathcal{N}$, then there exists n such that $|a_i|$ is independent of i for $i > n$. Let $(a_i) \in \mathcal{R}$ and define

$$|(a_i)| = \lim_{i \to \infty} |a_i| \text{ .}$$

It is obvious that this mapping, when restricted to constant sequences, coincides with the valuation on $\tilde{K}$ induced by the valuation on K. If $(b_i) \in \mathcal{N}$ and $(a_i) \in \mathcal{R}$ then $|(a_i + b_i)| = |(a_i)|$, so the valuation $|-|$ on sequences induces a valuation (again denoted by $|-|$) on the field $K' = \mathcal{R}/\mathcal{N}$ via any choice of representatives for equivalence classes. The valued field K' is called the *completion* of K.

As usual, the completion of the field $\mathbf{Q}$ with respect to the p-adic valuation will be denoted by $\mathbf{Q}_p$ and its valuation ring by $\mathbf{Z}_p$. The field $\mathbf{Q}_p$ is called the field of p-adic numbers, and the ring $\mathbf{Z}_p$ the ring of p-adic integers.

We leave as an exercise the proof of the following facts:

a) The field $\tilde{K}$ is *dense* in K', that is, every element of K' is the limit of a sequence of elements in $\tilde{K}$.

b) The field K' is complete.

c) If K is a complete field, a sequence (a_i) is a Cauchy sequence in K if and only if

$$\lim_{i \to \infty} |a_{i+1} - a_i| = 0 \text{ .}$$

As a consequence, the series $\sum_{i=0}^{\infty} a_i$ converges in K if and only if $\lim_{i \to \infty} a_i = 0$. If this is the case, then

$$\left| \sum_{i=0}^{\infty} a_i \right| \leq \sup_i \{|a_i|\} \text{ ,}$$

and, if there exist an index N such that $|a_i| < |a_N|$ for every $i \neq N$, then

$$\left| \sum_{i=0}^{\infty} a_i \right| = |a_N| \, .$$

3. Normed Vector Spaces.

We wish to study the problem of extending the valuation on a complete field K to an extension field of K. In order to do so, we need some general preliminaries on normed vector spaces.

Let K be a field with a non-trivial valuation $|-|$ and let V be a vector space over K. A *norm* on V is a map

$$V \longrightarrow \mathbf{R}_+$$
$$v \longmapsto \|v\|$$

satisfying the following properties

(N1) $\|v\| = 0$ if and only if $v = 0$,

(N2) $\|av\| = |a| \cdot \|v\|$ for all $a \in K$ and $v \in V$,

(N3) $\|v + w\| \leq \sup\{\|v\|, \|w\|\}$.

A K-vector space with a norm is called a *normed space* (*over* K). Associated to a norm $\|-\|$ on V there is the topology on V whose fundamental system of neighborhoods of a vector v consists of the family $\{D(v, \varepsilon^-)\}_{\varepsilon > 0}$ where $D(v, \varepsilon^-) = \{w \in V | \, \|w - v\| < \varepsilon\}$. We will also make use of the following notation: $V_\varepsilon = D(0, \varepsilon^+) = \{v \in V | \, \|v\| \leq \varepsilon\}$.

The definitions of Cauchy sequence, completeness, etc. are analogous to those given in Section 2.

Let $(V, \|-\|)$ and $(W, \|-\|)$ be normed spaces over K, and let $\theta : V \longrightarrow W$ be a K-linear map. We define

$$\|\theta\| = \sup_{\|v\| \leq 1} \|\theta(v)\| \in [0, \infty]$$

Proposition 3.1.   *The linear map θ is continuous if and only if*

$$\|\theta\| < \infty \, .$$

Proof.   Let $\tau \in G_K$ (the valuation group of K) and let $\alpha \in K$ be such that $|\alpha| = \tau$. Suppose that $\|\theta\| = c < \infty$. If $\|v\| \leq \tau$ then $\|\alpha^{-1}v\| \leq 1$

so that $\|\theta(\alpha^{-1}v)\| \le c$ and $\|\theta(v)\| \le c\tau$. Hence we have established that $\theta(V_\tau) \subseteq W_{c\tau}$, for every $\tau \in G_K$, so θ is continuous since G_K contains elements arbitrarily close to 0.

Conversely, let θ be continuous. Then for every $\varepsilon' > 0$ there exists $\varepsilon > 0$ such that $\theta(V_\varepsilon) \subseteq W_{\varepsilon'}$. Fix $\alpha \in K$, $\alpha \ne 0$ such that $|\alpha| \le \varepsilon$. If $v \in V_1$ we have $\|\alpha v\| \le \varepsilon$ and then $\|\theta(\alpha v)\| \le \varepsilon'$, so that $\|\theta(v)\| \le \varepsilon'/|\alpha|$. Hence, if we put $c = \varepsilon'/|\alpha|$ if follows that $\|\theta\| \le c$. **Q.E.D.**

Now let's suppose that V has finite dimension n over K. Let $\omega_1, \ldots, \omega_n$ be a base of V over K. Define, for $i = 1, 2, \ldots, n$, the K-linear form

$$\lambda_i : V \longrightarrow K$$
$$v \longmapsto \lambda_i(v) = x_i$$

where the x_i are defined by $v = \sum_{i=1}^{n} x_i \omega_i$. Now we can put

$$\|v\|' = \sup_{1 \le i \le n} (|\lambda_i(v)|).$$

It is easily checked that $\|\text{--}\|'$ is indeed a norm on V (note that this definition depends on the choice of basis for V). If the field K is complete, then obviously V is also complete with respect to this norm.

We say that two norms $\|\text{--}\|_1$ and $\|\text{--}\|_2$ on V are *equivalent* if there exist positive constants C_1 and C_2 such that

$$\|v\|_1 \le C_1 \|v\|_2 \qquad \text{and} \qquad \|v\|_2 \le C_2 \|v\|_1$$

for every $v \in V$.

Theorem 3.2.   *Let K be a complete field and $V = (V, \|\text{--}\|)$ a finite dimensional normed space over K. Then the norm $\|\text{--}\|$ and the norm $\|\text{--}\|'$ defined above are equivalent. As a consequence, V is complete.*

Proof.   We proceed by induction on n, the dimension of V. If $n = 1$, the result is clear since there exists $C > 0$ such that $\|v\| = C\|v\|'$ for every $v \in V$. We now suppose that $n > 1$ and the result true for dimensions less than n. If $v = \sum_{i=1}^{n} \lambda_i(v)\,\omega_i$, then

$$\|v\| \le \sup_i \|\lambda_i(v)\,\omega_i\| = \sup_i |\lambda_i(v)|\,\|\omega_i\|$$
$$\le \sup_i |\lambda_i(v)| \cdot \sup_i \|\omega_i\| = C_1\|v\|',$$

where $C_1 = \sup_i \|\omega_i\|$.

On the other hand, we have

$$\|v\|' = \sup_i |\lambda_i(v)| \leq \|v\| \cdot \sup_i \|\lambda_i\|.$$

Therefore it suffices to prove that λ_i is continuous for every i.

To fix ideas, we prove that λ_1 is continuous. Suppose the contrary. Then there exists a sequence $(v^{(j)})$ such that $v^{(j)} \to 0$ but $\lambda_1(v^{(j)}) \not\to 0$, so we can find a subsequence (which we still call $(v^{(j)})$) such that $v^{(j)} \to 0$ but $\lambda_1(v^{(j)}) \geq \varepsilon > 0$. Let

$$u^{(j)} = \frac{1}{\lambda_1(v^{(j)})} \cdot v^{(j)},$$

so that $u^{(j)} = \omega_1 + X_2^{(j)}\omega_2 + \cdots + X_n^{(j)}\omega_n$ for suitable $X_2^{(j)}, \ldots, X_n^{(j)} \in K$. Since $u^{(j)} \to 0$ it follows that

$$\lim_{j\to\infty} \left(X_2^{(j)}\omega_2 + \cdots + X_n^{(j)}\omega_n \right) = -\omega_1.$$

If V' is the subspace spanned by $\omega_2, \ldots, \omega_n$, then V' is complete by the induction hypothesis, so that $\omega_1 \in V'$ which gives a contradiction. **Q.E.D.**

4. Hensel's Lemma.

In this section we shall assume K to be a complete valued field. The ring of polynomials in one variable $K[X]$ is an infinite dimensional K-vector space and the set $\{1, X, X^2, \cdots\}$ forms a basis. We now define a norm on the K-vector space $K[X]$, called the *Gauss norm* by setting

$$\left| \sum_i a_i X^i \right|_{\text{gauss}} = \sup_i |a_i|.$$

Note that $K[X]$ is not complete with respect to the Gauss norm. Note also that the definition of the Gauss norm does not require that K be complete.

The following equality is known as Gauss's lemma: If $f, g \in K[X]$, then

$$|fg|_{\text{gauss}} = |f|_{\text{gauss}}|g|_{\text{gauss}}.$$

Let $W_n = \{f \in K[X] \mid \deg f \le n - 1\}$. Clearly the dimension of W_n over K is n, so that it is a complete K-space. Let g, h be fixed elements of $K[X]$ and let $\deg g = n$ and $\deg h = m$. Consider the linear map

$$\theta : W_n \oplus W_m \longrightarrow W_{n+m}$$
$$(\xi, \eta) \longmapsto g\eta + h\xi \tag{4.1}$$

In terms of the basis $\{1, X, X^2, \cdots, X^{s-1}\}$, for W_s, $s = 1, 2, \ldots$, one has

$$\det(\text{matrix}(\theta)) = \pm R(h, g),$$

where $R(h, g)$ is the Sylvester resultant of h and g.

Theorem 4.1 (Hensel's Lemma).   *Let K be a complete field with valuation $|{-}|$ and valuation ring $\mathcal{O}$. Let $f, g, h \in \mathcal{O}[X]$ with*

$$\deg g = n, \qquad \deg h = m, \qquad \deg f = n + m.$$

Let $0 < \varepsilon < 1$ and suppose
a) $|f - gh|_{\text{gauss}} \le \varepsilon |R(g, h)|^2$
b) $f - gh \in W_{n+m}$.
Then there exist $\tilde{g}, \tilde{h} \in \mathcal{O}[X]$ such that

$$f = \tilde{g}\tilde{h}, \qquad \tilde{g} - g \in W_n, \qquad \tilde{h} - h \in W_m$$

$$|\tilde{g} - g|_{\text{gauss}} \le \varepsilon |R(g, h)|, \quad |\tilde{h} - h|_{\text{gauss}} \le \varepsilon |R(g, h)|$$

Proof.   For brevity we put $|{-}|_{\text{gauss}} = |{-}|$. We obviously can assume $|R(g, h)| > 0$. Let

$$V = \{\xi \in W_n \mid |\xi| \le \varepsilon |R(g, h)|\},$$
$$W = \{\eta \in W_m \mid |\eta| \le \varepsilon |R(g, h)|\}$$

and define the norm

$$\|(\xi, \eta)\| = \sup\{|\xi|, |\eta|\} \tag{4.2}$$

on $V \oplus W$. Hypothesis (b) gives the map

$$\phi : V \oplus W \longrightarrow W_{n+m}$$
$$(\xi, \eta) \longmapsto f - gh - \xi\eta$$

so that we can consider the map $\Lambda = \theta^{-1}\phi$. We claim that

$$\Lambda(V \oplus W) \subseteq V \oplus W. \tag{4.3}$$

Indeed, from hypothesis (a) it follows that

$$
\begin{aligned}
|f - gh - \xi\eta| &\le \sup\{|f - gh|, |\xi\eta|\} \\
&\le \sup\{\varepsilon|R(g,h)|^2, \varepsilon^2|R(g,h)|^2\} \\
&= \varepsilon|R(g,h)|^2
\end{aligned}
$$

Since $g, h \in \mathcal{O}[X]$ the matrix of θ (with respect to the previously described basis) has elements in $\mathcal{O}$ so that

$$\|\theta^{-1}\| \le \frac{1}{|R(g,h)|}$$

and therefore

$$\|\theta^{-1}(f - gh - \xi\eta)\| \le \varepsilon|R(g,h)|$$

which, by (4.2), proves (4.3).

We now prove that Λ is a contractive map. Let $(\xi, \eta), (\xi', \eta') \in V \oplus W$. Since $\varepsilon < 1$, this follows from

$$
\begin{aligned}
\|\Lambda((\xi, \eta)) - \Lambda((\xi', \eta'))\| &= \|\theta^{-1}(\xi'\eta' - \xi\eta)\| \\
&= \|\theta^{-1}(\xi'(\eta' - \eta) + \eta(\xi' - \xi))\| \\
&= \|\theta^{-1}(\xi'(\eta' - \eta)) + \theta^{-1}(\eta(\xi' - \xi))\| \\
&\le \sup \frac{1}{|R(g,h)|}\{|\xi'|\|\eta' - \eta|, |\eta|\|\xi' - \xi|\} \\
&\le \varepsilon \sup\{|\eta' - \eta|, |\xi' - \xi|\}.
\end{aligned}
$$

Now, from the Fixed Point Theorem it follows that there exists an element $(\xi, \eta) \in V \oplus W$ such that $\Lambda((\xi, \eta)) = (\xi, \eta)$, that is $f - gh - \xi\eta = \theta((\xi, \eta)) = g\eta + h\xi$. Finally, take $\tilde{g} = g + \xi$, and $\tilde{h} = h + \eta$. **Q.E.D.**

Note that the role played by the resultant in the preceding discussion comes from the fact that $\pm R(g,h)$ is the Jacobian determinant of the map $(\xi, \eta) \to -f + (g + \xi)(h + \eta)$, i.e. Hensel's lemma is a special case of a multivariable implicit function theorem.

Before proceeding we make some remarks on Taylor's formula for polynomials with coefficients in a field of arbitrary characteristic. Let

$f(X) \in K[X]$, where K is a field (not necessarily valued) and let $\alpha \in K$. As in the classical case, when K has characteristic 0, one can write

$$f(X) = f(\alpha) + (Df)(\alpha)(X - \alpha) + \frac{1}{2!}(D^2 f)(\alpha)(X - \alpha)^2 + \cdots$$
$$+ \frac{1}{n!}(D^n f)(\alpha)(X - \alpha)^n$$

where n is the degree of f, and $D = \dfrac{d}{dX}$. If the characteristic of K is $p > 0$, then $n! = 0$ for $n \geq p$ so that the formula, as it stands, loses its meaning. However, we observe that for $m \geq 0$

$$\frac{1}{s!}D^s(X^m) = \begin{cases} \binom{m}{s} X^{m-s} & \text{if } m \geq s \\ \\ 0 & \text{otherwise.} \end{cases} \tag{4.4}$$

Furthermore, $\binom{m}{s} = \dfrac{1}{s!} m(m-1) \cdots (m-s+1)$ has a well defined image in K, and if K is a field with valuation then that image lies in $\mathcal{O}$. Hence we can take formula (4.4) as the definition of the symbol $\dfrac{1}{s!}D^s$. With this interpretation we are free to apply Taylor's formula in the general case. We shall also write

$$\frac{1}{s!}f^{(s)}(\alpha) = \left(\frac{1}{s!}D^s f\right)(\alpha).$$

In particular, if K is a valued field and $f(X) \in \mathcal{O}[X]$ then we have $\left(\dfrac{1}{s!}D^s f\right)(X) \in \mathcal{O}[X]$.

We can now give an application of Hensel's lemma which allows one to "refine" an approximate root α of a polynomial over a complete field to a true root $\tilde{\alpha}$ of the given polynomial.

Theorem 4.2 (Newton's Lemma). *Let K be a complete field with valuation $|\text{-}|$ and valuation ring $\mathcal{O}$. Let $f \in \mathcal{O}[X]$ and $\alpha \in \mathcal{O}$. Suppose that*

$$|f(\alpha)| \leq \varepsilon |f'(\alpha)|^2 \quad \text{with} \quad 0 < \varepsilon < 1.$$

Then there is an $\tilde{\alpha} \in \mathcal{O}$ such that

$$f(\tilde{\alpha}) = 0 \quad \text{and} \quad |\alpha - \tilde{\alpha}| \leq \varepsilon |f'(\alpha)|.$$

Proof. We apply Hensel's lemma to the given polynomial f and taking $g(X) = X - \alpha$,

$$h(X) = \frac{f(X) - f(\alpha)}{X - \alpha}$$

$$= f'(\alpha) + \frac{f''(\alpha)}{2!}(X - \alpha) + \cdots + \frac{f^{(n)}(\alpha)}{n!}(X - \alpha)^{n-1} \, ,$$

f, g and h being viewed as polynomials in the new variable $Y = X - \alpha$. We have g, $h \in \mathcal{O}[Y]$ and $\deg g = 1$, $\deg h = n - 1$. Since $f - gh = f(\alpha)$, condition (b) in Theorem 4.1 is obviously satisfied and so we need only verify that $|f(\alpha)| \le \varepsilon |f'(\alpha)|^2$ is equivalent to condition (a) of Theorem 4.1, where we take the Gauss norm with respect to Y. To prove this equivalence, it suffices to observe that

$$R(g,h) = \pm \det \begin{pmatrix} 0 & \cdots & 0 & f'(\alpha) \\ 1 & & & \frac{1}{2!}f''(\alpha) \\ & \ddots & & \vdots \\ 0 & \cdots & 1 & \frac{1}{n!}f^{(n)}(\alpha) \end{pmatrix} = \pm f'(\alpha) \, .$$

Thus, by Theorem 4.1, $f = (X - \tilde{\alpha})\tilde{h}(X)$ for some $\tilde{\alpha} \in \mathcal{O}$. **Q.E.D.**

As an explicit application of Newton's lemma we can consider $K = \mathbf{Q}_5 =$ the completion of $\mathbf{Q}$ with respect to the valuation $|-| = |-|_5$, and $f(X) = X^2 + 1$. Now $f(2) = 5$ so that $|f(2)| = 1/5 < 1$, and $f'(2) = 4$ so that $|f'(2)| = 1$ and therefore the hypotheses of Newton's lemma are satisfied. Thus we obtain $\alpha \in \mathbf{Q}_5$ with $\alpha^2 = -1$ and $|\alpha - 2| \le |5|$.

Indeed we can find an explicit formula for α. Since we can write $-4 = 1 - 5$ we obtain

$$2 \cdot (-1)^{1/2} = (-4)^{1/2} = (1 - 5)^{1/2} = \sum_{s=0}^{\infty} \binom{1/2}{s}(-5)^s. \qquad (4.5)$$

We recall that the binomial coefficient is defined for $a \in \mathbf{Q}$ and s a non-negative integer by

$$\binom{a}{s} = \begin{cases} \dfrac{1}{s!}a(a - 1) \cdots (a - s + 1) & \text{if } s \ne 0 \, , \\ 1 & \text{if } s = 0 \, . \end{cases}$$

We wish to prove that the series in (4.5) is convergent. To that end, it suffices to prove that $\left| \binom{1/2}{s} \right| \leq 1$. Observe that, for fixed $s \geq 0$, the map

$$\mathbf{Q} \longrightarrow \mathbf{Q}$$
$$a \longmapsto \binom{a}{s}$$

is continuous (with respect the $|-|_5$-topology), it being of polynomial type. Now let (a_N) be a sequence of positive integers converging in $\mathbf{Q}_5$ to $1/2$ (for example, since

$$\frac{1}{2} = 1 - \frac{1}{2} = 1 + 2\frac{1}{1-5} = 1 + 2\sum_{n=0}^{\infty} 5^n$$

we can take $a_N = 1 + 2\sum_{n=0}^{N} 5^n$). Then

$$\left| \binom{1/2}{s} \right| = \lim_{N \to \infty} \left| \binom{a_N}{s} \right| \leq 1,$$

as desired. Finally from (4.5) it follows that

$$(-1)^{1/2} = \frac{1}{2} \cdot \sum_{s=0}^{\infty} \binom{1/2}{s}(-5)^s.$$

The same reasoning we used to prove that $\binom{1/2}{s} \in \mathbf{Z}_5$ allows one to conclude more generally that

Proposition 4.3. *Let p be a rational prime. If $a \in \mathbf{Z}_p$ then $\binom{a}{s} \in \mathbf{Z}_p$ for all $s \geq 0$.*

We now offer a weaker version of Hensel's lemma. If $\alpha \in \mathcal{O}$, $f(X) \in \mathcal{O}[X]$ we will denote by $\bar{\alpha}$, $\bar{f}(X)$ the image of α, f in $\overline{K}, \overline{K}[X]$; respectively ($\overline{K} = \mathcal{O}/\mathfrak{p}$ is the residue field). Furthermore, we recall that two polynomials g, h, with coefficients in a field, are relatively prime if and only if $R(g,h) \neq 0$. In fact, we have $R(g,h) = 0$ if and only if ker θ is different from 0, where θ is the map defined in (4.1). This is equivalent to saying that the greatest common divisor $(g,h) \neq 1$. Indeed, if $\xi h + \eta g = 0$ then $h \mid g\eta$ and since $\deg h > \deg \eta$ this division is possible only if $(g,h) = r$ with $\deg r > 0$.

Theorem 4.4.   *Let K be a complete field with valuation $|{-}|$ and valuation ring $\mathcal{O}$. Let $f, g, h \in \mathcal{O}[X]$ with*

$$\deg g = n, \qquad \deg h = m, \qquad \deg f = m + n.$$

Suppose that

$$|f - gh|_{\text{gauss}} < 1 \qquad \text{and} \quad f - gh \in W_{n+m}$$

and suppose $\bar{g}, \bar{h}$ are relatively prime. Then there exist $\tilde{g}, \tilde{h} \in \mathcal{O}[X]$ such that

$$f = \tilde{g}\tilde{h}, \qquad \tilde{g} - g \in W_n, \qquad \tilde{h} - h \in W_m,$$

$$|\tilde{g} - g|_{\text{gauss}} < 1, \qquad |\tilde{h} - h|_{\text{gauss}} < 1.$$

Proof.   Since $g, h \in \mathcal{O}[X]$, then $R(g, h) \in \mathcal{O}$, and we can write

$$\overline{R(g, h)} = R(\bar{g}, \bar{h}) .$$

To conclude it suffices to observe that $R(\bar{g}, \bar{h}) \neq 0$ if and only if $|R(g, h)| = 1$, so that if $(\bar{g}, \bar{h}) = 1$ then the conditions (a) and (b) of Theorem 4.1 become the hypotheses of the present theorem.   **Q.E.D.**

As an application of Theorem 4.4 we prove the following reducibility criterion.

Corollary 4.5.   *Let*

$$f(X) = a_{n+m}X^{n+m} + \cdots + a_{n+1}X^{n+1} + X^n + a_{n-1}X^{n-1} + \cdots + a_0$$

with $|a_i| \leq 1$ and $|a_{n+j}| < 1$ for $0 \leq i \leq n - 1$ and $1 \leq j \leq m$. Then $f(X)$ is reducible in $K[X]$.

Proof.   Take $g = X^n + a_{n-1}X^{n-1} + \cdots + a_0$ and $h = 1 + a_{n+m}X^m$. Then

$$f - gh \in W_{n+m} \qquad \text{and} \qquad |f - gh| < 1.$$

Moreover $\bar{g} = X^n + \cdots$, $\bar{h} = 1$, and $(\bar{g}, \bar{h}) = 1$. It is now sufficient to apply Theorem 4.4.   **Q.E.D.**

5.   Extensions of Valuations.

We shall now discuss how to extend the valuation on the complete field K to a valuation on a finite extension field Ω of K. By an *extension to* Ω of the valuation $|-|$ on K we mean a valuation θ on Ω such that

$$\theta(a) = |a| \qquad \text{for every } a \in K.$$

Theorem 5.1. *Let K be a complete field with respect to the valuation $|-|$ and let Ω be a finite extension field of K. Then there exists a unique extension of $|-|$ to Ω.*

Proof.   Uniqueness.

(a) The valuation $|-|$ is trivial: $|a| = 1$ for every $a \in K^{\times}$. Let $|-|_1$ be an extension of $|-|$ on Ω. If $\alpha \in \Omega^{\times}$ there exist $a_i \in K$ such that

$$\alpha^m + a_1 \alpha^{m-1} + \cdots + a_m = 0 .$$

Then there must exist i, i' with $i \neq i'$ such that

$$|a_i \alpha^{m-i}|_1 = |a_{i'} \alpha^{m-i'}|_1 \neq 0 .$$

It follows that $|\alpha^{i-i'}|_1 = |a_i/a_{i'}|_1 = |a_i/a_{i'}| = 1$ so that $|\alpha|_1^{i-i'} = 1$ and finally $|\alpha|_1 = 1$.

(b) The valuation $|-|$ is not trivial. Let $|-|_1$ and $|-|_2$ be two extensions. From Theorem 3.2 and Lemma 1.2 it follows that there exists $\gamma > 0$ such that

$$|\alpha|_1 = |\alpha|_2^{\gamma} \qquad \text{for every} \quad \alpha \in \Omega .$$

Since $|a|_1 = |a|_2 = |a|$ for every $a \in K$ we have

$$|a|^{\gamma} = |a| \qquad \text{for every} \quad a \in K .$$

Now, taking $a \in K$ with $|a| \neq 0, 1$, we obtain $\gamma = 1$.

Existence.

Assume for a moment that Ω/K is a Galois extension and let $\mathcal{G} = \{\sigma_1, \cdots, \sigma_n\}$ be the Galois group of Ω/K. Let

$$\theta : \Omega \longrightarrow \mathbf{R}_+$$

be an extension of $|-|$ to Ω. The map θ satisfies:

(a) $\theta(\alpha) = 0$ if and only if $\alpha = 0$,
(b) $\theta(\alpha\beta) = \theta(\alpha)\theta(\beta)$,
(c) $\theta(\alpha) \leq 1$ implies $\theta(1 + \alpha) \leq 1$,
(d) $\theta(a) = |a|$ for every $a \in K$.

Now if $\sigma \in G$ one readily checks that the map $\alpha \mapsto \theta(\sigma\alpha)$ satisfies the four properties just listed. For example, if $\theta(\sigma(\alpha)) \leq 1$ then $\theta(\sigma(1 + \alpha)) = \theta(1 + \sigma(\alpha)) \leq 1$ by property (c) of θ with $\sigma(\alpha)$ playing the role of α. Uniqueness of the extension then implies that $\theta \circ \sigma = \theta$, and so

$$\prod_{i=1}^{n} \theta(\sigma_i(\alpha)) = \theta(\alpha)^n ,$$

that is,

$$\theta\left(\prod_{i=1}^{n} \sigma_i(\alpha)\right) = \theta(N_{\Omega/K}(\alpha)) = |N_{\Omega/K}(\alpha)| = \theta(\alpha)^n ,$$

where $N_{\Omega/K}$ denotes the norm from Ω to K.

We now return to the general case. Let n be the degree of Ω/K. The last equation suggests that we define θ as follows

$$\theta(\alpha) = |N_{\Omega/K}(\alpha)|^{1/n} .$$

We now verify that θ satisfies conditions (a)–(d) above. The first condition is obvious, the second follows from the identity

$$N_{\Omega/K}(\alpha\beta) = N_{\Omega/K}(\alpha)N_{\Omega/K}(\beta),$$

and the fourth is also obvious since for $a \in K$ we have

$$N_{\Omega/K}(a) = a^n.$$

It remains only to establish condition (c). So, suppose that $\alpha \in \Omega$ satisfies $|N_{\Omega/K}(\alpha)| \leq 1$. Let

$$f(X) = X^m + a_1 X^{m-1} + \cdots + a_m$$

be the minimal polynomial of α over K; note that m divides n and that $N_{\Omega/K}(\alpha) = \pm(a_m)^{n/m}$, so that $|a_m| \leq 1$. The polynomial $g(X) = f(X - 1)$ is the minimal polynomial of $1 + \alpha$ over K, so that

$$N_{\Omega/K}(1 + \alpha) = \pm g(0)^{n/m} = \pm f(-1)^{n/m} .$$

We wish to prove that $|f(-1)^{n/m}| \leq 1$. Suppose the contrary. Then $|f(-1)| > 1$ and so $|\pm 1 \mp a_1 \pm \cdots \pm a_m| > 1$. Hence there is an $i < m$ with $|a_i| > 1$ and we may choose the minimum i such that $|a_i| \geq |a_j|$ for all j. Now we have

$$\frac{1}{a_i}f(X) = \frac{1}{a_i}X^m + \frac{a_1}{a_i}X^{m-1} + \cdots + X^{m-i} + \frac{a_{i+1}}{a_i}X^{m-i-1} + \cdots + \frac{a_m}{a_i}.$$

By Corollary 4.5 we conclude that f is reducible, which is a contradiction. **Q.E.D.**

In the sequel we will denote the valuation θ defined above by the same symbol used for the valuation on K. Then, for every $\alpha \in \Omega$, we have

$$|\alpha| = |N_{\Omega/K}(\alpha)|^{1/[\Omega:K]} \tag{5.1}$$

where $[\Omega : K]$ is the degree of the field extension.

Let Ω be a finite extension of K. We recall that $\overline{\Omega} = \mathcal{O}_\Omega/\mathfrak{p}_\Omega$ and $\overline{K} = \mathcal{O}_K/\mathfrak{p}_K$. Then since $\mathfrak{p}_\Omega \cap \mathcal{O}_K = \mathfrak{p}_K$ we have $\overline{\Omega} \supseteq \overline{K}$, and so we may define

$$f = f(\Omega/K)$$
$$= \text{the } \textit{relative degree} \text{ of } |\text{-}| = [\overline{\Omega} : \overline{K}]\,.$$

Similarly G_K is a subgroup of G_Ω, so that we can define

$$e = e(\Omega/K) = \text{the } \textit{ramification index} \text{ of } |\text{-}| = (G_\Omega : G_K)$$

where $(G_\Omega : G_K)$ is the index of the subgroup G_K in G_Ω.

We now wish to prove that e and f are finite numbers. We actually have a more precise result.

Theorem 5.2. *Let K be a complete valued field and Ω a finite extension of K. Then*

$$ef \leq [\Omega : K]\,.$$

Proof. Let $\omega_1, \cdots, \omega_\mu \in \mathcal{O}_\Omega$ be such that their residues $\overline{\omega}_1, \cdots, \overline{\omega}_\mu$ in $\overline{\Omega}$ are linearly independent over $\overline{K}$ (here μ is a positive integer). Furthermore, let $\pi_1, \cdots, \pi_\nu$ elements of $\Omega^\times$ such that the cosets $|\pi_1|G_K$, $\ldots, |\pi_\nu|G_K$ in G_Ω/G_K are distinct. We claim that the elements $\pi_j\omega_i$, for $1 \leq j \leq \nu$ and $1 \leq i \leq \mu$, are linearly independent over K. This will give the desired result since both μ and ν must be bounded by $n = [\Omega : K]$, so that e and f are finite and $ef \leq n$.

To prove the claim, let us prove first that if

$$\theta = y_1\omega_1 + \cdots + y_\mu\omega_\mu$$

with the $y_i \in K$ and not all zero, then

$$|\theta| = \max_{1 \le i \le \mu} |y_i| \in G_K.$$

Clearly we may assume that $|y_1|$ is the maximum of the $|y_i|$, so that, in particular, $y_1 \ne 0$. Now the element

$$\omega_1 + \frac{y_2}{y_1}\,\omega_2 + \cdots + \frac{y_\mu}{y_1}\,\omega_\mu$$

belongs to $\mathcal{O}_\Omega$ and its residue in $\overline{\Omega}$ is not 0 since the $\overline{\omega}_i$ are linearly independent over $\overline{K}$. Hence

$$\left| \omega_1 + \frac{y_2}{y_1}\,\omega_2 + \cdots + \frac{y_\mu}{y_1}\,\omega_\mu \right| = 1$$

whence $|\theta| = |y_1|$ as desired.

If the $\pi_j\omega_i$ were linearly dependent we would have

$$\sum_{j=1}^{\nu} \pi_j \sum_{i=1}^{\mu} X_{ij}\omega_i = 0$$

with $X_{ij} \in K$ and not all zero. Let us define $Y_j = \sum_{i=1}^{\mu} X_{ij}\omega_i$. Now

$$\left| \sum_{j=1}^{\nu} \pi_j Y_j \right| = \sup_{1 \le j \le \nu} \{|\pi_j Y_j|\}$$

since the $|\pi_j Y_j|$ are in distinct cosets by what we have seen above. This gives the desired contradiction.   **Q.E.D.**

The inequality $ef \le [\Omega : K]$ of Theorem 5.2 also holds if K is not complete, and in that case one should not expect equality to hold. However even in the complete case (which we consider) the inequality can be strict. We cite without proof the following theorem, called

Ostrowski's Defect Theorem. *With the same hypotheses and notation as in Theorem 5.1, there exists an integer δ, called the* defect *of Ω over K, such that*

$$[\Omega : K] = ef\delta \ .$$

Moreover, $\delta = 1$ unless the residue class field $\overline{K}$ has characteristic $p > 0$ in which case $\delta = p^l$ for some $\in \mathbf{N}$.

For a proof see Artin [1], p. 60.

There are examples of fields for which the integer l is positive. However, in the particular case which we will now discuss the equality $ef = [\Omega : K]$ holds. We say that the valuation on K is *discrete* if G_K is a discrete subgroup of the topological group $\mathbf{R}^\times$. If π is an element of K such that $|\pi| = \max\{z \in G_K \mid z < 1\}$ then $G_K = \langle|\pi|\rangle$, the multiplicative cyclic group generated by $|\pi|$ and $\mathfrak{p}_K = \pi\mathcal{O}_K$. If $e = (G_\Omega : G_K) < \infty$ then G_Ω is also a cyclic group, so that there exists a $\Pi \in \mathcal{O}_\Omega$ such that $|\Pi|^e = |\pi|$ and $G_\Omega = \langle|\Pi|\rangle$, $\mathfrak{p}_\Omega = \Pi\mathcal{O}_\Omega$.

Theorem 5.3. *Let K be a complete field with respect to a discrete valuation and let Ω be a finite extension of K. Then*

$$ef = [\Omega : K].$$

Proof. Let $\omega_1, \ldots, \omega_f \in \mathcal{O}_\Omega$ be a set of representatives for a basis of $\overline{\Omega}$ over $\overline{K}$. We prove that the set $\{\omega_i\Pi^j\}_{1 \leq i \leq f, 0 \leq j \leq e-1}$ is a basis of Ω over K . We observe that for every $\alpha \in \Omega$ there exists a power π^t such that $\pi^t\alpha \in \mathcal{O}_\Omega$ since $|\pi| < 1$ and the group G_Ω is archimedean. Then it suffices to prove that

$$\mathcal{O}_\Omega = \sum_{i=1}^{f} \sum_{j=0}^{e-1} \mathcal{O}_K\omega_i\Pi^j.$$

We first prove that if $\alpha \in \mathcal{O}_\Omega$ there exist $X_{1,0}, \ldots, X_{f,0} \in \mathcal{O}_K$ and $\alpha_1 \in \mathcal{O}_\Omega$ such that

$$\alpha = X_{1,0}\omega_1 + \cdots + X_{f,0}\omega_f + \Pi\alpha_1.$$

This follows immediately from the fact that we can write

$$\overline{\alpha} = \overline{X}_{1,0}\overline{\omega}_1 + \cdots + \overline{X}_{f,0}\overline{\omega}_f$$

for suitable $\overline{X}_{i,0} \in \overline{K}$. Now we iterate this procedure and write

$$
\begin{aligned}
\alpha_1 \quad &= \quad X_{1,1}\omega_1 + \cdots + X_{f,1}\omega_f + \Pi\alpha_2 \\
&\vdots \qquad\qquad\qquad\qquad\qquad \vdots \\
\alpha_{e-1} &= \quad X_{1,e-1}\omega_1 + \cdots + X_{f,e-1}\omega_f + \Pi\alpha_e.
\end{aligned}
$$

with suitable $X_{i,j} \in \mathcal{O}_K$ and $\alpha_h \in \mathcal{O}_\Omega$. Hence we have

$$
\alpha = \sum_{i=1}^{f}\sum_{j=0}^{e-1} X_{i,j}\omega_i \Pi^j + \Pi^e \alpha_e.
$$

But $\Pi^e = \pi u$ where $u \in \mathcal{O}_\Omega$ satisfies $|u| = 1$; so if we put $\beta_1 = u\alpha_e \in \mathcal{O}_\Omega$, we can write

$$
\alpha = \sum_{i=1}^{f}\sum_{j=0}^{e-1} X_{i,j,0}\omega_i \Pi^j + \pi\beta_1
$$

where $X_{i,j,0} \in \mathcal{O}_K$, and then, iterating,

$$
\begin{aligned}
\beta_1 \quad &= \quad \sum_{i=1}^{f}\sum_{j=0}^{e-1} X_{i,j,1}\omega_i \Pi^j + \pi\beta_2 \\
&\vdots \qquad\qquad\qquad\qquad \vdots \\
\beta_s \quad &= \quad \sum_{i=1}^{f}\sum_{j=0}^{e-1} X_{i,j,s}\omega_i \Pi^j + \pi\beta_{s+1}
\end{aligned}
$$

for suitable $X_{i,j,\lambda} \in \mathcal{O}_K$ and $\beta_\mu \in \mathcal{O}_\Omega$. Thus

$$
\alpha = \sum_{i=1}^{f}\sum_{j=0}^{e-1} \omega_i \Pi^j \left(\sum_{\lambda=0}^{s} X_{i,j,\lambda}\pi^\lambda \right) + \pi^{s+1}\beta_{s+1} \ .
$$

We now take the limit for $s \to \infty$ and obtain

$$
\alpha = \sum_{i=1}^{f}\sum_{j=0}^{e-1} \omega_i \Pi^j \left(\sum_{\lambda=0}^{\infty} X_{i,j,\lambda}\pi^\lambda \right)
$$

because $\pi^{s+1}\beta_{s+1} \to 0$ since $\beta_{s+1} \in \mathcal{O}_\Omega$ and the series

$$
\sum_{\lambda=0}^{\infty} X_{i,j,\lambda}\pi^\lambda
$$

is convergent in $\mathcal{O}_K$ since $X_{i,j,\lambda}\pi^\lambda$ tends to 0 for $\lambda \to \infty$.   **Q.E.D.**

Let K be a field complete with respect to the valuation $|-|$ and let K^{alg} its algebraic closure. From Theorem 5.1 it easily follows that $|-|$ admits a unique extension to K^{alg}. However K^{alg} need not be complete with respect to this valuation. We call its completion Ω and we will still use $|-|$ to denote the unique extension of $|-|$ to Ω.

Theorem 5.4.   *Let K be a complete field and let Ω be the completion of its algebraic closure. Then Ω is algebraically closed.*

Proof.   Assume not. Then there exists a monic irreducible polynomial f in $\Omega[X]$ of degree $n > 1$ such that $|f|_{\mathrm{gauss}} \leq 1$. We begin with the case of a separable polynomial, that is we assume that $f'(X) \neq 0$ (the inseparable case can occur only in positive characteristic). We can write

$$f(X) = \prod_{i=1}^{n}(X - \beta_i)$$

with $\beta_i \neq \beta_j$ for $i \neq j$. We again use $|-|$ to denote its unique extension to $\Omega(\beta_1, \ldots, \beta_n)$. Let

$$R_i = \min_{j \neq i}\{|\beta_i - \beta_j|\} \ .$$

Let $\alpha \in K^{\mathrm{alg}}$; since $f(\alpha) = \prod_i(\alpha - \beta_i)$, from $|\alpha - \beta_i| \geq R_i$ for every i we deduce $|f(\alpha)| \geq R_1 R_2 \cdots R_n$. On the other hand, if $|\alpha - \beta_i| < R_i$ for at least one index i, we claim that

$$|f(\alpha)| = |\alpha - \beta_i||f'(\beta_i)| \ .$$

To fix notation assume $i = 1$. It suffices to show that $|\alpha - \beta_j| = |\beta_1 - \beta_j|$ for every $j \neq 1$, and this follows from the identity $\alpha - \beta_j = \alpha - \beta_1 + \beta_1 - \beta_j$ since $|\beta_1 - \beta_j| \geq R_1$.

Since K^{alg} is dense in Ω there exists a sequence (g_s) of monic elements from $K^{\mathrm{alg}}[X]$ with $\deg g_s = n$ and $|g_s|_{\mathrm{gauss}} \leq 1$ such that

$$\lim_{s \to \infty} |g_s - f|_{\mathrm{gauss}} = 0 \ .$$

Let $\alpha_s \in K^{\mathrm{alg}}$ be a root of g_s. Then we have $|\alpha_s| \leq 1$, and

$$\lim_{s \to \infty} |f(\alpha_s)| = 0$$

since

$$|f(\alpha_s)| = |g_s(\alpha_s) - f(\alpha_s)| \le |g_s - f|_{\text{gauss}} \ .$$

We can conclude that there exists an index j with $1 \le j \le n$, and a subsequence (α_{s_μ}) of (α_s) such that $|\alpha_{s_\mu} - \beta_j| \to 0$. Hence, the sequence (α_{s_μ}) converges to β_j, so that $\beta_j \in \Omega$, which gives a contradiction.

Finally, if K has characteristic p and f is inseparable then $f(X) = g(X^q)$ with q a power of p and $g(t)$ separable, monic and irreducible. By the case just discussed we see that if α is a root of g then $\alpha = \lim_n a_n$ with $a_n \in K^{\text{alg}}$. But then $(a_n^{1/q})$ is a Cauchy sequence in K^{alg} which tends to a root of f.   **Q.E.D.**

The completion of the algebraic closure of the field $\mathbf{Q}_p$ of p-adic numbers will be denoted by $\mathbf{C}_p$.

6.   Newton Polygons.

Given a valuation $|-|$ on a field K, we define the *order function* associated to $|-|$ by setting

$$\text{ord } a = -\log |a| \ .$$

If K has characteristic 0, and $\overline{K}$ is of positive characteristic p then one frequently takes

$$\text{ord } a = -\frac{\log |a|}{\log p} \ .$$

The following are the defining properties of the ord function

(O1) $\text{ord } a = +\infty$ if and only if $a = 0$,

(O2) $\text{ord }(ab) = \text{ord } a + \text{ord } b$,

(O3) $\text{ord }(a + b) \ge \min\{\text{ord } a, \text{ord } b\}$ with equality when $\text{ord } a \ne \text{ord } b$.

Now let $f \in K[X]$ be of degree n, say $f(X) = a_n X^n + a_{n-1} X^{n-1} + \cdots + a_0$. We define the *Newton polygon* of $f(X)$ as the convex hull of the set of points

$$\{(j, \text{ord } a_j) \mid j \ge 0\} \cup \{Y_{+\infty}\}$$

where $Y_{+\infty}$ denotes the point at infinity of the positive vertical axis; if $a_j = 0$, we define $(j, \text{ord } a_j) = Y_{+\infty}$.

The following is the Newton polygon of the polynomial $f(X) = 5X^{10} + 5^2X^6 + 5^{-1}X^5 + X^4 + 5^2X + 5^7$ over the field $\mathbf{Q}_5$.

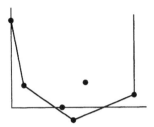

Theorem 6.1. *Let K be field complete with respect to a valuation, and let $f \in K[X]$. Then to each finite side of the Newton polygon of f there corresponds at least one root of f. The number (counting multiplicities) of roots corresponding to a given side is equal to the length of the projection of that side on the x-axis, and all roots α corresponding to the same side have ord $\alpha = -\lambda$ where λ is the slope of the side. If*

$$f_\lambda(X) = \prod_{\substack{f(\alpha)=0 \\ \text{ord } \alpha = -\lambda}} (X - \alpha)$$

then $f_\lambda(X) \in K[X]$.

Proof. We first prove the last statement. Let f be of degree n and let the roots of f (counted with multiplicities) be $\alpha_1, \dots, \alpha_n$. Let us first assume that $K(\alpha_1, \dots, \alpha_n)$ is a Galois extension of K (this is always the case if K has characteristic zero). As we have seen in the proof of Theorem 5.1, if σ is any automorphism of $K(\alpha_1, \cdots, \alpha_n)$ over K then $|\sigma(\alpha_i)| = |\alpha_i|$, that is the values of the roots are invariant under σ. If we set $R_\lambda = \{\alpha \mid f(\alpha) = 0 \text{ and ord } \alpha = -\lambda\}$, we have $\sigma R_\lambda = R_\lambda$ and then $f_\lambda = \prod_{\alpha \in R_\lambda}(X - \alpha)$ is invariant for σ so that it is in $K[X]$.

If K has positive characteristic p we proceed as follows. We first observe that if f is irreducible, then all roots of f have the same magnitude: this is certainly true if f is also separable, since in that case $K(\alpha_1, \dots, \alpha_n)$ is a Galois extension; otherwise it suffices to observe that $f(X) = g(X^{p^\nu})$ with $g \in K[X]$ irreducible and separable, and ν a positive integer. If f is reducible we proceed by induction on n, the case $n = 1$ being obvious. Let $h(X) \in K[X]$ be the minimal polynomial of α_1 over K, and put $f(X) = g(X)h(X)$, with $g(X) \in K[X]$. Then, if we define

$$g_\lambda(X) = \prod_{\substack{g(\alpha)=0 \\ \text{ord } \alpha = -\lambda}} (X - \alpha),$$

then by the inductive hypothesis $g_\lambda(X) \in K[X]$ for every λ. Let $\lambda_1 = -\text{ord } \alpha_1$. Then since $h(X)$ is irreducible, $f_{\lambda_1}(X) = g_{\lambda_1}(X)h(X)$ and $f_\lambda(X) = g_\lambda(X)$ for all $\lambda \neq \lambda_1$. Therefore, $f_\lambda(X) \in K[X]$ for all λ, as desired.

We now prove the first part of the theorem. It is not restrictive to assume, as we do, that $a_0 = 1$ (multiplying the polynomial f by a non-zero constant or by X has the effect of a translation on the Newton polygon); changing the preceding notation we write

$$f = \prod_{i=1}^{n}(1 + \alpha_i X) = 1 + a_1 X + a_2 X^2 + \ldots + a_n X^n$$

and suppose

$$\text{ord } \alpha_1 \leq \text{ord } \alpha_2 \leq \ldots \leq \text{ord } \alpha_n \ . \qquad (6.1)$$

Suppose $\{\text{ord } \alpha_1, \ldots, \text{ord } \alpha_n\} = \{\nu_1, \ldots, \nu_l\}$ with $\nu_1 < \nu_2 < \ldots < \nu_l$ and let the order ν_i appear with multiplicity κ_i, for $1 \leq i \leq l$ (that is, $\text{ord } \alpha_1 = \text{ord } \alpha_2 = \ldots = \text{ord } \alpha_{\kappa_1} = \nu_1$, $\text{ord } \alpha_{\kappa_1+1} = \text{ord } \alpha_{\kappa_1+2} = \ldots = \text{ord } \alpha_{\kappa_1+\kappa_2} = \nu_2, \ldots$). We must prove that the Newton polygon of f has l (non-vertical) sides $P_0P_1, P_1P_2, \ldots, P_{l-1}P_l$ where

$$P_0 = (0,0), \ P_1 = (\kappa_1, \nu_1\kappa_1), \ P_2 = (\kappa_1 + \kappa_2, \nu_1\kappa_1 + \nu_2\kappa_2), \ \ldots,$$

$$P_\rho = (\sum_{i=1}^{\rho} \kappa_i, \sum_{i=1}^{\rho} \nu_i\kappa_i), \ \ldots \ .$$

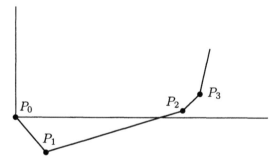

This is equivalent to proving that

$$\text{ord } a_{\kappa_1} = \nu_1\kappa_1, \quad \text{ord } a_{\kappa_2} = \nu_1\kappa_1 + \nu_2\kappa_2, \ldots,$$

$$\text{ord } a_{\kappa_1+\ldots+\kappa_\rho} = \sum_{i=1}^{\rho} \nu_i\kappa_i, \quad \ldots$$

and that, if the integer s is such that $\kappa_0 + \kappa_1 + \ldots + \kappa_\rho < s < \kappa_1 + \ldots + \kappa_{\rho+1}$, then

$$\operatorname{ord} a_s \geq \sum_{i=0}^{\rho} \nu_i \kappa_i + \nu_{\rho+1}(s - \kappa_1 - \ldots - \kappa_\rho) , \qquad (6.2)$$

for $\rho = 0, \ldots, l-1$ (where $\kappa_0 = \nu_0 = 0$). This last condition means that the points $(s, \operatorname{ord} a_s)$ lie on or above the side $P_\rho P_{\rho+1}$. Clearly, we have

$$a_s = \sum_{1 \leq i_1 \leq i_2 \leq \ldots \leq i_s \leq n} \alpha_{i_1} \alpha_{i_2} \cdots \alpha_{i_s} , \qquad (6.3)$$

so that, by (6.1)

$$\operatorname{ord} a_s \geq \operatorname{ord}(\alpha_1 \alpha_2 \cdots \alpha_s) = \operatorname{ord} \alpha_1 + \ldots + \operatorname{ord} \alpha_s .$$

which, by the hypotheses, is the same as (6.2).

Finally, if $s = \kappa_1 + \ldots + \kappa_\rho$, then $\operatorname{ord} \alpha_{s+1} > \operatorname{ord} \alpha_s$; therefore from (6.3) it follows that

$$\operatorname{ord} a_s = \operatorname{ord}(\alpha_1 \alpha_2 \cdots \alpha_s) = \sum_{i=1}^{s} \operatorname{ord} \alpha_i = \sum_{i=1}^{s} \nu_i \kappa_i . \qquad (6.4)$$

Q.E.D.

Example 6.2. Let $K = \mathbf{Q}_5$ and $f = X^2 + 1$.

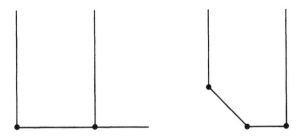

There are two roots, each of which has order 0. If we make the change of variable $X = Y + 2$, then $f(X) = g(Y) = Y^2 + 4Y + 5$. Now the two roots have different orders; one is of order 0 and the other of order 1.

7.   The y-intercept Method.

Let

$$a_0 + a_1X + \ldots + a_nX^n = f(X) = A(X - \alpha_1)\ldots(X - \alpha_n) \ .$$

For $\lambda \in \mathbf{R}$, the *line of support*, of slope λ, for the Newton polygon of f is the line $\ell : Y = \lambda X + b_\lambda$ with b_λ maximal such that each of the points $(j, \mathrm{ord}\, a_j)$ lies on or above ℓ.

Trivially $b_\lambda = \inf\limits_j(\mathrm{ord}\, a_j - j\lambda)$, a continuous function of λ.

If $\beta \in \Omega$, $\mathrm{ord}\, \beta = -\lambda$ then $\mathrm{ord}\,(a_j\beta^j) \geq b_\lambda$ for all j and hence $\mathrm{ord}\, f(\beta) \geq b_\lambda$.

We distinguish two cases

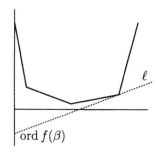

ord $f(\beta)$

1. General Case.

If no side of the polygon is of slope λ then there exists precisely one subscript j such that $b_\lambda = \mathrm{ord}\, a_j - \lambda j$ and hence $\mathrm{ord}\, f(\beta) = b_\lambda$, i.e.

$$\mathrm{ord}\, A + \sum_{i=1}^{n} \mathrm{ord}\,(\beta - \alpha_i) = b_\lambda \ .$$

But in this case $\mathrm{ord}\, \beta$ and $\mathrm{ord}\, \alpha_i$ are distinct for all i and hence

$$b_\lambda = \mathrm{ord}\, A + \sum_{i=1}^{n} \inf(\mathrm{ord}\, \alpha_i, -\lambda)$$

provided λ is not the slope of a side. But both sides of this equation are continuous functions of λ and hence the relation is valid also if λ is the slope of a side.

2. Special case.

For $\operatorname{ord}\beta = -\lambda$, λ a slope of a side, we cannot hope to determine $\operatorname{ord} f(\beta)$ in terms of λ alone since $\operatorname{ord} f(\beta)$ may assume arbitrarily large values. To assist in a geometric presentation of the situation we use a multiplicative notation. Let $|\beta| = R$ correspond to $\operatorname{ord}\beta = -\lambda$. Let $\Gamma_R = \{\beta \mid |\beta| = R\}, D(\alpha, R^-) = \{\beta \mid |\alpha - \beta| < R\}$. Let $\{\alpha_1, \ldots, \alpha_h\}$ be the zeros of f in Γ_R.

If $\beta \in \Gamma_R \backslash \bigcup\limits_{i=1}^{h} D(\alpha_i, R^-)$ then, for $1 \le i \le n$,

$$|\alpha_i - \beta| = \ \sup\left(|\alpha_i|, |\beta|\right)$$

and hence

$$|f(\beta)| = |A| \prod_{i=1}^{n} \ \sup\left(|\alpha_i|, |\beta|\right)$$

i.e. once again

$$\operatorname{ord} f(\beta) = \operatorname{ord} A + \sum_{i=1}^{n} \inf\left(\operatorname{ord} a_i, -\lambda\right) = b_\lambda \ .$$

Summarizing:

Let $\operatorname{ord}\beta = -\lambda$; then $\operatorname{ord} f(\beta) \ge b_\lambda$ in all cases. Equality holds if $\operatorname{ord}(\beta - \alpha_i) = -\lambda$ for each root α_i with $\operatorname{ord}\alpha_i = -\lambda$. In particular equality holds if the polygon has no side of slope λ.

We invite the reader to note that in "ultrametric" geometry a "circumference" of radius R (like Γ_R above) contains open disks of radius R centered at points of the circumference. Thus, if one wishes to make an analogy with "classical" geometry, one should view the circumference as an annulus rather than a circle.

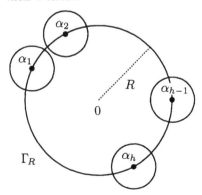

8. Ramification Theory

Let K be a complete field and let Ω be a finite extension of K. We say that Ω/K is an *unramified* extension if

a) $\overline{\Omega}/\overline{K}$ is a separable extension,
b) $f(\Omega/K) = [\Omega : K]$.

Clearly, by Theorem 5.2, if Ω/K is an unramified extension then $e(\Omega/K) = 1$. Conversely, if the valuation of K is discrete, this last condition, by Theorem 5.3, implies that Ω/K is unramified (subject to condition (a)). If $f(\Omega/K) = 1$ we say that the extension Ω/K is *totally ramified*.

Theorem 8.1. *Let K be a complete field and Ω be a finite extension of K. Suppose that $\overline{\Omega}/\overline{K}$ is a separable extension. Then Ω contains a unique field extension T of K such that Ω/T is totally ramified and T/K is unramified.*

Proof. We remind the reader that $\overline{\Omega}/\overline{K}$ is a separable extension if and only if $\overline{\Omega} = \overline{K}(\overline{\alpha})$ where $\overline{\alpha}$ is a root of a monic irreducible polynomial $\overline{g}(X) \in \overline{K}[X]$ (of degree f) such that $\overline{g}'(\overline{\alpha}) \neq 0$. There exists $g(X) \in \mathcal{O}_\Omega[X]$ with $\deg g = f$ and whose reduction modulo $\mathfrak{p}_K[X]$ is equal to $\overline{g}$. We claim that g is irreducible. Indeed, we have $\overline{g}(0) \neq 0$ and hence the Newton polygon of g consists of one side of slope zero and length f. Thus the roots of g (in an extension field of Ω) have valuation one. If g factored in $K[X]$, it would have monic factors in $\mathcal{O}_K[X]$ and by reduction mod $\mathfrak{p}_K[X]$, the polynomial $\overline{g}$ would be reducible in $\overline{K}[X]$. Let $\alpha \in \mathcal{O}_\Omega$ satisfy $(\alpha \mod \mathfrak{p}_\Omega) = \overline{\alpha}$; then we have $\overline{g(\alpha)} = \overline{g}(\overline{\alpha}) = 0$, $\overline{g'(\alpha)} = \overline{g}'(\overline{\alpha}) \neq 0$, so that $|g(\alpha)| < 1$ and $|g'(\alpha)| = 1$. Hence by Newton's lemma there exists $\beta \in \Omega$ such that $g(\beta) = 0$ and $|\beta - \alpha| < 1$, that is $\overline{\beta} = \overline{\alpha}$. We define $T = K(\beta)$. Now we have $\overline{T} = \overline{K(\beta)} \supseteq \overline{K}(\overline{\beta}) = \overline{\Omega}$, so that Ω/T is totally ramified. On the other hand, $[T : K] = [K(\beta) : K] = \deg g = f = [\overline{\Omega} : \overline{K}] = [\overline{T} : \overline{K}]$ so that T/K is unramified.

To prove that T is unique with these properties, assume that E is another such field. Then $\overline{E} = \overline{\Omega}$ and so we can find $\gamma \in E$ such that $(\gamma \mod \mathfrak{p}_E) = \overline{\alpha}$. This implies that $|g(\gamma)| < 1$ and $|g'(\gamma)| = 1$ since $\overline{g}$ is a separable polynomial. Repeating the preceding argument, we can assume that γ is actually a root of g and so it must be equal to β since distinct roots of g have distinct reductions.   **Q.E.D.**

The field T is called the *inertial subfield* of Ω/K.

Theorem 8.2. *Let K be a complete field. Let Ω/K be a finite Galois extension, and assume $\overline{\Omega}/\overline{K}$ to be a separable extension. Then $\overline{\Omega}/\overline{K}$ is a Galois extension and there is a group epimorphism*

$$\mathcal{G}(\Omega/K) \longrightarrow \mathcal{G}(\overline{\Omega}/\overline{K})$$

whose kernel is $\mathcal{G}(\Omega/T)$, where T is the inertial subfield of Ω/K. Therefore there is a natural group isomorphism

$$\mathcal{G}(T/K) \longleftrightarrow \mathcal{G}(\overline{\Omega}/\overline{K}) .$$

Proof. We maintain the notations from the proof of Theorem 8.1. We can write

$$g(X) = \prod_{i=1}^{f}(X - \alpha_i) \quad \text{with the } \alpha_i \in \Omega \text{ and } \alpha_1 = \alpha .$$

We recall that the polynomial $g(X)$ is irreducible over K. Since $\overline{g}$ is separable the $\overline{\alpha}_i$ are distinct elements of $\overline{\Omega}$ so that

$$\overline{g}(X) = (X - \overline{\alpha}_1) \cdots (X - \overline{\alpha}_f)$$

in $\overline{\Omega} = \overline{K}(\overline{\alpha})$. Thus $\overline{\Omega} = \overline{K}(\overline{\alpha})$ is the splitting field of a separable polynomial, and so is a Galois extension.

If $\sigma \in \mathcal{G}(\Omega/K)$ then $\sigma\mathcal{O}_\Omega \subseteq \mathcal{O}_\Omega$ and $\sigma\mathfrak{p}_\Omega \subseteq \mathfrak{p}_\Omega$ since $|\sigma\beta| = |\beta|$ for every $\beta \in \Omega$. Let

$$\overline{\sigma} : \mathcal{O}_\Omega/\mathfrak{p}_\Omega \longrightarrow \mathcal{O}_\Omega/\mathfrak{p}_\Omega = \overline{\Omega}$$

be defined by

$$\overline{\sigma}\overline{\beta} = \overline{\sigma\beta}$$

for every $\beta \in \mathcal{O}_\Omega$. It is an immediate verification that $\overline{\sigma}$ is an automorphism of $\overline{\Omega}/\overline{K}$ and that

$$\mathcal{G}(\Omega/K) \longrightarrow \mathcal{G}(\overline{\Omega}/\overline{K})$$
$$\sigma \longmapsto \overline{\sigma}$$

is a group homomorphism. To prove surjectivity, let $\sigma^* \in \mathcal{G}(\overline{\Omega}/\overline{K})$ be defined by $\sigma^*\overline{\alpha} = \overline{\alpha}_i$ and let $\sigma \in \mathcal{G}(\Omega/K)$ be defined by $\sigma\alpha = \alpha_i$. It is clear that $\overline{\sigma} = \sigma^*$. To conclude, observe that by varying i from 1 to f we obtain all the elements of $\mathcal{G}(\overline{\Omega}/\overline{K})$.

If $\sigma \in \mathcal{G}(\Omega/K)$ is such that $\overline{\sigma} = 1$ then $\overline{\sigma\alpha} = \overline{\alpha}$ so that we have $|\sigma\alpha - \alpha| < 1$. The roots of g being all in distinct classes modulo $\mathfrak{p}_\Omega$, we conclude that $\sigma\alpha = \alpha$ so that $\sigma \in \mathcal{G}(\Omega/T)$ (recall that $T = K(\alpha)$, by definition). The natural isomorphism follows from the fact that the two groups $\mathcal{G}(\Omega/K)/\mathcal{G}(\Omega/T)$ and $\mathcal{G}(\overline{\Omega}/\overline{K})$ have the same order and the fact that $\mathcal{G}(\Omega/K)/\mathcal{G}(\Omega/T)$ is naturally isomorphic to $\mathcal{G}(T/K)$ (Galois theory). **Q.E.D.**

Exercise. In the notation of the preceding theorem, but with the hypothesis "Ω/K galois" replaced by "$\overline{\Omega}/\overline{K}$ galois", prove that T/K is a galois extension.

Suppose that $\overline{K} = \mathbf{F}_q$ is a finite field. Let $[\overline{\Omega} : \overline{K}] = f$ so that $\overline{\Omega} = \mathbf{F}_{q^f}$. Then $\overline{\Omega}/\overline{K}$ is a Galois extension and $\mathcal{G}(\overline{\Omega}/\overline{K})$ is a cyclic group generated by the map

$$\overline{\sigma} : \overline{\alpha} \longmapsto \overline{\alpha}^q .$$

Then T/K is also a Galois extension with cyclic Galois group; this group is generated by the lifting of $\overline{\sigma}$ to T/K (the so-called *Frobenius automorphism*) that is the unique automorphism Frob of T/K such that

$$\text{Frob } \alpha \equiv \alpha^q \mod \mathfrak{p}_T \quad \text{for every } \alpha \in \mathcal{O}_T . \tag{8.1}$$

The field $\overline{\Omega} = \mathbf{F}_{q^f}$ is the splitting field of the polynomial $X^{q^f} - X$ (coefficients in $\overline{K} = \mathbf{F}_q$) and therefore the same polynomial (coefficients in K) splits in $T = K(\zeta_{q^f-1})$ where ζ_{q^f-1} is a primitive $(q^f - 1)$-th root of unity. Since in each class modulo $\mathfrak{p}_T$ there is only one $(q^f - 1)$-th root of unity, from (8.1) it follows that

$$\text{Frob } \zeta_{q^f-1} = \zeta_{q^f-1}^q , \tag{8.2}$$

which gives a more precise description of the automorphism Frob.

Finally, the following result is easily obtained.

Theorem 8.3. *Let K be a finite extension of $\mathbf{Q}_p$ with $\#\overline{K} = q$ and let f be a positive integer. Then the unique unramified extension of K of degree f is $K(\zeta)$ where ζ is a primitive $q^f - 1$ root of unity.*

9. Totally Ramified Extensions.

In this section we shall be concerned with the following situation: K is a field complete with respect to a discrete valuation; the residue field $\overline{K}$ is of characteristic $p \neq 0$, and Ω is a totally ramified extension of K of degree $n = e(\Omega/K)$, where $n = n_0 p^l$ with $(n_0, p) = 1$. We note that since K has a discrete valuation, we have $G_K = \langle |\pi| \rangle$ with $|\pi| < 1$ and therefore for any $\alpha \in K$ the relation $|\pi| \leq |\alpha| < 1$ implies that $|\alpha| = |\pi|$. Since $f(\Omega/K) = 1$, we have $\overline{\Omega} = \overline{K}$. Furthermore if Π is a generator for $\mathfrak{p}_\Omega$ then, as we have seen (cf. proof of Theorem 5.3), $\Omega = K(\Pi)$ and

$$\mathcal{O}_\Omega = \mathcal{O}_K + \mathcal{O}_K \Pi + \mathcal{O}_K \Pi^2 + \cdots + \mathcal{O}_K \Pi^{e-1} . \tag{9.1}$$

Lemma 9.1. *With the above assumptions, suppose that $z \in \Omega$ satisfies $z^{n_0} = 1$. Then in fact $z \in K$.*

Proof. Let $f(X) = X^{n_0} - 1$, and choose $y \in K$ such that $|y - z| < 1$ (such a choice is certainly possible in view of equation (9.1)). Since $|y| = |y - z + z|$ and $|z| = 1$ we see that $|y| = 1$. Moreover $|f(y)| = |f(y) - f(z)| = |y^{n_0} - z^{n_0}| < 1$. Now $f'(y) = n_0 y^{n_0 - 1}$ so that $|f'(y)| = 1$ since $(n_0, p) = 1$. Thus we may apply Newton's lemma to find a $y_0 \in K$ with $y_0^{n_0} = 1$, and $|y_0 - y| < 1$. Hence $|y_0 - z| < 1$ so that $y_0 = z$ since the reduced polynomial is separable while y_0 and z are roots of $f(X)$ with the same reduction. **Q.E.D.**

Theorem 9.2. *Let K be a field complete with respect to a discrete valuation, and let Ω/K be a totally ramified extension of degree $n = e(\Omega/K)$. Let the characteristic of the residue field $\overline{K}$ be $p > 0$ and suppose that $n = n_0\, p^l$ with $(n_0, p) = 1$. Then there exists a unique extension V of K with $\Omega \supseteq V \supseteq K$ such that $[V : K] = n_0$. Moreover, $V = K(\sqrt[n_0]{\pi})$ where π is an element of K such that $\mathfrak{p}_K = \pi \mathcal{O}_K$.*

Proof. As in the discussion preceding Lemma 9.1, we have $\overline{K} = \overline{\Omega}$, and if $G_K = \langle |\pi| \rangle$, $G_\Omega = \langle |\Pi| \rangle$ then $\Omega = K(\Pi)$ and

$$\mathcal{O}_\Omega = \mathcal{O}_K + \mathcal{O}_K \Pi + \cdots \mathcal{O}_K \Pi^{e-1}, \quad \text{with } e = n = n_0 p^l .$$

It follows that $\Pi^{n_0 p^l} = \pi U$ with $U \in \mathcal{O}_\Omega$ satisfying $|U| = 1$. Since $\overline{\Omega} = \overline{K}$ we may write $U = uZ$ where $u \in K$ satisfies $|u| = 1$ and $Z \in \mathcal{O}_\Omega$ satisfies $|Z - 1| < 1$. Then $\Pi^{n_0 p^l} = u\pi Z$. We now consider the polynomial $f(X) = X^{n_0} - Z$ in $\Omega[X]$. Since $|Z - 1| < 1$ we have $|f(1)| < 1$, and clearly $|f'(1)| = |n_0| = 1$. Hence by Newton's lemma we find a root w of f in Ω, that is we have $w^{n_0} = Z$. Therefore

$$\Pi^{n_0 p^l} = u\pi Z = u\pi w^{n_0},$$

which we write in the form

$$\left(\frac{\Pi^{p^l}}{w}\right)^{n_0} = \pi u \in K \ .$$

Thus if we set $\Pi' = \Pi^{p^l}/w$, then Π' is a root of $g(X) = X^{n_0} - \pi u$, a polynomial lying in $K[X]$. Let $V = K(\Pi')$; clearly $[V : K] \leq n_0$ but $e(V/K) \geq n_0$ (since $|\Pi'|^{n_0} = |\pi|$) whence $[V : K] = n_0$. It remains only to prove that V is unique.

Assume to the contrary that there are two such extensions, say $K(\Pi')$ and $K(\Pi'')$ so that $\Pi'^{n_0} = \pi u'$ and $\Pi''^{n_0} = \pi u''$ with u', $u'' \in K$ such that $|u'| = |u''| = 1$. Consider the quotient $(\Pi'/\Pi'')^{n_0} = u'/u''$ and let $Z = \Pi'/\Pi''$. Then Z satisfies the equation $g(X) = X^{n_0} - u'/u'' = 0$, and since $\overline{Z} \in \overline{\Omega} = \overline{K}$ there exists a $z \in K$ such that $|z - Z| < 1$. Thus we have $|g(z)| < 1$ and $|g'(z)| = |n_0| = 1$. Again invoking Newton's lemma over K we obtain an exact solution $z_0 \in K$, that is we find $z_0 \in K$ such that $z_0^{n_0} = u'/u''$. Thus we have $(\Pi'/z_0\Pi'')^{n_0} = 1$; now Lemma 9.1 allows us to conclude that $\Pi'/z_0\Pi'' \in K$, whence $\Pi'/\Pi'' \in K$ so that $K(\Pi') = K(\Pi'')$.   **Q.E.D.**

Assume that Ω/K is a finite (not necessarily totally ramified) extension. Suppose that char $\overline{K} = p > 0$ and that $\overline{\Omega}/\overline{K}$ is a separable extension. We say that the extension is

tamely ramified if $p \nmid e(\Omega/K)$,
wildly ramified if $p \mid e(\Omega/K)$.

With this terminology we have proved that, under the hypotheses of Theorem 9.2, every totally ramified extension Ω/K contains a unique maximal tamely ramified subfield V, and that when $\Omega \neq V$ then the extension Ω/V is wildly ramified.

We can summarize what we have seen in the last two sections. For every finite extension Ω of the complete discrete valued field K with $\overline{\Omega}/\overline{K}$ a separable extension, there exist unique fields T and V such that

$$K \subseteq T \subseteq V \subseteq \Omega$$

where T/K is unramified, Ω/T is totally ramified, V/T is tamely ramified, and Ω/V is wildly ramified.

$$\begin{array}{ll} \text{Wild} & \left. \begin{array}{c} \Omega \\ | \\ V \\ | \\ T \end{array} \right] \quad \begin{array}{c} \text{Totally} \\ \text{ramified} \end{array} \\ \text{Tame} & \\ & \left. \begin{array}{c} | \\ K \end{array} \right. \quad \text{Unramified} \end{array}$$

Example 9.3. Consider the polynomial

$$\frac{X^p - 1}{X - 1} = X^{p-1} + X^{p-2} + \cdots + 1 \quad \text{over} \quad \mathbf{Q}_p.$$

If we set $Y = X - 1$ we obtain the Eisenstein polynomial

$$g(Y) = \frac{(1 + Y)^p - 1}{Y} = Y^{p-1} + \binom{p}{1} Y^{p-2} + \cdots + \binom{p-1}{p}.$$

A glance at the Newton polygon of g (*cf.* below where the case $p = 7$ is illustrated) shows that all its roots λ_1 have $\operatorname{ord} \lambda_1 = 1/(p - 1)$ (we suppose the ord function normalized by the condition $\operatorname{ord} p = 1$).

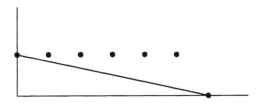

If $\zeta_p = 1 + \lambda_1$, the extension $\mathbf{Q}_p(\zeta_p)/\mathbf{Q}_p$ is tamely ramified since clearly here we have $f = 1$, and $(p - 1, p) = 1$. From Theorem 9.2 we know that $\mathbf{Q}_p(\zeta_p)$ is generated over $\mathbf{Q}_p$ by a root of a polynomial of the type $X^{p-1} - pu$ with u a unit in $\mathbf{Z}_p$. We can actually show that we can take $u = -1$, that is,

$$\mathbf{Q}_p(\zeta_p) = \mathbf{Q}_p(\pi) \quad \text{where} \quad \pi^{p-1} = -p.$$

To see that this is indeed the case it suffices to show that $\zeta_p \in \mathbf{Q}_p(\pi)$ since the two fields have the same degree over $\mathbf{Q}_p$. Taking Z as a new indeterminate, the polynomial

$$h(Z) = \frac{-1}{p} g(\pi Z)$$

$$= \frac{-1}{p} \left(\pi^{p-1} Z^{p-1} + \binom{p}{1} \pi^{p-2} Z^{p-2} + \cdots + \binom{p}{p-1} \right)$$

$$= Z^{p-1} - \pi^{p-2} Z^{p-2} - \cdots - \frac{p-1}{2} \pi Z - 1$$

has its coefficients in $\mathbf{Q}_p(\pi)$. Since $|\pi| < 1$ and $h'(1) = (p-1) +$ (things divisible by π), we have $|h(1)| < 1$ and $|h'(1)| = 1$ so we can appeal to Newton's lemma to find an exact root of h in $\mathbf{Q}_p(\pi)$. This shows that at least one, and hence all ζ_p do indeed belong to $\mathbf{Q}_p(\pi)$, as desired. (Since, by Hensel's lemma, the polynomial $X^{p-1} - 1$ splits over $\mathbf{Q}_p$ it follows that $X^{p-1} - \pi$ splits in $\mathbf{Q}_p(\pi)$, and so any choice of the root π works equally well.)

Continuing this example, we now consider $\mathbf{Q}_p(\zeta_{p^2})$, where ζ_{p^2} is a root of

$$\frac{X^{p^2} - 1}{X^p - 1} = X^{p(p-1)} + X^{p(p-2)} + \cdots + 1$$

(that is, a primitive p^2-th root of unity) and we prove that this is a wildly ramified extension of $\mathbf{Q}_p(\zeta_p)$. If α is a root of the polynomial $X^p - \zeta_p$ in $\mathbf{Q}_p(\zeta_p)[X]$ then $\mathbf{Q}_p(\zeta_{p^2}) = \mathbf{Q}_p(\zeta_p)(\alpha)$. We set $\alpha = 1 + \lambda_2$ so that $(1 + \lambda_2)^p = 1 + \lambda_1$, that is

$$\lambda_2^p + \binom{p}{1} \lambda_2^{p-1} + \cdots + \binom{p}{p-1} \lambda_2 - \lambda_1 = 0.$$

Hence $[\mathbf{Q}_p(\zeta_{p^2}) : \mathbf{Q}_p(\zeta_p)] \geq p$. The Newton polygon of this last polynomial in λ_2 consists of a single finite side between the points $(0, 1/(p-1))$ and $(p, 0)$ so the order of any root λ_2 must be $1/p(p-1)$. Now, $G_{\mathbf{Q}_p(\zeta_p)} = \langle |\lambda_1| \rangle = \langle |p|^{1/(p-1)} \rangle$ and $G_{\mathbf{Q}_p(\zeta_{p^2})} \supset \langle |\lambda_2| \rangle = \langle |p|^{1/p(p-1)} \rangle$ so that the index is at least p. Hence $[\mathbf{Q}_p(\zeta_{p^2}) : \mathbf{Q}_p(\zeta_p)] = p$.

We conclude by observing that, using the same reasoning, we can inductively define a sequence

$$\lambda_0, \lambda_1, \ldots, \lambda_s, \ldots,$$

by the conditions that $\lambda_0 = 0$ and that, for $s \geq 1$,

$$(1 + \lambda_s)^p = 1 + \lambda_{s-1} \; ;$$

then, for $s \geq 1$ we have

$$\operatorname{ord} \lambda_s = \frac{1}{p^{s-1}(p-1)} \, .$$

The element $1 + \lambda_s$ lies among the $p^{s-1}(p-1)$ roots of the polynomial

$$\frac{X^{p^s} - 1}{X^{p^{s-1}} - 1} = X^{p^{s-1}(p-1)} + X^{p^{s-1}(p-2)} + \cdots + 1 \, .$$

This shows that the set of pth power roots of unity consists of points inside the disk $D(1, 1^-) = \{x \in \mathbf{C}_p \mid |x - 1| < 1\}$; in this set $p - 1$ elements are of order $1/(p-1)$, $p(p-1)$ elements are of order $1/p(p-1)$, $\ldots$, $p^{s-1}(p-1)$ elements are of order $1/p^{s-1}(p-1), \ldots$.

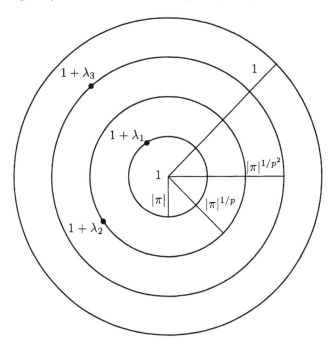

CHAPTER II

ZETA FUNCTIONS

1. Logarithms.

We recall that $\mathbf{C}_p$ denotes the completion of the algebraic closure of $\mathbf{Q}_p$ with respect to the p-adic valuation which we denote $|-|$ for simplicity. Moreover we have $|p| = p^{-1}$, and we normalize the associated order function as follows:

$$\operatorname{ord} x = -\frac{\log |x|}{\log p} = -\log_p |x| \ .$$

We find it convenient to fix an element $\pi \in \mathbf{C}_p$ satisfying

$$\pi^{p-1} = -p \ .$$

Consider the formal power series

$$\log (1 + X) = X - \frac{X^2}{2} + \frac{X^3}{3} + \cdots = \sum_{n=1}^{\infty} (-1)^{n-1} \frac{X^n}{n} \ ; \qquad (1.1)$$

this series is convergent for values $x \in \Omega$ such that

$$\lim_{n \to \infty} \frac{|x|^n}{|n|} = 0 \ .$$

Since $\operatorname{ord} n \leq \log n / \log p$ we have

$$\operatorname{ord} \left(\frac{x^n}{n}\right) = n \operatorname{ord} x - \operatorname{ord} n \geq n \operatorname{ord} x - \frac{\log n}{\log p} \ ,$$

and so if $\operatorname{ord} x > 0$ then $\operatorname{ord} (x^n/n) \to +\infty$ as $n \to \infty$, that is the series converges. If $\operatorname{ord} x \leq 0$ it is clear that the series is not convergent.

The following formal identity for power series in two variables X, Y holds

$$\log((1 + X)(1 + Y)) = \log(1 + X) + \log(1 + Y) . \qquad (1.2)$$

In fact, to prove this we must verify the following congruences for polynomials over $\mathbf{Q}$:

$$\sum_{i=1}^{N} \frac{(-1)^{i-1}}{i}(X + Y + XY)^i \equiv \sum_{i=1}^{N} \frac{(-1)^{i-1}}{i}(X^i + Y^i) \quad \mathrm{mod} \ \deg(N + 1)$$

for every integer $N \geq 1$. These formal congruences can be obtained by observing that for x, $y \in \mathbf{C}$ with say $|x| < 1/4$, $|y| < 1/4$ we have

$$\log\left((1 + x)(1 + y)\right) = \log(1 + x) + \log(1 + y) ; \qquad (1.3)$$

then use the series representation of both sides, and finally invoke the uniqueness of series representations.

If $|x| < 1$ and $|y| < 1$ then clearly both sides of (1.2) are convergent and the identity (1.3) also holds p-adically.

We wish to discuss the Newton polygon for the series (1.1), which is defined, in strict analogy with the polynomial case, to be the convex hull of the set

$$\{(n, \mathrm{ord} \ ((-1)^{n-1}/n)\}_{n \geq 1} \cup \{Y_{+\infty}\} .$$

Since for $p^j < n < p^{j+1}$ we have $\mathrm{ord}\, n \leq j$, while clearly $\mathrm{ord}\, p^j = j$, we see that the vertices are precisely the points $(p^j, -j)$ for $j = 0, 1, \ldots$. Thus the slopes of the finite sides are

$$\frac{-1}{p - 1}, \quad \frac{-1}{p(p - 1)}, \quad \frac{-1}{p^2(p - 1)}, \quad \cdots, \quad \frac{-1}{p^j(p - 1)}, \cdots$$

which obviously tend to 0 as $j \to \infty$. The following diagram illustrates the case $p = 2$.

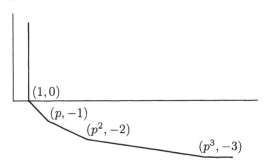

The validity of the y-intercept method (Section I.7) for λ not the slope of a side is trivial.

Proposition 1.1.   If ord $x > 1/(p-1)$ then ord $\log(1+x) = $ ord x. If

$$\frac{1}{p^s(p-1)} < \text{ord } x < \frac{1}{p^{s-1}(p-1)}$$

for some $s \geq 1$, then

$$\text{ord } \log(1+x) = p^s \text{ord } x - s \; .$$

If a theorem like Theorem I.6.1 also holds for power series then one expects that for each side of this polygon there will be as many roots x of $\log(1+X)$ as the length of the horizontal projection of the side, and that for all such x we will have ord $x = -\beta$ where β is the slope of the side. To verify that this is the case fix $s \geq 0$. We must then find $p^s(p-1)$ elements $x \in \mathbf{C}_p$ such that $\log(1+x) = 0$ and that ord $x = 1/p^s(p-1)$, that is $|x| = |\pi|^{1/p^s}$. This is a consequence of the discussion in Example I.9.3 and the following Proposition.

Proposition 1.2.   If $x \in \mathbf{C}_p$, with $|x| < 1$ and $\log(1+x) = 0$ then there exists an integer $l \geq 0$ such that

$$(1+x)^{p^l} = 1.$$

Conversely, if $x \in \mathbf{C}_p$ is such that $(1+x)^{p^l} = 1$ then $|x| < 1$ and $\log(1+x) = 0$.

Proof.   We first prove that if $|\pi| \leq |x| < 1$ then there exists a positive integer l such that

$$\left| z^{p^l} - 1 \right| < |\pi|$$

where $z = 1 + x$. We have

$$z^p - 1 = (1+x)^p - 1 = x^p + \binom{p}{1} x^{p-1} + \cdots + \binom{p}{p-1} x \; .$$

From $|x| \geq |\pi|$ it follows that $|x|^i \geq |\pi|^i \geq |\pi|^{p-1} = |p|$, for every $i = 1, \ldots, p-1$, so that

$$|x|^p \geq |p||x|^{p-i} = \left| \binom{p}{i} x^{p-i} \right| \; .$$

We have proved that if $|\pi| \leq |z - 1| < 1$ then

$$|z^p - 1| \leq |z - 1|^p .$$

Hence by iteration the assertion holds if we choose l such that $|z-1|^{p^l} < |\pi|$.

Assume now $|x| < 1$ and $\log(1 + x) = 0$. If $|x| < |\pi|$ then $x = 0$ since by Proposition 1.1 we have $|\log(1 + x)| = |x|$. On the other hand, if $|\pi| \leq |x| < 1$ take $l > 0$ such that

$$\left| (1 + x)^{p^l} - 1 \right| < |\pi| .$$

Since we surely have $p^l \log(1 + x) = 0$ and so, by (1.3), $\log(1 + x)^{p^l} = 0$ we can again conclude that $(1 + x)^{p^l} = 1$.

Conversely, assume that $(1 + x)^{p^l} = 1$. Then $|x| < 1$ (cf. Example I.9.3) and $p^l \log (1 + x) = \log (1 + x)^{p^l} = \log 1 = 0$ so that $\log(1+x) = 0$. **Q.E.D.**

2. Newton Polygons for Power Series.

Our goal in this Section will be to generalize Theorem I.6.1 to power series as illustrated in the preceding Section.

Let K be a complete field and let Ω be an algebraically closed extension of K, complete with respect to a valuation extending that of K.

We recall that if

$$f(X) = \sum_{n=0}^{\infty} a_n X^n \qquad \text{with} \quad a_n \in K ,$$

the *radius of convergence* $R \in [0, +\infty]$ of f is defined by

$$\frac{1}{R} = \limsup_{n \to \infty} |a_n|^{1/n} .$$

The following characterization of R is easily checked (using the fact that G_Ω is dense in $\mathbf{R}_+$): for every $x \in \Omega$ such that $|x| < R$, the series $f(x)$ is convergent, while for $|x| > R$ it is not convergent. The additive version of this criterion is the following: let

$$-M = \liminf_{n \to \infty} \frac{\operatorname{ord} a_n}{n} ;$$

then $M \in [-\infty, +\infty]$ and might be called the *additive radius of convergence*. If $x \in \Omega$ satisfies ord $x > M$ then $f(x)$ is convergent, if ord $x < M$ then $f(x)$ is not convergent. We recall also that the series $f(x)$ converges if and and only if $|a_n||x|^n \to 0$ (or equivalently, ord $a_n + n$ ord $x \to +\infty$) as $n \to \infty$.

As in the case of polynomials, the Newton polygon of f is defined to be the convex hull of the set $\{(j, \text{ord } a_j)\}_{j \geq 0} \cup \{Y_{+\infty}\}$. If $a_j = 0$ then we set $(j, +\infty) = Y_{+\infty}$.

There is in any case a first infinite vertical side. There are three possible cases:

(i) there is no other side (e.g. $\sum_n p^{-n^2} X^n$),

(ii) there are $l + 1$ other sides, $l \geq 0$,

(iii) there is an infinite set of other sides.

In case (ii), l of the other sides are finite in length, the $(l+1)$-st is of infinite length. Each finite side has a first point and a last point. These are vertices of the polygon. The initial point $(j_1, \text{ord } a_{j_1})$ of the last side is again a vertex, but there need not be any other point $(j, \text{ord } a_j)$ on the infinite side. We must have ord $a_j \geq$ ord $a_{j_1} + \lambda(j - j_1)$ for all $j > j_1$, and

$$\liminf_{j \to \infty} \frac{\text{ord } a_j - \text{ord } a_{j_1}}{j - j_1} = \lambda .$$

In case (iii) the slopes of the sides form a strictly monotonic sequence. An example is given by $\log(1 + X)$.

At this point it is not difficult to prove the following useful formula for the additive radius of convergence of f:

$$M = -\sup\{\lambda \mid \lambda \text{ is a slope of the Newton polygon of } f\}.$$

As in the classical case, on the "circle of convergence" the series may or may not converge, as we see from the following two examples.

• Let
$$f(X) = 1 + X + X^2 + \cdots + X^n + \cdots .$$

The Newton polygon of f is the positive x-axis so the additive radius of convergence is 0, but for x a unit the series $f(x)$ does not converge.

• Let
$$f(X) = \sum_{n=0}^{\infty} p^n X^{2^n} .$$

The Newton polygon of this f is again the positive x-axis, but in this case when ord $x = 0$ we have convergence. More generally we have the same behavior for any series of the type $\displaystyle\sum_{n=0}^{\infty} p^{g(n)} X^n$ where $g(n)$ is a sequence satisfying $\displaystyle\lim_{n\to\infty} g(n) = +\infty$ and $\displaystyle\lim_{n\to\infty} \frac{g(n)}{n} = 0$.

As promised, we now prove the analogue of Theorem I.6.1 for power series. In the proof we use the spaces W_N introduced in the proof of Hensel's lemma (Theorem I.4.1).

Theorem 2.1. *Let K be a complete field and let $f = \displaystyle\sum_{n=0}^{\infty} a_n X^n \in K[[X]]$. Then to each finite side of the Newton polygon of f there correspond l zeros α (counting multiplicities) of f where l is the length of the horizontal projection of the side. Moreover, if λ is the slope of the side, then $\operatorname{ord}\alpha = -\lambda$. Conversely if α is a root then $-\operatorname{ord}\alpha$ is the slope of a (possibly infinite) side.*

Proof. If $\operatorname{ord}\alpha = -\lambda$ and λ is not the slope of a side then the slope intercept method shows that $\operatorname{ord} f(\alpha)$ is the y-intercept of the line of support of slope λ and hence $f(\alpha) \neq 0$. This proves the converse part of the theorem.

As to the direct part of the theorem, it suffices to prove the assertion in the following special situation: the Newton polygon of f contains a finite side with extreme vertices $(m, 0)$ and $(m + n, 0)$.

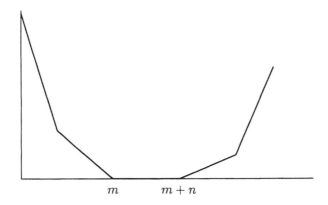

$$m \qquad m+n$$

In fact we can easily reduce to this situation as follows: We start from a side of slope λ whose horizontal projection has length l. We take a $\beta \in \Omega$ with $\operatorname{ord}\beta = -\lambda$. Then the series $f(\beta X)$ has a horizontal side of length l. Now the series $\delta f(\beta X)$, for a suitable $\delta \neq 0$, has a side on

the X-axis of length l. So, assume that we are in the special situation.
There are two cases to consider.

Case 1. The number of sides is infinite, so that we can find an infinite
sequence of vertices $(N, \operatorname{ord} a_N)$ from the Newton polygon of f with
$N > m + n$. Let $(N, \operatorname{ord} a_N)$ be one of these vertices, and take

$$f_N(X) = \sum_{n=0}^{N} a_n X^n$$

. Then in view of Theorem I.6.1 we can write $f_N(X) = P_N Q_N$ where

$$P_N = X^n + b_{n-1}^{(N)} X^{n-1} + \cdots + b_0^{(N)} \quad \text{with} \quad |b_i^{(N)}| \le 1 \quad \text{and} \quad |b_0^{(N)}| = 1 ,$$

$$Q_N = \gamma_N \left(X^m + \sum_{j=1}^{m} c_{m-j}^{(N)} X^{m-j} \right) \left(1 + \sum_{h=1}^{N-m-n} d_h^{(N)} X^h \right)$$

with $|c_j^{(N)}| < 1$ and $|d_h^{(N)}| < 1$. The polynomial P_N corresponds to
the horizontal side, the first factor of Q_N corresponds to the sides of
negative slope, and the second factor to sides of positive slope. Here γ_N
is a unit in $\mathcal{O}_K$ since $1 = |f_N|_{\text{gauss}} = |P_N|_{\text{gauss}} |Q_N|_{\text{gauss}} = |\gamma_N|$. Let
$(N', \operatorname{ord} a_{N'})$ be another vertex with $N' > N$. Then $f_{N'} = P_{N'} Q_{N'}$ as
above. We also have

$$f_{N'} - P_N(Q_N + a_{N'} X^{N'-n}) \in W_{N'} ,$$

where $W_{N'} = \{ f \in K[X] \mid \deg f \le N' - 1 \}$. The reduction mod
$\mathfrak{p}_K$ of $Q_N + a_{N'} X^{N'-n}$ is the polynomial $\overline{Q}_N = \overline{\gamma}_N X^m$ while $\overline{P}_N =$
$X^n + \cdots + b_0^{(N)}$. Since $b_0^{(N)} \neq 0$ we see that $\overline{P}_N$ and $\overline{Q}_N$ are relatively
prime. Now let us define

$$\varepsilon_N = \sup_{j \ge N} \{|a_j|\} < 1.$$

Then

$$|f_{N'} - P_N(Q_N + a_{N'} X^{N'-n})|_{\text{gauss}} \le \varepsilon_N .$$

Hensel's lemma gives a factorization $f_{N'} = \tilde{P}_{N'} \tilde{Q}_{N'}$ with $\tilde{P}_{N'}, \tilde{Q}_{N'} \in$
$\mathcal{O}_K[X]$ such that $\deg \tilde{P}_{N'} = n$, $\deg \tilde{Q}_{N'} = N' - n$, that $\tilde{P}_{N'} - P_N \in W_n$,
$\tilde{Q}_{N'} - (Q_N + a_{N'} X^{N'-n}) \in W_{N'-n}$ and that $|\tilde{P}_{N'} - P_N|_{\text{gauss}} \le \varepsilon_N$,
$|\tilde{Q}_{N'} - (Q_N + a_{N'} X^{N'-n})|_{\text{gauss}} \le \varepsilon_N$. From this we can deduce first

that $\tilde{P}_{N'}$ is monic and its roots are all units. Since the unit roots of $f_{N'}$ are the roots of $P_{N'}$ we obtain $\tilde{P}_{N'} = P_{N'}$ and then $\tilde{Q}_{N'} = Q_{N'}$. Since $\varepsilon_N \to 0$ for $N \to \infty$ we have that $|P_{N'} - P_N|_{\text{gauss}} \to 0$. Similarly

$$|Q_{N'} - (Q_N + a_{N'} X^{N'-n})|_{\text{gauss}} \leq \varepsilon_N,$$

so that $|Q_{N'} - Q_N|_{\text{gauss}} \leq \varepsilon_N$. We can now take

$$P = \lim_{N \to \infty} P_N \qquad Q = \lim_{N \to \infty} Q_N$$

with respect to the topology on $K[[X]]$ induced by the Gauss norm. Since the P_N are all monic polynomials of degree n, the same holds for P; on the other hand Q is a power series which is surely different from zero since the coefficient of X^m is a unit, this being the case for all of the Q_N. Taking the limit in $f_N = P_N Q_N$ we find that $f = PQ$. Observe that the Newton polygon of Q is obtained by "removing" the horizontal side from the Newton polygon of f, and so Q has the same radius of convergence as f. In particular Q is convergent for $|x| = 1$. Thus we have found at least n unit roots of f, namely the roots of P (counting multiplicities, if necessary). Now let α satisfy $|\alpha| = 1$ and $f(\alpha) = 0$. We wish to prove that $Q(\alpha) \neq 0$ so that $P(\alpha) = 0$. To see that $Q(\alpha) \neq 0$ it suffices to observe that $\overline{Q} = \overline{\gamma} X^m$ with $\gamma = \lim_{N \to \infty} \gamma_N$ which is a unit.

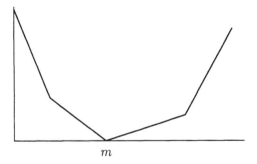

$$m$$

Case 2.   The number of sides is finite, so that there is a last side. Let λ be the slope of the last side, $(N_0, \text{ord } a_{N_0})$ the initial point of the last side, and λ' the slope of the next to last side. For the points $(j, \text{ord } a_j)$ lying on or above the Newton polygon, we have $\text{ord } a_j \geq \text{ord } a_{N_0} + \lambda(j - N_0)$ for all j. We now consider the partial sums f_N with $N > \max\{N_0, m + n\}$ and, to fix ideas, we may insist that $a_N \neq 0$ so that $\deg f_N = N$. The Newton polygon of f_N is not, as in Case 1, the truncation of the

Newton polygon of f at N. The Newton polygon of f_N can have several sides of slopes $> \lambda'$, but the Newton polygon of f_N lies on or above the truncation at N of the Newton polygon of f. Bearing this in mind, we can once again follow the same steps as in case 1 to obtain the polynomial P and the series Q. The Newton polygon of Q lies on or above the Newton polygon of f with its horizontal side removed. Then sup {slopes of the Newton polygon of Q} $\geq \lambda$ and we can conclude that the radius of convergence of Q is at least $-\lambda$ so that $f = PQ$, as in case 1. Then the radius of convergence of Q must be equal to $-\lambda$ and the Newton polygon of Q must be equal to the Newton polygon of f with its horizontal side removed.

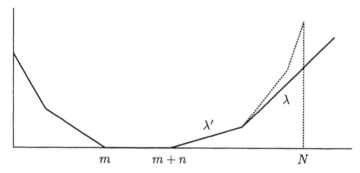

As a final remark we observe that if α is a root of P of multiplicity s, then it is of the same multiplicity as a zero of f (that is $f(\alpha) = f'(\alpha) = \cdots = f^{(s-1)}(\alpha) = 0$, while $f^{(s)}(\alpha) \neq 0$). This follows immediately from Leibniz's formula for the successive derivatives of a product and the fact that the various derived series have better convergence than their primitives. **Q.E.D.**

3. Newton Polygons for Laurent Series.

We now wish to extend our considerations on Newton polygons to the case of *Laurent series*, that is series of the form

$$f(X) = \sum_{n=-\infty}^{\infty} a_n X^n \quad \text{with} \quad a_n \in K \ .$$

We can write $f(X) = f^+(X) + f^-(1/X)$ where

$$f^+(X) = \sum_{n=0}^{+\infty} a_n X^n \quad \text{and} \quad f^-(X) = \sum_{n=1}^{+\infty} a_{-n} X^n .$$

We say that $f(X)$ converges at $x \neq 0$, $x \in \Omega$ if f^+ converges at x and f^- converges at $1/x$. This is is equivalent to saying that

$$\lim_{|n| \to \infty} (\text{ord } a_n + n \text{ ord } x) = +\infty .$$

If M^+ and M^- are the additive radii of convergence of f^+ and f^- respectively, then f converges for $\text{ord } x$ in $[-\infty, -M^-) \cap (M^+, +\infty]$. If $M^+ < -M^-$ then we have an *annulus of convergence*, and we will consider only this case.

The Newton polygon of a Laurent series has the same definition as in the case of a power series. In order to have an annulus of convergence we assume that $\lambda_1 = \inf\{\text{slopes}\} < \lambda_2 = \sup\{\text{slopes}\}$ and then we have convergence for $-\lambda_1 > \text{ord } x > -\lambda_2$.

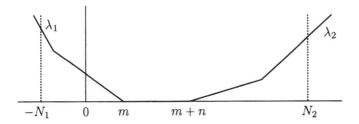

We now generalize Theorem 2.1 to the case of Laurent series. We give a sketch of the proof. As in the case of a power series, we first reduce to the case of a finite side on the x-axis. Using the notation introduced above, we can consider the truncated Laurent polynomial

$$f_{N_1, N_2} = \sum_{s=-N_1}^{N_2} a_s X^s$$

where N_1 and N_2 are positive integers with $N_1, N_2 \to \infty$. Now we write

$$X^{N_1} f_{N_1, N_2} = P_{N_1, N_2} \cdot Q_{N_1, N_2}$$

where

$$P_{N_1, N_2} = X^n + b_{n-1} X^{n-1} + \cdots + b_0$$

with $|b_0| = 1$ and $|b_i| \leq 1$ for $i = 1, 2, \ldots, n-1$, and

$$Q_{N_1, N_2} = \gamma_{N_1, N_2} R_{N_1, N_2} S_{N_1, N_2}$$

where

$$R_{N_1,N_2} = 1 + c_1 X + \cdots + c_{N_2-n-m} X^{N_2-n-m}$$

and

$$S_{N_1,N_2} = X^{N_1+m} + d_{N_1+m-1} X^{N_1+m-1} + \cdots + d_0 \, ,$$

with $|c_j| < 1$, $|d_l| < 1$. These conditions ensure that $|\gamma_{N_1,N_2}| = 1$, and that the reductions $\overline{P}_{N_1,N_2}$ and $\overline{Q}_{N_1,N_2}$ mod $\mathfrak{p}_K$ are relatively prime. Now take $N_1' > N_1$ and $N_2' > N_2$ and write, as above,

$$X^{N_1'} f_{N_1',N_2'} = P_{N_1',N_2'} \cdot Q_{N_1',N_2'}.$$

We then have

$$X^{N_1'} f_{N_1',N_2'} - P_{N_1,N_2} \left(Q_{N_1,N_2} X^{N_1'-N_1} + a_{N_2'} X^{N_1'+N_2'-n} \right) \in W_{N_1'+N_2'-n} \, ,$$

and

$$\left| X^{N_1'} f_{N_1',N_2'} - P_{N_1,N_2} \left(Q_{N_1,N_2} X^{N_1'-N_1} + a_{N_2'} X^{N_1'+N_2'-n} \right) \right| \le \varepsilon_{N_1,N_2} \, ,$$

where

$$\varepsilon_{N_1,N_2} \stackrel{\text{def}}{=} \sup_{\substack{j \ge N_2 \\ j \le -N_1}} \{|a_j|\} \to 0$$

as $N_1, N_2 \to \infty$. Now, the reduction modulo $\mathfrak{p}_K$ of the two polynomials P_{N_1,N_2} and $Q_{N_1,N_2} X^{N_1'-N_1} + a_{N_2'} X^{N_1'+N_2'-n}$ are the polynomials $\overline{P}_{N_1,N_2}$ and $X^{N_1'-N_1} \overline{Q}_{N_1,N_2}$ respectively, so that they are relatively prime (since $\overline{b}_0 \ne 0$). Thus, Hensel's lemma furnishes us a $\tilde{P}_{N_1',N_2'}$ and a $\tilde{Q}_{N_1',N_2'}$ such that $\deg \tilde{P}_{N_1',N_2'} = n$,

$$X^{N_1'} f_{N_1',N_2'} = \tilde{P}_{N_1',N_2'} \tilde{Q}_{N_1',N_2'} \, ,$$

and

$$|\tilde{P}_{N_1',N_2'} - P_{N_1,N_2}| \le \varepsilon_{N_1,N_2} \, ,$$

$$|\tilde{Q}_{N_1',N_2'} - X^{N_1'-N_1} Q_{N_1,N_2}| \le \varepsilon_{N_1,N_2}$$

so that $\tilde{P}_{N_1',N_2'}$ is monic and has only unit roots. Therefore $\tilde{P}_{N_1',N_2'} = P_{N_1',N_2'}$ and

$$|P_{N_1',N_2'} - P_{N_1,N_2}| \le \varepsilon_{N_1,N_2} \, .$$

We obtain similar results for the $X^{-N_1} Q_{N_1,N_2}$. Passing to the limit as N_1 and $N_2 \to \infty$ we obtain a polynomial P and a Laurent series Q such that $f = PQ$, and then one proceeds as in the proof of Theorem 2.1.

Remark 3.1. The restriction to finite sides is actually unnecessary. Suppose $\operatorname{ord} a_j \geq 0$ for all j, $\operatorname{ord} a_0 = 0 = \operatorname{ord} a_m$ and $\operatorname{ord} a_j > 0$ both for $j < 0$ and for $j > m$. Further suppose $|a_j| \to 0$ as $|j| \to \infty$. Then $f(X) = \sum_{j=-\infty}^{\infty} a_j X^j$ has precisely m unit roots.

The proof is as before except for one point. The Newton polygon of $X^{-N_1} Q_{N_1, N_2}$ coincides with that of f_{N_1, N_2} except that the horizontal side is excised. From this we deduce $Q(X) = \sum_s b_s X^s$ with $|b_m| = 1$, $|b_s| \to 0$ as $|s| \to \infty$ and $|b_s| < 1$ for all $s \neq m$.

4. The Binomial and Exponential Series.

Let $\alpha \in \mathbf{C}_p$. We define

$$(1 + X)^\alpha = \sum_{n=0}^{\infty} \binom{\alpha}{n} X^n , \tag{4.1}$$

where $\binom{\alpha}{0} = 1$ and

$$\binom{\alpha}{n} = \frac{\alpha(\alpha - 1) \cdots (\alpha - n + 1)}{n!} .$$

The problem of convergence for general α is rather delicate and will be treated in Section IV.7. However, by Proposition I.4.3,

$$\text{if} \quad \alpha \in \mathbf{Z}_p \quad \text{then} \quad \binom{\alpha}{n} \in \mathbf{Z}_p ,$$

so that in this case the series is convergent if $|x| < 1$, with $x \in \mathbf{C}_p$. Moreover

$$(1 + x)^\alpha (1 + x)^\beta = (1 + x)^{\alpha + \beta} \tag{4.2}$$

for every $\alpha, \beta \in \mathbf{Z}_p$ and every $x \in \mathbf{C}_p$ with $|x| < 1$. This identity follows from

$$\binom{\alpha + \beta}{s} = \sum_{i+j=s} \binom{\alpha}{i} \binom{\beta}{j}$$

which holds since (4.2) holds over the complex field. From (4.2), taking $\beta = n$ a positive integer, we immediately obtain

$$(1 + x)^{\alpha n} = ((1 + x)^\alpha)^n \tag{4.3}$$

for every $\alpha \in \mathbf{Z}_p$ and every $x \in \mathbf{C}_p$ with $|x| < 1$. More generally, using Lemma IV.7.1, it can be established that

$$((1+x)^\alpha)^\beta = (1+x)^{\alpha\beta} \tag{4.4}$$

for every $\alpha, \beta \in \mathbf{Z}_p$ and every $x \in \mathbf{C}_p$ with $|x| < 1$.

We have already studied $\log(1+X)$, and so we turn our attention to the exponential series. As usual we define $\exp X$ by

$$\exp X = \sum_{n=0}^{\infty} \frac{X^n}{n!} . \tag{4.5}$$

To clarify the question of convergence we must estimate $\operatorname{ord} n!$. As usual, for $x \in \mathbf{R}$ we let $[x]$ denote the greatest integer less than or equal to x, and if

$$n = a_0 + a_1 p + \cdots + a_l p^l$$

with $0 \le a_i \le p - 1$ for $i = 0, 1, \ldots, l$, is the expansion of the positive integer n in the base p we set

$$S_n = a_0 + a_1 + \cdots + a_l .$$

Then we see immediately that

$$[n/p] = a_1 + a_2 p + \cdots + a_l p^{l-1}$$
$$[n/p^2] = a_2 + a_3 p + \cdots + a_l p^{l-2}$$
$$\cdots$$
$$[n/p^l] = a_l$$
$$[n/p^{l+1}] = [n/p^{l+2}] = \cdots = 0 .$$

Clearly the number $[n/p^j]$ counts the number of multiples of p^j which are less than or equal to n, so that the sum $[n/p] + [n/p^2] + \cdots [n/p^l]$ counts all multiples of p which are less than or equal to n once, all multiples of p^2 twice, etc. Thus

$$\operatorname{ord} n! = [n/p] + [n/p^2] + \cdots + [n/p^l] .$$

Collecting the coefficients of $a_1, a_2, \ldots a_l$ now gives

$$\operatorname{ord} n! = a_1 + a_2(1 + p) + a_3(1 + p + p^2) + \cdots + a_l(1 + p + \cdots + p^{l-1})$$
$$= a_1 + a_2 \left(\frac{p^2 - 1}{p - 1} \right) + \cdots + a_l \frac{p^l - 1}{p - 1} ,$$

so that $(p-1)\text{ord}\,n! = a_1(p-1) + a_2(p^2-1) + \cdots + a_l(p^l-1)$. Now adding and subtracting a_0 gives

$$\text{ord}\,n! = \frac{n - S_n}{p-1}\,. \tag{4.6}$$

In particular, $\text{ord}\,p^s! = (p^s-1)/(p-1)$.

Now, for $x \in \mathbf{C}_p$ the series $\exp x$ is convergent if $\text{ord}\,(x^n/n!) \to \infty$ as $n \to \infty$, that is, if

$$n\,\text{ord}\,x - \frac{n}{p-1} + \frac{S_n}{p-1} = n\left(\text{ord}\,x - \frac{1}{p-1}\right) + \frac{S_n}{p-1} \to \infty\,.$$

Thus we have convergence for $\text{ord}\,x > 1/(p-1)$, or in other words, if $|x| < |\pi|$ where, as in Section 1, $\pi^{p-1} = -p$. We note that if $\text{ord}\,x = 1/(p-1)$ the series does not converge, since, for instance, the terms corresponding to $n = p^s$ for $s = 1, 2, \ldots$ all have order $1/(p-1)$.

The following lemma is an easy consequence of the preceding discussion.

Lemma 4.1. *Let $n \geq 2$ be an integer and let $x \in \mathbf{C}_p$ with $|x| < |\pi|$. Then*

$$\left|\frac{x^n}{n!}\right| < |x|\,.$$

Proof. Indeed, if $\text{ord}\,x > 1/(p-1)$, we have

$$\text{ord}\,(x^{n-1}/n!) = (n-1)\text{ord}\,x - (n - S_n)/(p-1) > (S_n - 1)/(p-1) \geq 0.$$

Q.E.D.

It follows that, for $|x| < |\pi|$,

$$\exp x = 1 + x + o(x) \qquad \text{where} \qquad \frac{|o(x)|}{|x|} < 1 \ .$$

Thus exp sends the disk $D(0, |\pi|^-)$ to the disk $D(1, |\pi|^-)$. On the other hand we have already seen that

$$\log(1 + X) = \sum_{n=0}^{\infty} (-1)^{n-1} \frac{X^n}{n}$$

converges for $|x| < |\pi|$, and since for $n \geq 2$ we have

$$\left| \frac{x^n}{n} \right| \leq \left| \frac{x^n}{n!} \right|$$

we see that log maps $D(1, |\pi|^-)$ into $D(0, |\pi|^-)$. Furthermore, reducing to formal polynomial identities as in the case of log one can establish that

$$\exp(x + y) = \exp x \exp y \qquad \text{for} \quad |x| < |\pi|, \ |y| < |\pi| \ ,$$

and that for $|x| < |\pi|$

$$\log \exp x = x \qquad \text{and} \quad \exp \log(1 + x) = 1 + x \ .$$

Hence the functions $\exp x$ and $\log(1 + x)$ are mutually inverse isomorphisms between the additive group $D(0, |\pi|^-)$ and the multiplicative group $D(1, |\pi|^-)$.

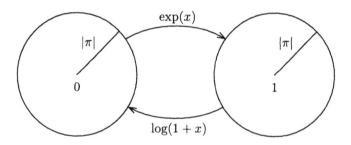

5. Dieudonné's Theorem.

We consider the completion T of the field $\mathbf{Q}_p(\zeta_{p^2-1}, \zeta_{p^3-1}, \ldots)$. The field T is an infinite unramified extension of $\mathbf{Q}_p$, and we have the Frobenius automorphism

$$\sigma : T \longrightarrow T$$

defined by the following property

$$\sigma\alpha \equiv \alpha^p \mod p\mathcal{O}_T$$

for every $\alpha \in T$ with $|\alpha| \leq 1$; recall also that $\sigma\zeta_{p^s-1} = \zeta_{p^s-1}^p$ (see (I.8.2)). The valuation on T remains discrete since T is unramified over $\mathbf{Q}_p$. Thus the maximal ideal of $\mathcal{O}_T$ is $p\mathcal{O}_T$.

Let us now consider

$$f(X) = 1 + a_1 X + a_2 X^2 + \cdots \in 1 + XT[[X]] \tag{5.1}$$

and define

$$f^\sigma(X) = 1 + (\sigma a_1)X + (\sigma a_2)X^2 + \cdots .$$

We can write

$$f(X) = \prod_{j=1}^{\infty}(1 + b_j X^j) , \tag{5.2}$$

where the $b_j \in \mathbf{Z}[a_1, \ldots, a_j]$. In fact we have the following lemma:

Lemma 5.1. *Assume that $a_1, a_2, a_3, \ldots$ in (5.1) are indeterminates over $\mathbf{Z}$. Then b_j in equation (5.2) is a uniquely determined polynomial in $\mathbf{Z}[a_1, a_2, \ldots, a_j]$ for $j = 1, 2, \ldots$. Conversely, if the b's in (5.2) are indeterminates then each a_j in (5.1) is a uniquely determined polynomial in $\mathbf{Z}[b_1, b_2, \ldots, b_j]$.*

Proof. To calculate a_n we need only the first n factors in the product of (5.2), so the result in this direction is clear. Conversely, we proceed by induction, and suppose that $b_1, \ldots, b_{n-1}$ have already been determined. Then $b_n = $ coefficient of X^n in

$$(1 + b_1 X)^{-1} \cdots (1 + b_{n-1}X^{n-1})^{-1}(1 + a_1 X + a_2 X^2 + \cdots) ,$$

that is in the series

$$(\sum_{k=0}^{\infty}(-b_1 X)^k) \cdots (\sum_{k=0}^{\infty}(-b_{n-1}X^{n-1})^k)(1 + a_1 X + a_2 X^2 + \cdots),$$

so that $b_n \in \mathbf{Z}[b_1, \ldots, b_{n-1}, a_1, \ldots, a_n]$ which, by the inductive hypothesis, equals $\mathbf{Z}[a_1, \ldots, a_n]$. **Q.E.D.**

The following result is due to Dieudonné:

Theorem 5.2. Let $f(X) = 1 + a_1 X + a_2 X^2 + \cdots \in 1 + XT[[X]]$. Then

$$f(X) \in 1 + X\mathcal{O}_T[[X]]$$

if and only if

$$\frac{f(X)^p}{f^{\sigma}(X^p)} \in 1 + pX\mathcal{O}_T[[X]]. \tag{5.3}$$

Proof. Assume $a_i \in \mathcal{O}_T$ for all i. Then by Lemma 5.1 the b_j in (5.2) all belong to $\mathcal{O}_T$, so to verify (5.3) it suffices to prove that each factor

$$\frac{(1 + b_n X^n)^p}{1 + \sigma b_n X^{pn}} \in 1 + pX\mathcal{O}_T[[X]] .$$

But

$$\frac{(1 + b_n X^n)^p}{1 + \sigma b_n X^{pn}} = \frac{1 + b_n^p X^{pn} + \sum_{i=1}^{p-1} \binom{p}{i}(b_n X^n)^i}{1 + \sigma b_n X^{pn}}$$

$$= \frac{1 + \sigma b_n X^{pn} + pX(\cdots)}{1 + \sigma b_n X^{pn}} ,$$

since $b_n \in \mathcal{O}_T$ so that $|b_n^p - \sigma b_n| \le |p|$. Clearly the last term in the preceding equations is an element of $1 + pX\mathcal{O}_T[[X]]$.

Conversely, assuming (5.3), we wish to prove that $|b_n| \le 1$ for all the b_n appearing in (5.2). We proceed by induction, and suppose that $b_1, \ldots, b_{n-1} \in \mathcal{O}_T$. Then

$$f(X) = (1 + b_1 X)(1 + b_2 X^2) \cdots (1 + b_{n-1} X^{n-1})g(X)$$

where $g(X) = (1 + b_n X^n)(1 + b_{n+1} X^{n+1}) \cdots$. Then we have

$$\frac{f(X)^p}{f^{\sigma}(X^p)} = \left[\prod_{j=1}^{n-1} \frac{(1 + b_j X^j)^p}{1 + \sigma b_j X^{pj}} \right] \frac{g(X)^p}{g^{\sigma}(X^p)} .$$

The left-hand side of this equation, and the first factor on the right are both elements of $1 + pX\mathcal{O}_T[[X]]$ which is a group under multiplication, and so also the last factor is in $1 + pX\mathcal{O}_T[[X]]$. Now

$$\frac{g(X)^p}{g^{\sigma}(X^p)} = \frac{((1 + b_n X^n)(1 + b_{n+1} X^{n+1}) \cdots)^p}{(1 + \sigma b_n X^{pn})(1 + \sigma b_{n+1} X^{p(n+1)}) \cdots}.$$

$$\equiv 1 \mod pX\mathcal{O}_T[[X]]$$

Hence the coefficient of X^n in $g(X)^p/g^\sigma(X^p)$ is congruent to $0 \bmod p$. But the only term contributing to this coefficient is $(1+b_nX^n)^p$ and this gives pb_nX^n. Therefore $pb_n \equiv 0 \mod p\mathcal{O}_T$, which is to say $b_n \in \mathcal{O}_T$, as desired. **Q.E.D.**

Here is an important application of Theorem 5.2.

Corollary 5.3. *Let*

$$E(X) = \exp(X + \frac{X^p}{p} + \frac{X^{p^2}}{p^2} + \cdots) \tag{5.4}$$

be the Artin-Hasse exponential *series. Then*

$$E(X) \in 1 + X\mathbf{Z}_p[[X]] \ .$$

Proof. Clearly $E(X) \in 1 + X\mathbf{Q}_p[[X]]$. Since the Frobenius automorphism σ is the identity on $\mathbf{Q}_p$ it suffices to prove that

$$\frac{E(X)^p}{E(X^p)} = \frac{\exp(pX + X^p + X^{p^2}/p + \cdots)}{\exp(X^p + X^{p^2}/p + \cdots)} = \exp(pX) = \sum_{n=0}^\infty \frac{(pX)^n}{n!}$$

is an element of $1+pX\mathbf{Z}_p[[X]]$. This is the result of a calculation: indeed, it suffices to prove that $\mathrm{ord}\,(p^n/n!) \geq 1$ for $n \geq 1$. Since the valuation is discrete, it suffices to prove that $\mathrm{ord}\,(p^n/n!) > 0$ for $n \geq 1$. Now, by (4.6), we have

$$\mathrm{ord}\,(p^n/n!) = n - \mathrm{ord}\,n! = n - \frac{n - S_n}{p - 1} > 0$$

since $S_n > 0$ and $n\left(1 - \dfrac{1}{p-1}\right) \geq 0$ for every positive prime. **Q.E.D.**

It follows that $E(x)$ is defined as a function for $|x| < 1$. One should note that if we let

$$f_1(x) = x + \frac{x^p}{p} + \frac{x^{p^2}}{p^2} + \cdots$$

and

$$f_2(y) = \exp y \ ,$$

then it is NOT true that $E(x) = (f_2 \circ f_1)(x)$ for all x with $|x| < 1$; that is, E is not the composite of f_2 and f_1 on the entire disk $|x| < 1$. However, on the smaller disk $|x| < |\pi|$ the composition is valid. The question of convergence of composition of power series will be discussed again in Lemma IV.7.1.

6. Analytic Representation of Additive Characters.

In this Section we give an analytic construction for an additive character on $\mathbf{F}_p$, or more generally, on $\mathbf{F}_{p^s}$. We recall that if Ω is a field, then an *additive character* of $\mathbf{F}_{p^s}$ in Ω is a map

$$\chi : \mathbf{F}_{p^s} \longrightarrow \Omega$$

such that

$$\chi(x + y) = \chi(x)\chi(y) \qquad \text{for all} \quad x, y \in \mathbf{F}_{p^s}.$$

We shall insist that Ω be a field of characteristic zero. It is well known that the set of additive characters of $\mathbf{F}_{p^s}$ is a multiplicative group of order p^s. The identity of the group is the *trivial character* which maps all elements to 1. If χ is a non-trivial character, then all other characters are of the form χ_α for $\alpha \in \mathbf{F}_{p^s}$, where $\chi_\alpha(z) = \chi(\alpha z)$.

There is a simple way to obtain a non-trivial character. Let $q = p^s$. We define the *Teichmüller representative* (or *lifting*) of an element $\overline{x} \in \mathbf{F}_q$: it is the unique element $x = \text{Teich}\,\overline{x} \in \mathcal{O}_{\mathbf{C}_p}$ whose reduction mod $\mathfrak{p}_{\mathbf{C}_p}$ is $\overline{x}$ and such that $x^q = x$. (Observe that if $\overline{x} \neq 0$ then this x is a $(q-1)^{\text{st}}$ root of unity.) We list some properties of Teich.
First,

$$\text{Teich}\,(\overline{xy}) = (\text{Teich}\,\overline{x})(\text{Teich}\,\overline{y}) \tag{6.1}$$

for all $x, y \in \mathbf{F}_q$.

Next, $\text{Teich}\,\overline{x} \in T$ where T is the maximal unramified extension of $\mathbf{Q}_p$ in $\mathbf{C}_p$, and, in particular, if $\overline{x} \in \mathbf{F}_p$ then $\text{Teich}\,\overline{x} \in \mathbf{Z}_p$.
We also have that

$$\text{Teich}\,(\overline{x} + \overline{y}) \equiv \text{Teich}\,\overline{x} + \text{Teich}\,\overline{y} \mod p\mathcal{O}_T. \tag{6.2}$$

Finally, if σ is the Frobenius automorphism of T over $\mathbf{Q}_p$, then from (I.8.2) it follows that

$$\sigma\,\text{Teich}\,\overline{x} = (\text{Teich}\,\overline{x})^p. \tag{6.3}$$

We now define

$$\chi : \mathbf{F}_p \longrightarrow \mathbf{C}_p$$

by setting

$$\chi(\overline{x}) = \zeta_p^{\text{Teich}\,\overline{x}}, \tag{6.4}$$

where ζ_p is a primitive p-th root of unity. The right-hand side is a well-defined element of $\mathbf{C}_p$ since $\zeta_p = 1 + \lambda_1$ with $|\lambda_1| < 1$ and Teich $\overline{x} \in \mathbf{Z}_p$. In this way we obtain an additive character since Teich $\overline{x}$ + Teich $\overline{y}$ = Teich$(\overline{x} + \overline{y}) + pz$ with $z \in \mathbf{Z}_p$, so that

$$\chi(\overline{x})\chi(\overline{y}) = \zeta_p^{\text{Teich }\overline{x} + \text{Teich }\overline{y}} = \zeta_p^{\text{Teich}(\overline{x}+\overline{y})+pz}$$
$$= \chi(\overline{x} + \overline{y})\zeta_p^{pz} = \chi(\overline{x} + \overline{y}) \ .$$

The last equality follows from $\zeta_p^{pz} = (\zeta_p^p)^z = 1$ (see (4.4)). It is clear that χ is non-trivial since Teich $\overline{x} \notin p\mathbf{Z}_p$ for $\overline{x} \neq 0$.

We now extend this construction to the field $\mathbf{F}_q$ with $q = p^s$. We define

$$\chi_s : \mathbf{F}_q \longrightarrow \mathbf{C}_p$$

by setting

$$\chi_s(\overline{x}) = \chi\left(\text{Teich Tr}_{\mathbf{F}_q/\mathbf{F}_p}\overline{x}\right) , \tag{6.5}$$

for every $\overline{x} \in \mathbf{F}_q$, where

$$\text{Tr}_{\mathbf{F}_q/\mathbf{F}_p}\overline{x} = \overline{x} + \overline{x}^p + \cdots \overline{x}^{p^{s-1}}$$

is the trace mapping. Since the trace mapping for finite fields is surjective we surely obtain a non-trivial character in this manner.

Since

$$\text{Teich Tr}_{\mathbf{F}_q/\mathbf{F}_p}\overline{x} = \sum_{i=0}^{s-1}(\text{Teich }\overline{x})^{p^i} + pz$$

with $z \in \mathbf{Z}_p$, we can write

$$\chi_s(\overline{x}) = \zeta_p^{\sum_{i=0}^{s-1}(\text{Teich }\overline{x})^{p^i}} \ .$$

Remark 6.1. The desirable equality

$$\zeta_p^{\sum_{i=0}^{s-1}x^{p^i}} = \prod_{i=0}^{s-1}\zeta_p^{x^{p^i}} ,$$

where $x = \text{Teich }\overline{x}$ does NOT hold, since x is not in general an element of $\mathbf{Z}_p$.

In view of the preceding Remark we now look for a function θ such that

- $\theta(x)$ is defined for $|x| \leq 1$,
- $\overline{x} \longrightarrow \prod_{i=0}^{s-1}\theta\left((\text{Teich }\overline{x})^{p^i}\right)$ is an additive character of $\mathbf{F}_q$.

There are many such functions but we seek the simplest. We recall that π is a fixed element in $\mathbf{C}_p$ satisfying $\pi^{p-1} = -p$.

Proposition 6.2. *Let*

$$\theta(X) = \exp\left(\pi(X - X^p)\right) . \tag{6.6}$$

(i) The series $\theta(x)$ is convergent if $|x| \leq 1 + \varepsilon$ for a suitable $\varepsilon > 0$.
(ii) $\theta(1)$ is a primitive p-th root of unity.
(iii) If $x \in \mathbf{C}_p$ is such that $x^{p^s} = x$, then

$$\theta(1)^{x + x^p + \cdots + x^{p^{s-1}}} = \theta(x)\theta(x^p) \cdots \theta(x^{p^{s-1}}) .$$

Proof. Since $\pi^p/p = -\pi$ we can write

$$E(\pi X) = \exp\left(\pi X + \frac{(\pi X)^p}{p} + \frac{(\pi X)^{p^2}}{p^2} + \cdots\right)$$

$$= \theta(X) \cdot \prod_{j=2}^{\infty} \exp\left(\frac{(\pi X)^{p^j}}{p^j}\right) ,$$

so that

$$\theta(X) = E(\pi X) \prod_{j=2}^{\infty} \exp\left(-\frac{(\pi X)^{p^j}}{p^j}\right) . \tag{6.7}$$

Let $b \in \mathbf{R}$, $b > 0$; we define

$$\mathcal{G}(b) = \left\{ f(X) \;\middle|\; \begin{array}{c} f(X) \in 1 + X\mathbf{C}_p[[X]] \\ f \text{ converges for ord } x > -b \\ |f(x) - 1| < 1 \end{array} \right\} ; \tag{6.8}$$

clearly, if $f(X) = 1 + a_1 X + a_2 X^2 + \cdots \in \mathbf{C}_p[[X]]$ then $f(X) \in \mathcal{G}(b)$ if and only if ord $a_j \geq jb$ for every $j \geq 1$.

We now prove that $\theta(X) \in \mathcal{G}\left(\frac{p-1}{p^2}\right)$. Since the product in (6.7) is convergent in the X-adic sense, it suffices to show that each factor is an element of $\mathcal{G}\left(\frac{p-1}{p^2}\right)$. Clearly $E(\pi X) \in \mathcal{G}\left(\frac{1}{p-1}\right) \subseteq \mathcal{G}\left(\frac{p-1}{p^2}\right)$ and $\exp\left(\frac{(\pi X)^{p^j}}{p^j}\right)$ converges for $p^j \text{ord}\,(\pi X) - j > 1/(p-1)$, that is for

$$\text{ord } x > -\frac{1}{p-1} + \frac{j + \dfrac{1}{p-1}}{p^j} .$$

The left-hand side is a decreasing function of $j \geq 2$, so that

$$\frac{1}{p-1} - \frac{j + \dfrac{1}{p-1}}{p^j} \geq \frac{1}{p-1} - \frac{2p-1}{p^2(p-1)} = \frac{(p-1)^2}{p^2(p-1)} = \frac{p-1}{p^2} ;$$

therefore $\exp\left(-\dfrac{(\pi X)^{p^j}}{p^j}\right) \in \mathcal{G}\left(\frac{p-1}{p^2}\right)$ for $j = 2, 3, \ldots$.

If we write

$$\theta(X) = 1 + C_1 X + C_2 X^2 + \cdots ,$$

then

$$\text{ord } C_j \geq j\frac{p-1}{p^2} \quad \text{for every} \quad j \geq 2 .$$

On the other hand, since

$$C_j = \frac{\pi^j}{j!} \quad \text{for} \quad 1 \leq j \leq p-1 ,$$

we have

$$\text{ord } C_j = \frac{j}{p-1} \quad \text{for} \quad 1 \leq j \leq p-1 ,$$

and, since

$$\theta(X) \equiv E(\pi X) \mod X^{p^2} ,$$

then

$$\text{ord } C_j \geq \frac{j}{p-1} \quad \text{for} \quad j \leq p^2 - 1 .$$

This proves statement (i).

We now claim that $\theta(1) \neq 1$. Since

$$\theta(1) = 1 + \pi + \sum_{j=2}^{\infty} C_j ,$$

it suffices to prove that $\text{ord } C_j > \dfrac{1}{p-1}$ if $j \geq 2$. This is clear for $j < p^2$. If $j \geq p^2$ we have

$$\text{ord } C_j \geq j\frac{p-1}{p^2} \geq p^2\frac{p-1}{p^2} = p - 1$$

which is $> \dfrac{1}{p-1}$ if $p \geq 3$. In the case of $p = 2$ we have only to check what happens for $j = 4$. Since, in this case, $\pi = -2$, we have

$$\exp\left(-2(X - X^2)\right) = \left(1 - 2X + \frac{4X^2}{2!} - \cdots\right)\left(1 + 2X^2 + \frac{4X^4}{2} + \cdots\right),$$

so that

$$C_4 = \frac{4}{2} + \frac{4}{2}\cdot 2 + \frac{16}{4!} = 4 + 2\cdot\frac{4}{3}$$

and ord $C_4 \geq 2 > 1$.

Similarly to what we have seen for the Artin-Hasse exponential function, although $\theta(x)$ converges for $|x| \leq 1 + \varepsilon$, it is not true that $\theta(x) = \exp\left(\pi(x - x^p)\right)$ when $|x| \geq 1$. However, this last equality does hold for $|x| < 1$ since for such x the series

$$\sum_{j=0}^{\infty} (\pi(X - X^p))^j\,\frac{1}{j!}$$

is convergent (see Lemma IV.7.1). If $|x| = 1$, it is possible to have $|x - x^p| = 1$ and for such x the series is not convergent. Therefore we have

$$\log\theta(x) = \pi(x - x^p) \quad \text{for} \quad |x| < 1\,.$$

On the other hand $|\theta(x) - 1| < 1$ if ord $x > -\dfrac{p-1}{p^2}$, so that $\log\theta(X)$ is a series which converges for ord $x > -\dfrac{p-1}{p^2}$. It follows that $\log\theta(x) = \pi(x - x^p)$ for $|x| < 1 + \varepsilon$, with $\varepsilon > 0$; from this equality we obtain $\log\theta(1) = 0$ so that $\theta(1)$ is a primitive p^l-th root of unity for some $l > 0$. Since $|\theta(1) - 1| = |\pi|$ we can conclude that $l = 1$. This proves statement (ii).

Now let $x \in \mathbf{C}_p$ be such that $x^{p^s} = x$. We prove that the equality in (iii) holds. We define the series

$$H(T) = \theta(Tx)\theta(Tx^p)\cdots\theta(Tx^{p^{s-1}})$$

which converges for $|t| < 1 + \varepsilon$. If $|t| < 1$ we have

$$\theta(tx^{p^i}) = \exp\left(\pi(tx^{p^i} - t^p x^{p^{i+1}})\right)$$

so that

$$
\begin{aligned}
H(t) &= \exp\left\{\pi\left[(tx) - (tx)^p + (tx^p) - (tx^p)^p + \cdots - (tx^{p^{s-1}})^p\right]\right\} \\
&= \exp\left(\pi(t - t^p)(x + x^p + \cdots + x^{p^{s-1}})\right) \\
&= \exp\left(\pi(t - t^p)\right)^{x + x^p + \cdots + x^{p^{s-1}}} \quad .
\end{aligned}
$$

The last equality follows from the fact that $x + x^p + \cdots + x^{p^{s-1}}$, being the trace over $\mathbf{Q}_p$ of x, is an element of $\mathbf{Z}_p$. Hence

$$
H(T) = \theta(T)^{x + x^p + \cdots + x^{p^{s-1}}}
$$

and, since both series converge for $|t| < 1 + \varepsilon$, this equality holds when specialized at $T = 1$. **Q.E.D.**

We now generalize the preceding construction. Let $q = p^l$. We define

$$
\theta_l(X) = \exp\left(\pi(X - X^q)\right) = \theta(X)\theta(X^p)\cdots\theta(X^{p^{l-1}}) \ .
$$

Since $\theta(X^{p^i}) \in \mathcal{G}\left((p-1)/p^2 p^i\right)$ for all i, we have $\theta_l \in \mathcal{G}\left((p-1)/pq\right)$. We can now define a non-trivial additive character

$$
\chi_s : \mathbf{F}_{q^s} \longrightarrow \mathbf{C}_p
$$

by setting

$$
\chi_s(\bar{z}) = \theta_l(z)\theta_l(z^q)\cdots\theta_l(z^{q^{s-1}}) \tag{6.9}
$$

where $z = \text{Teich}\,\bar{z}$.

7. Meromorphy of the Zeta Function of a Variety.

Let V be an algebraic set defined over $\mathbf{F}_q$, with $q = p^l$. For $s \in \mathbf{N}$, we denote by $V_s = V(\mathbf{F}_{q^s})$ the set of points of V rational over $\mathbf{F}_{q^s}$ and let $N_s = \# V_s$. We define the *zeta function* of V to be

$$
\zeta(V, t) = \exp\left(\sum_{s=1}^{\infty} N_s \frac{t^s}{s}\right) \ . \tag{7.1}
$$

We will consider the special case of

$$V = \{x \in \mathbf{F}_{q^\infty}^n \mid f(x) = 0, \ x_1 x_2 \cdots x_n \neq 0\},$$

where $f \in \mathbf{F}_q[X_1, \ldots, X_n]$ and $\mathbf{F}_{q^\infty}$ denotes the algebraic closure of $\mathbf{F}_q$. The algebraic set V is an open set of the affine hypersurface $W = \{x \in \mathbf{F}_{q^\infty}^n \mid f(x) = 0\}$, precisely the set of the points of W which do not belong to the union of the coordinate hyperplanes. Clearly $V_s = \{x \in \mathbf{F}_{q^s} \mid f(x) = 0, \ x_1 x_2 \cdots x_n \neq 0\}$.

We will show that $\zeta(V, t) \in \mathbf{Q}(t)$, i.e., it is a rational function. From this, using a combinatorial argument, one can obtain the same result for W and then for an arbitrary algebraic set over $\mathbf{F}_q$, see Dwork [1].

We begin with a simple observation. As a consequence of Galois theory we have

Proposition 7.1.   *With the notation given above we have*

$$\zeta(V, t) \in \mathbf{Z}[[t]] .$$

Proof.   Let

$$\sigma : x \longrightarrow x^q$$

be the Frobenius automorphism of $\mathbf{F}_{q^\infty}/\mathbf{F}_q$. The automorphism σ acts naturally on V: if $x = (x_1, \ldots, x_n)$ is a point of V then $\sigma x = x^q = (x_1^q, \ldots, x_n^q)$ is again a point of V. The orbits of this action are finite sets and give a partition of V. Let $x = (x_1, \ldots, x_n)$ be a point of V and let l be the length of the orbit $\{x, x^q, \ldots, x^{q^{l-1}}\}$ of x. Then $\mathbf{F}_q(x_1, \ldots, x_n) = \mathbf{F}_{q^l}$ and $x \in V_s$ if and only if $l \mid s$. If we let $D_l = \#$ of orbits of length l, then,

$$N_s = \#V_s = \sum_{l \mid s} l D_l .$$

Now

$$\sum_{s=1}^{\infty} N_s \frac{t^s}{s} = \sum_{s=1}^{\infty} \frac{t^s}{s} \left(\sum_{l \mid s} l D_l \right) = \sum_{l=1}^{\infty} D_l \left(\sum_{j=1}^{\infty} l \frac{t^{lj}}{lj} \right)$$

$$= \sum_{l=1}^{\infty} D_l \left(-\log(1 - t^l) \right) = \log \prod_{l=1}^{\infty} \left(\frac{1}{1 - t^l} \right)^{D_l} ,$$

so that

$$\zeta(V,t) = \prod_{l=1}^{\infty} \left(\frac{1}{1-t^l}\right)^{D_l} = 1 + a_1 t + a_2 t^2 + \cdots$$

where the a_i are positive integers.   **Q.E.D.**

Remark 7.2.   Since $N_s \le q^{sn} = \#\mathbf{F}_{q^s}^n$, the series $\zeta(V,t)$ is dominated (in the sense of Cauchy, cf. Appendix III) by the series

$$\exp\left(\sum_{s=1}^{\infty} \frac{(q^n t)^s}{s}\right) = \exp\left(\log\frac{1}{1-q^n t}\right) = \frac{1}{1-q^n t} = \sum_{j=0}^{\infty} q^{nj} t^j \ ,$$

so that

$$a_j \le q^{nj}$$

for every $j \ge 0$ where a_j is as in the proof of the preceding proposition.

Our aim is now to find a useful formula for the numbers N_s. In order to do so we note first that if χ_s is the additive character defined in (6.9), we have

$$\sum_{x_0 \in \mathbf{F}_{q^s}} \chi_s\left(x_0 f(x_1, x_2, \ldots, x_n)\right) = \begin{cases} 0 & \text{if} \quad f(x_1, \ldots, x_n) \ne 0 \\ q^s & \text{if} \quad f(x_1, \ldots, x_n) = 0. \end{cases}$$

Thus

$$q^s N_s = \sum_{\substack{x_0 \in \mathbf{F}_{q^s} \\ (x_1, \ldots, x_n) \in \left(\mathbf{F}_{q^s}^{\times}\right)^n}} \chi_s\left((x_0 f(x_1, x_2, \ldots, x_n)\right)$$

$$= (q^s - 1)^n + \sum_{(x_0, \ldots, x_n) \in \left(\mathbf{F}_{q^s}^{\times}\right)^{n+1}} \chi_s\left(x_0 f(x_1, \ldots, x_n)\right) \ . \tag{7.2}$$

Now we suppose that f has degree d and set

$$X_0 f(X_1, \ldots, X_n) = \sum_u \overline{a}_u X^u \tag{7.3}$$

where $\overline{a}_u \in \mathbf{F}_q$ and the $u = (u_0, u_1, \ldots, u_n)$ are multi-indices with $u_0 = 1$ and $d \ge u_1 + \cdots + u_n$. Moreover we set $X^u = X_0^{u_0} X_1^{u_1} \cdots X_n^{u_n}$. If $\overline{x} = (\overline{x}_0, \ldots, \overline{x}_n) \in \left(\mathbf{F}_{q^s}^{\times}\right)^{n+1}$ we then have

$$\chi_s(\overline{x}_0 f(\overline{x})) = \chi_s\left(\sum_u \overline{a}_u \overline{x}^u\right) = \prod_u \chi_s(\overline{a}_u \overline{x}^u) \ .$$

If we now put $a_u = \text{Teich}\,\bar{a}_u$ and

$$x = (x_0, x_1, \ldots, x_n) = \text{Teich}\,\bar{x} = (\text{Teich}\,\bar{x}_0, \text{Teich}\,\bar{x}_1, \ldots, \text{Teich}\,\bar{x}_n)$$

then from (7.3) and (6.9) we obtain

$$\chi_s(\bar{x}_0 f(\bar{x})) = \prod_u \left(\theta_l(a_u x^u) \theta_l(a_u^q x^{qu}) \cdots \theta_l(a_u^{q^{s-1}} x^{q^{s-1}u}) \right)$$

$$= \prod_u \theta_l(a_u x^u) \prod_u \theta_l(a_u x^{qu}) \cdots \prod_u \theta_l(a_u x^{q^{s-1}u})$$

since $a_u^q = a_u$.

We define

$$F(X) = \prod_u \theta_l(a_u X^u) = \sum_{\substack{v = (v_0, \ldots, v_n) \\ d v_0 \geq v_1 + \cdots v_n}} A_v X^v .$$

Since $\theta_l \in \mathcal{G}\left(\frac{p-1}{pq}\right)$, whenever $v_1 + \cdots + v_n \leq d v_0$ we find, after a moment's reflection, that

$$\text{ord}\, A_v \geq v_0 \kappa \qquad (7.4)$$

where

$$\kappa = \frac{p-1}{pq} .$$

Therefore $F(X)$ and also $F(X)F(X^q) \cdots F(X^{q^{s-1}})$ converge for

$$|x| \overset{\text{def}}{=} \sup\{|x_1|, \cdots, |x_n|\} < 1 + \varepsilon$$

with a suitable positive ε. Then

$$\chi_s\left(\bar{x}_0 f(\bar{x})\right) = F(x)F(x^q) \cdots F(x^{q^{s-1}})$$

and finally from (7.2) it follows that

$$q^s N_s = (q^s - 1)^n + \sum_{x^{q^s-1}=1} F(x)F(x^q) \cdots F(x^{q^{s-1}}) . \qquad (7.5)$$

Before proceeding we need some algebraic considerations. Let T be a field of characteristic 0, and consider the T-vector space $W =$

$T[X_0, X_1, \ldots, X_n]$. For $H \in W$ we indicate by the same letter H the multiplication by H in W:

$$H : W \longrightarrow W$$
$$\xi \longmapsto \xi H \; ;$$

obviously H is a T-linear map.

Let

$$\psi_q : W \longrightarrow W$$

be the unique T-linear map such that

$$\psi_q(X^v) = \begin{cases} 0 & \text{if } q \nmid v \\ X^{v/q} & \text{otherwise,} \end{cases}$$

where $v = (v_0, v_1, \ldots, v_n) \in \mathbf{N}^{n+1}$ and $q \mid v$ means $q \mid v_i$ for all i. A more intrinsic definition of ψ_q is the following: if $\xi \in W$

$$(\psi_q \xi)(x) = \frac{1}{q^{n+1}} \sum_{y^q = x} \xi(y).$$

Let R be positive real number and define

$$W_R' = \{\xi \in W \mid \xi = \sum_{\substack{u_0 \le R \\ d u_0 \ge u_1 + \cdots + u_n}} A_u X^u\}. \tag{7.6}$$

Now we fix a positive integer N and suppose that $H \in W_N'$. The composed map $\psi_q \circ H$

$$W_R' \longrightarrow W_{R+N}' \longrightarrow W_{\frac{R+N}{q}}'$$

is an endomorphism of the T-vector space W_R' if R is sufficiently large. In fact if we assume $R > N/(q-1)$ then we have $(R+N)/q < R$ and $W_{(R+N)/q}' \subseteq W_R'$. We denote by $\operatorname{Tr} \phi$ the trace of a vector-space endomorphism ϕ.

Lemma 7.3. *In the preceding notations we have*

$$(q-1)^{n+1}\mathrm{Tr}\,(\psi_q \circ H)|_{W'_R} = \sum_{x^{q-1}=1} H(x)\,.$$

Proof. Since both members are linear in H we can assume H to be a monomial. Let $H = X^u$ with $u = (u_0, u_1, \ldots, u_n)$ such that $u_0 \le R$, $du_0 \ge u_1 + \cdots + u_n$; then we have

$$\sum_{x^{q-1}=1} x^u = \begin{cases} 0 & \text{if } (q-1)\nmid u \\ (q-1)^{n+1} & \text{otherwise} \end{cases}$$

since $x \longmapsto x^u$ is a non-trivial multiplicative character if $(q-1)\nmid u$. On the other hand, for fixed u we have that $(\psi_q \circ X^u)X^v = \psi_q X^{u+v}$ is zero or a monomial for every v. To calculate $\mathrm{Tr}\,(\psi_q \circ X^u)$ we need only the v's such that $(\psi_q \circ X^u)X^v = X^v$, that is such that $u + v = qv$. It follows that

$$\mathrm{Tr}\,(\psi_q \circ X^u) = \begin{cases} 0 & \text{if } (q-1)\nmid u \\ 1 & \text{otherwise}\,, \end{cases}$$

as desired. **Q.E.D.**

From this Lemma it follows in particular that $\mathrm{Tr}\,(\psi_q \circ H)|_{W'_R}$ does not depend on R for large R. (More generally it is possible to prove that the characteristic polynomial of $(\psi_q \circ H)|_{W'_R}$ does not depend on R.)

Now, let s be a positive integer; since $H(X)H(X^q)\cdots H(X^{q^{s-1}})$ is an element of $W'_{N(1+q+\cdots+q^{s-1})}$ and $\left(R + N(1+q+\cdots+q^{s-1})\right)/q^s <$ R provided that $R > N/(q-1)$, the map $\psi_{q^s}\circ H(X)H(X^q)\cdots H(X^{q^{s-1}})$ is an endomorphism of W'_R for R sufficiently large independently of s. Since

$$\psi_q \circ H(X^q) = H(X) \circ \psi_q\,,$$

as is easily verified, we have

$$\begin{aligned} \psi_{q^2} \circ H(X)H(X^q) &= \psi_q \circ \psi_q \circ H(X^q) \circ H(X) \\ &= \psi_q \circ H(X) \circ \psi_q \circ H(X) \\ &= (\psi_q \circ H(X))^2 \end{aligned}$$

and, inductively, for $s > 2$

$$\psi_{q^s} \circ H(X)H(X^q)\cdots H(X^{q^{s-1}}) = (\psi_q \circ H(X))^s\,.$$

Then, from Lemma 7.3, it follows that

$$(q^s - 1)^{n+1} \mathrm{Tr}\, (\psi_q \circ H(X))\big|^s_{W'_R} = \sum_{x^{q^s-1}=1} H(x)H(x^q)\cdots H(x^{q^{s-1}}) \quad (7.7)$$

and again $\mathrm{Tr}\, (\psi_q \circ H(X))\big|^s_{W'_R}$ is independent of R.

We define

$$Z(t) = \exp\left(\sum_{s=1}^{\infty} \frac{t^s}{s} \sum_{x^{q^s-1}=1} F(x)F(x^q)\cdots F(x^{q^{s-1}})\right). \quad (7.8)$$

Let

$$F_N(X) = \sum_{\substack{v_0 \leq N \\ d v_0 \geq v_1 + \cdots v_n}} A_v X^v.$$

Then, since the series $F(X)F(X^q)\cdots F(X^{q^{s-1}})$ converges for $|x| < 1+\varepsilon$, by (7.7) we obtain

$$\sum_{x^{q^s-1}=1} F(x)F(x^q)\cdots F(x^{q^{s-1}})$$

$$= \lim_{N\to\infty} \sum_{x^{q^s-1}=1} F_N(x)F_N(x^q)\cdots F_N(x^{q^{s-1}}) \quad (7.9)$$

$$= \lim_{N\to\infty} (q^s - 1)^{n+1} \mathrm{Tr}\, (\psi_q \circ F_N)^s,$$

so that

$$Z(t) = \lim_{N\to\infty} \exp\left(\sum_{s=1}^{\infty} \frac{t^s}{s}(q^s - 1)^{n+1} \mathrm{Tr}\, (\psi_q \circ F_N)^s\right),$$

where the limit is taken "coefficient by coefficient" (i.e., identifying $C_p[[t]]$ with C_p^N endowed with the product topology).

Now, for any $h(t) \in C_p[[t]]$, we define

$$h^\phi(t) = h(qt).$$

Consider

$$h_N(t) = \exp\left(\sum_{s=1}^{\infty} \frac{t^s}{s} \mathrm{Tr}\, (\psi_q \circ F_N)^s\right);$$

then we have

$$\exp\left(\sum_{s=1}^{\infty}\frac{t^s}{s}(q^s-1)\mathrm{Tr}\,(\psi_q\circ F_N)^s\right) = \frac{h_N(qt)}{h_N(t)} = h_N^{\phi-1}(t)$$

and

$$\exp\left(\sum_{s=1}^{\infty}\frac{t^s}{s}(q^s-1)^{n+1}\mathrm{Tr}\,(\psi_q\circ F_N)^s\right) = h_N^{(\phi-1)^{n+1}}(t)\ .$$

On the other hand, we recall the following elementary fact:

Proposition 7.4. *If f is an endomorphism of a finite dimensional vector space over a field then*

$$\exp\left(\sum_{s=1}^{\infty}\frac{t^s}{s}\mathrm{Tr}\,(f^s)\right) = (\det(I - tf))^{-1}\ ,$$

where I is the identity map.

Proof.   Consider the Jordan canonical form for f.   **Q.E.D.**

Returning to our situation, we have

$$h_N(t) = \exp\left(\sum_{s=1}^{\infty}\frac{t^s}{s}\mathrm{Tr}\,(\psi_q\circ F_N)^s\right) = (\det(I - t(\psi_q\circ F_N)))^{-1}\ ,$$

so that

$$Z(t) = \Delta(t)^{-(\phi-1)^{n+1}} = \prod_{i=0}^{n+1}\Delta(q^i t)^{(-1)^{n-i}\binom{n+1}{i}} \tag{7.10}$$

where

$$\Delta(t) = \lim_{N\to\infty}\det(I - t(\psi_q\circ F_N))\ . \tag{7.11}$$

Our next step will be to prove that $\Delta(t)$ is a p-adic *entire* function, that is an everywhere convergent power series. From this result and (7.10) it will follow that $Z(t)$ is a p-adic *meromorphic* function, that is a quotient of two p-adic entire functions. Now, from (7.7) and (7.9) it follows that

$$\zeta(V, qt) = \exp\left(\sum_{s}q^s\frac{N_s t^s}{s}\right)$$

$$= \exp\left(\sum_{s}(q^s-1)^n\frac{t^s}{s}\right)\lim_{N\to\infty}\exp\left(\sum_{s=1}^{N}(q^s-1)^{n+1}\frac{t^s}{s}\mathrm{Tr}\,(\psi_q\circ F_N)^s\right)$$

$$= \exp\left(\sum_{s}(q^s-1)^n\frac{t^s}{s}\right)Z(t)\ .$$

Moreover, since

$$\exp\left(\sum_{s=1}^{\infty}(q^s-1)^n\frac{t^s}{s}\right) = \left(\exp\left(\sum_{s=1}^{\infty}\frac{t^s}{s}\right)\right)^{(\phi-1)^n} = \frac{1}{(1-t)^{(\phi-1)^n}},$$

we obtain

$$\zeta(V,qt) = \frac{1}{(1-t)^{(\phi-1)^n}}Z(t),$$

so that also $\zeta(V,t)$ will be a p-adic meromorphic function.

We now prove that $\Delta(t)$ is a p-adic entire function. We set

$$\Delta_N(t) = \det(I - t(\psi \circ F_N)).$$

We then have

$$\Delta(t) = \lim_{N\to\infty}\Delta_N(t).$$

Let L_s be the number of points $u \in \mathbf{N}^{n+1}$ such that $du_0 \geq u_1 + \cdots + u_n$ and consider the polygon with vertices at $(L_0, 0)$, $(L_0+L_1, (q-1)\kappa L_1)$, $\ldots$, $(\sum_{i=0}^{t}L_i, (q-1)\kappa\sum_{i=0}^{t}iL_i)$, $\ldots$.

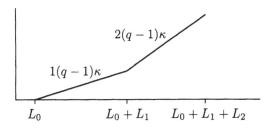

We claim that the Newton polygon of $\Delta(t)$ lies on or above the polygon just constructed. It suffices to establish this for each $\Delta_N(t)$. We consider the endomorphism $\psi_q \circ F_N$ of the space $W'_{N/(q-1)}$ defined in (7.6). We take a basis for $W'_{N/(q-1)}$ of the form $\{\alpha_{v_0}X^v\}_{dv_0 \geq v_1+\cdots v_n}$ for a suitable choice of $\alpha_{v_0} \in \mathbf{C}_p$. We then have

$$\psi_q F_N(a_{u_0}X^u) = \alpha_{u_0}\sum_{v}\psi_q(A_vX^{v+u})$$

$$= \alpha_{u_0}\sum_{w}X^wA_{qw-u}$$

$$= \sum_{w}(X^w\alpha_{w_0})A_{qw-u}\frac{\alpha_{u_0}}{\alpha_{w_0}}.$$

Since by (7.4) we have ord $A_{qw-u} \geq (qw_0 - u_0)\kappa$, if we put

$$\alpha_{u_0} = p^{\kappa u_0}$$

we obtain

$$\text{ord}\left(A_{qw-u}\frac{\alpha_{u_0}}{\alpha_{w_0}}\right) \geq (qw_0 - u_0)\kappa + \kappa u_0 - \kappa w_0 = (q-1)w_0\kappa.$$

Thus the matrix B of $\psi_q \circ F_N$ has entries $B_{u,w}$ satisfying

$$\text{ord } B_{u,w} \geq (q-1)w_0\kappa.$$

We now consider the coefficient β_l of t^l in $\det(I - tB)$. Let $L_0 + \ldots + L_\nu \leq l \leq L_0 + \ldots + L_{\nu+1}$. The coefficient β_l is the sum of all products of the form

$$(\text{sgn } \sigma) \, B_{u^{(1)}, u(\sigma(1))} B_{u^{(2)}, u(\sigma(2))} \cdots B_{u^{(l)}, u(\sigma(l))}$$

where $u^{(1)}, \ldots, u^{(l)}$ is any choice of l distinct indices u and σ is any permutation of $(1, 2, \ldots, l)$. It follows that

$$\text{ord } \beta_l \geq (q-1)\kappa(u_0^{(1)} + \cdots + u_0^{(l)})$$
$$\geq (q-1)\kappa\left(\sum_{s=0}^{\nu} sL_s + (l - L_0 - L_1 - \cdots - L_\nu)(\nu+1)\right),$$

since in the set of indices there are exactly L_0 elements with $u_0 = 0$, L_1 with $u_0 = 1$, etc. This proves our assertion about the Newton polygon of $\Delta_N(t)$. Since, however, this estimate holds for all N it also holds for the limit $\Delta(t)$. From this we can conclude that the supremum of the slopes of the sides of the Newton polygon of $\Delta(t)$ is $+\infty$ so that $\Delta(t)$ is a p-adic entire function as desired.

Remark 7.5. From the relation

$$\zeta(qt) = \frac{1}{(1-t)^{(\phi-1)^n}} \frac{1}{\Delta^{(\phi-1)^{n+1}}}$$

we obtain

$$\zeta^\phi \cdot (1-t)^{(\phi-1)^n} = \Delta^{(1-\phi)^{n+1}(-1)^n}$$

whence

$$\Delta^{(1-\phi)^{n+1}} = \zeta^{(-1)^n\phi} \cdot (1-t)^{(1-\phi)^n},$$

so that

$$\Delta(t) = \zeta(t)^{(-1)^n \phi(1-\phi)^{-n-1}} \cdot (1-t)^{(1-\phi)^{-1}}$$
$$= \zeta(t)^{(-1)^n \phi \sum \binom{-n-1}{s} \phi^s (-1)^s} \cdot (1-t)^{1+\phi+\phi^2+\cdots}.$$

From the preceding discussion we can conclude that the right side of this equality is an entire function (clearly $(1-t)^{1+\phi+\phi^2+\cdots}$ is a p-adic entire function). One might try to invert the reasoning and take the last equality as a definition of the function $\Delta(t)$, and ask whether it is possible to derive the entireness of $\Delta(t)$ from the properties of $\zeta(t)$. For example, following the approach of Grothendieck and Deligne, do the properties of $\zeta(t)$ embodied in the Weil conjectures imply that $\Delta(t)$ is entire?

In the particular case in which V' is the projective hypersurface defined by $g(y) = 0$ with $g(y_0, y_1, \ldots, y_n) = y_0^d \overline{f}(y_1/y_0, y_2/y_0, \ldots, y_n/y_0)$, if we make the assumptions that V' is smooth and in general position then Δ entire follows from the Weil conjectures. For more details see Dwork [2].

8. Condition for Rationality.

Theorem 8.1. *Let* $f(t) = \sum_j A_j t^j$ *be a power series with coefficients in an arbitrary field* Ω. *Let*

$$\Lambda_{s,k} = \det \begin{pmatrix} A_s & A_{s+1} & \cdots & A_{s+k} \\ A_{s+1} & A_{s+2} & \cdots & A_{s+k+1} \\ \vdots & \vdots & & \vdots \\ A_{s+k} & A_{s+k+1} & \cdots & A_{s+2k} \end{pmatrix}$$

Then the series $f(t)$ *represents a rational function if and only if there is a positive integer* k *such that* $\Lambda_{s,k} = 0$ *for all* $s \geq s_0$.

Proof. Suppose first that

$$\left(\sum_{i=0}^{\infty} A_i t^i \right) \left(\sum_{j=0}^{k} B_j t^j \right)$$

is a polynomial, and at least one $B_j \neq 0$. Then

$$A_s B_k + A_{s+1} B_{k-1} + \cdots + A_{s+k} B_0 = 0$$

for s sufficiently large, which gives the desired necessity of our condition. To prove sufficiency, we first establish a lemma in linear algebra.

Lemma 8.2.   *Let U and V be vector spaces over a field k and let $v_1, v_2, \ldots, v_n$ be linearly dependent elements of V. Suppose further that ϕ and ψ are linear maps of V into U such that $\psi v_{i-1} = \phi v_i$ for $i = 2, 3, \ldots, n$. If $\phi(v_1), \phi(v_2), \ldots, \phi(v_{n-1})$ are linearly dependent, then $\phi(v_2), \phi(v_3), \ldots, \phi(v_n)$ are also linearly dependent.*

Proof.   Suppose that $\phi(v_2), \phi(v_3), \ldots, \phi(v_n)$ are linearly independent. Then surely $\phi(v_2), \phi(v_3), \ldots, \phi(v_{n-1})$ are linearly independent, and so in the assumed relation of linear dependence for $\phi(v_1), \phi(v_2), \ldots, \phi(v_{n-1})$ the coefficient of $\phi(v_1)$ must be non-zero. Hence

$$\phi(v_1) \in \langle \phi(v_2), \phi(v_3), \ldots, \phi(v_{n-1}) \rangle \ .$$

Thus if

$$a_1 v_1 + a_2 v_2 + \cdots + a_n v_n = 0 \tag{8.1}$$

is any non-trivial dependence relation between the v's we find that

$$a_1 \phi(v_1) + \cdots + a_{n-1} \phi(v_{n-1}) = -a_n \phi(v_n)$$

and the left side lies in $\langle \phi(v_2), \ldots, \phi(v_{n-1}) \rangle$. Hence, since we have assumed that the vectors $\phi(v_2), \phi(v_3), \ldots, \phi(v_n)$ are linearly independent we must have $a_n = 0$, and so the non-trivial dependence relation (8.1) must take the form

$$a_1 v_1 + a_2 v_2 + \cdots + a_{n-1} v_{n-1} = 0 \ .$$

If we now apply ψ and use the condition $\psi(v_i) = \phi(v_{i+1})$ we find the non-trivial relation

$$a_1 \phi(v_2) + a_2 \phi(v_3) + \cdots + a_{n-1} \phi(v_n) = 0 \ ,$$

which gives a contradiction.   **Q.E.D.**

We now return to the proof of sufficiency. Assume that k is minimal with respect to the condition that there exists an s_0 such that $\Lambda_{s,k} = 0$ for all $s \geq s_0$. For $s \geq 0$, let

$$v_s^{(k)} = (A_s, A_{s+1}, \ldots, A_{s+k}) \in \Omega^{k+1} \ ,$$

$$W_s = \langle v_s^{(k)}, v_{s+1}^{(k)}, \ldots, v_{s+k}^{(k)} \rangle \ ,$$

$$U_s = \langle v_s^{(k)}, v_{s+1}^{(k)}, \ldots, v_{s+k-1}^{(k)} \rangle \ ,$$

so that $U_s \subseteq W_s \subseteq \Omega^{k+1}$ but note that $U_{s+1} \subseteq W_s$. By hypothesis $\dim W_s \leq k$ for all $s \geq s_0$ and k is minimal for which such a relation of linear dependence holds for all s sufficiently large.

We will show that

$$\dim U_s = k \qquad \text{for } s \geq s_0 . \tag{8.2}$$

From this we will deduce that

$$U_s \quad \text{is independent of } s \text{ for } s \geq s_0 . \tag{8.3}$$

We first show that (8.3) implies the rationality of f. Indeed, the $(k+1)$-tuple $z = (z_k, z_{k-1}, \ldots, z_0) \in \Omega^{k+1}$ acts on Ω^{k+1} by

$$(x_0, x_1, \ldots, x_k) \longmapsto z_k x_0 + z_{k-1} x_1 + \cdots + z_0 x_k .$$

Certainly there exists z which annihilates U_s for all $s \geq s_0$. Hence z annihilates $v_s^{(k)}$ for all $s \geq s_0$ and so

$$z_k A_s + z_{k-1} A_{s+1} + \cdots + z_0 A_{s+k} = 0$$

for every $s \geq s_0$ which shows that

$$f(t)(z_0 + z_1 t + \cdots + z_k t^k)$$

is a polynomial as asserted.

We now show that (8.2) implies (8.3).

Indeed, for $s \geq s_0$, U_s and U_{s+1} are k dimensional subspaces of W_s which is of dimension bounded by k. Hence $U_s = W_s$, $U_{s+1} = W_s$ which shows that $U_s = U_{s+1}$ for $s \geq s_0$.

Let $s' \geq s_0$. We will show that $\dim U_{s'} < k$ implies $\Lambda_{s,k-1} = 0$ for all $s \geq s'$, contradicting the minimality of k.

Indeed let

$$\phi v_s^{(k)} = v_s^{(k-1)} ,$$

$$\psi v_s^{(k)} = v_{s+1}^{(k-1)}$$

for every $s \geq 0$. If $\dim U_{s'} < k$ then $\left\{ v_{s'}^{(k)}, \ldots, v_{s'+k-1}^{(k)} \right\}$ is linearly dependent; hence $\left\{ \phi v_{s'}^{(k)}, \ldots, \phi v_{s'+k-1}^{(k)} \right\}$ is linearly dependent and so, by Lemma 8.2, $\left\{ \phi v_{s'+1}^{(k)}, \ldots, \phi v_{s'+k}^{(k)} \right\}$ is linearly dependent. Therefore, by recursion, $\left\{ \phi v_s^{(k)}, \ldots, \phi v_{s+k-1}^{(k)} \right\}$ is linearly dependent for $s \geq s'$, i.e.

$\left\{v_s^{(k-1)}, \ldots, v_{s+k-1}^{(k-1)}\right\}$ is linearly dependent for $s \geq s'$, contradicting the minimality of k.   **Q.E.D.**

9. Rationality of the Zeta Function.

In view of the results of Sections 6 and 7, we know that $\zeta(t) = \sum_j A_j t^j$ satisfies

a) $|A_j|_\infty \leq q^{\alpha j}$ for a suitable α.

b) $A_j \in \mathbf{Z}$ for all j.

c) $\zeta(t) = h(t)/g(t)$, where g, h are p-adic entire functions and $g(0) \neq 0$.

To prove the rationality of $\zeta(t) = \sum_j A_j t^j$, by Theorem 8.1 it suffices to prove that there exist positive integers s_0 and k such that $\Lambda_{s,k} = 0$ for all $s \geq s_0$. We shall prove that

$$|\Lambda_{s,k}|_\infty |\Lambda_{s,k}|_p < 1 \tag{9.1}$$

(which, since $\Lambda_{s,k} \in \mathbf{Z}$, immediately implies that $\Lambda_{s,k} = 0$). Since

$$|\Lambda_{s,k}|_\infty \leq (k+1)! q^{\alpha(s+2k)(k+1)}$$

there remains only to estimate $|\Lambda_{s,k}|_p$.

Let $R > 1$ and write

$$g(t) = f(t) g_R(t)$$

where

$$f(t) = 1 + b_1 t + \cdots + b_l t^l$$

and $g_R(t)$ is a p-adic entire function whose zeros are exactly the zeros ω of f with $|\omega|_p > R$ (the existence of such a factorization follows immediately from the proof of Theorem 2.1; clearly, since the Newton polygon of a non-polynomial entire function has an infinite sequence of finite sides with slopes growing to $+\infty$, f has only a finite number of zeros in the disk $D(0, R^+)$). Then $f(t)\zeta(t) = h(t)/g_R(t) = \sum_j C_j t^j$ is bounded in the disk $|t|_p \leq R$, so that by the Cauchy estimates (see the Appendix at the end of the Chapter) we have

$$|C_j|_p \leq \frac{M}{R^j} \tag{9.2}$$

for a suitable constant M depending only on R. Let $k > l$; we have

$$C_{s+l} = A_s b_l + A_{s+1} b_{l-1} + \cdots + A_{s+l} b_0$$

so that

$$\Lambda_{s,k} = \det \begin{pmatrix} A_s & A_{s+1} & \cdots & A_{s+l-1} & C_{s+l} & \cdots & C_{s+k} \\ A_{s+1} & A_{s+2} & \cdots & A_{s+l} & C_{s+l+1} & \cdots & C_{s+k+1} \\ \vdots & \vdots & & \vdots & \vdots & & \vdots \\ A_{s+k} & A_{s+k+1} & \cdots & A_{s+k+l-1} & C_{s+k+l} & \cdots & C_{s+2k} \end{pmatrix}.$$

(The idea of substituting the A_j's with the C_j's is due to E. Borel.)
From (b) and (9.2) an easy calculation gives

$$|\Lambda_{s,k}|_p \le M^{k-l+1} \frac{1}{R^{(k-l+1)s + (k-l+1)\frac{l+k}{2}}}$$

so that

$$\begin{aligned} |\Lambda_{s,k}|_\infty |\Lambda_{s,k}|_p &\le \frac{M^{k+1-l}}{R^{(k+1-l)s} R^{(k-l+1)\frac{l+k}{2}}} (k+1)! q^{\alpha(k+1)s} q^{2\alpha k(k+1)} \\ &= \frac{\xi_{k,l}}{(R^{k+1-l}/q^{\alpha(k+1)})^s} \end{aligned}$$

where

$$\xi_{k,l} = \frac{M^{k+1-l}(k+1)! q^{2\alpha k(k+1)}}{R^{(k-l+1)\frac{l+k}{2}}}$$

does not depend on s. We now fix R such that $R/q^\alpha > 1$. Once R is
fixed then l is also determined. Finally we take k such that

$$\theta = \left(\frac{R}{q^\alpha}\right)^{k+1} R^{1-l} > 1 \, .$$

Since R, l, k and $\xi_{k,l}$ are now all fixed, we can find an s_0 such that

$$\frac{\xi_{k,l}}{\theta^s} < 1$$

for all $s > s_0$. It follows that for such s the desired inequality (9.1)
holds. **Q.E.D.**

Appendix to Chapter II

If $h(t) = \sum_{j \geq 0} A_j t^j \in \mathbf{C}_p[[t]]$ converges for $|t| \leq R$ then there exists an $M > 0$ such that $|A_j| \leq M/R^j$ for all j.

Proof.   The series $h(t)$ converges for $|t| \leq R$ if and only if

$$\lim_{j \to +\infty} |A_j| R^j = 0 \, ,$$

so that $M = \max_{0 \leq j < \infty} |A_j| R^j$ exists, and $|A_j| \leq M/R^j$ for all j. **Q.E.D.**

If $g(t) = \sum_{j \geq 0} B_j t^j \in \mathbf{C}_p[[t]]$ converges for $|t| \leq R + \varepsilon$ with $\varepsilon > 0$ and all its zeros z satisfy $|z| > R$ then $1/g(t)$ converges in the disk $|t| \leq R$.

Proof.   We may assume without loss of generality that $B_0 = 1$. Let μ be the first slope of the Newton polygon of g. Then we have $p^\mu > R$ and $\operatorname{ord} B_j \geq \mu j$. Thus $g \in \mathcal{G}(\mu)$ (cf. (6.8)), whence $1/g \in \mathcal{G}(\mu)$.   **Q.E.D.**

CHAPTER III

DIFFERENTIAL EQUATIONS

1. Differential Equations in Characteristic p.

We now turn our attention to differential operators over arbitrary fields. Let H be a *differential field*, that is a field endowed with a map

$$D : H \longrightarrow H$$

satisfying

$$\begin{aligned} D(x + y) &= Dx + Dy \\ D(xy) &= xDy + (Dx)y \end{aligned} \tag{1.1}$$

for all $x, y \in H$. (Such a map is called a *derivation* of H.) We assume that D is *non-trivial*, that is $D \neq 0$. Let $H_0 = \ker(D, H)$ be the kernel of D in H. One checks easily that H_0 is a subfield of H. If $u_1, u_2, \ldots, u_n$ are elements of H we define the *Wronskian matrix* $W(u_1, u_2, \ldots, u_n)$ of $u_1, u_2, \ldots, u_n$ with respect to D by

$$W(u_1, u_2, \ldots, u_n) = \begin{pmatrix} u_1 & u_2 & \cdots & u_n \\ Du_1 & Du_2 & \ldots & Du_n \\ \vdots & \vdots & & \vdots \\ D^{n-1}u_1 & D^{n-1}u_2 & \ldots & D^{n-1}u_n \end{pmatrix} \tag{1.2}$$

and the *Wronskian* (determinant) of $u_1, u_2, \ldots, u_n$ by

$$w(u_1, u_2, \ldots, u_n) = \det W(u_1, u_2, \ldots, u_n) . \tag{1.3}$$

Lemma 1.1. *The elements $u_1, u_2, \ldots, u_n \in H$ are linearly dependent over H_0 if and only if $w(u_1, u_2, \ldots, u_n) = 0$.*

Proof. If the u's are linearly dependent then there exist elements $a_1, a_2, \ldots, a_n \in H_0$, not all zero, such that

$$a_1 u_1 + a_2 u_2 + \cdots + a_n u_n = 0$$

so that on applying iterates of D we obtain

$$W(u_1, u_2, \ldots, u_n) \begin{pmatrix} a_1 \\ a_2 \\ \vdots \\ a_n \end{pmatrix} = 0$$

which proves $w(u_1, \ldots, u_n) = 0$.

Conversely, assume $w(u_1, \ldots, u_n) = 0$ and $w(u_1, \ldots, u_{n-1}) \neq 0$ (otherwise we proceed inductively), so that the columns of the Wronskian $W(u_1, \ldots, u_n)$ are linearly dependent over H while the first $n-1$ are independent over H. Then there exist $z_1, \ldots, z_{n-1} \in H$ such that

$$\begin{pmatrix} u_n \\ Du_n \\ \vdots \\ D^{n-1}u_n \end{pmatrix} = \begin{pmatrix} u_1 & \cdots & u_{n-1} \\ Du_1 & \cdots & Du_{n-1} \\ \vdots & & \vdots \\ D^{n-1}u_1 & \cdots & D^{n-1}u_{n-1} \end{pmatrix} \begin{pmatrix} z_1 \\ z_2 \\ \vdots \\ z_{n-1} \end{pmatrix}. \qquad (1.4)$$

Applying D to both sides of this identity we obtain

$$\begin{pmatrix} Du_n \\ \vdots \\ D^{n-1}u_n \\ D^n u_n \end{pmatrix} = \begin{pmatrix} Du_1 & \cdots & Du_{n-1} \\ \vdots & & \vdots \\ D^{n-1}u_1 & \cdots & D^{n-1}u_{n-1} \\ D^n u_1 & \cdots & D^n u_{n-1} \end{pmatrix} \begin{pmatrix} z_1 \\ z_2 \\ \vdots \\ z_{n-1} \end{pmatrix}$$

$$+ \begin{pmatrix} u_1 & \cdots & u_{n-1} \\ \vdots & & \vdots \\ D^{n-1}u_1 & \cdots & D^{n-1}u_{n-1} \end{pmatrix} \begin{pmatrix} Dz_1 \\ Dz_2 \\ \vdots \\ Dz_{n-1} \end{pmatrix}. \qquad (1.5)$$

Now removing the last row in (1.5) and comparing with (1.4), from which the first row has been removed, we obtain $W(u_1, \ldots, u_{n-1})D\vec{z} = 0$ whence $\vec{z} \in H_0^{n-1}$ in virtue of the invertibility of $W(u_1, \ldots, u_{n-1})$. **Q.E.D.**

We define the ring $H[D]$ of *(ordinary) differential operators* on H by

$$H[D] = \{A_0 D^n + A_1 D^{n-1} + \cdots + A_n \mid A_i \in H\}.$$

We shall denote the product of L_1 and L_2 in $H[D]$ by $L_1 \circ L_2$ or simply (with due attention to the meaning) $L_1 L_2$. The elements of $H[D]$ are to be thought of as non-commutative polynomials with the usual rules of multiplication, except that, for $a \in H$,

$$D \circ a = D(a) + a \circ D .$$

Let $L = A_0 D^n + A_1 D^{n-1} + \cdots + A_n \in H[D]$, then if $A_0 \neq 0$ we say that L has *order* n, and we write $n = \text{ord}\, L$. Clearly $\text{ord}\,(L_1 \circ L_2) = \text{ord}\, L_1 + \text{ord}\, L_2$. One easily verifies that the ring $H[D]$ is a left and right euclidean ring with respect to the function ord, so that, in particular, each (right or left) ideal is a principal ideal.

Let (H', D') be another differential field. We say that H' is a *differential extension* of H if $H' \supseteq H$ and $D'_{|H} = D$. In this case, for simplicity, we will write D for D'. We observe that for each differential extension E of H, any element $L \in H[D]$ can be interpreted as an endomorphism of E.

Lemma 1.2. *Let* $L = A_0 D^n + A_1 D^{n-1} + \cdots + A_n \in H[D]$, *with* $A_0 \neq 0$. *Then* $\ker(L, H)$ *is an* H_0-*vector space and*

$$\dim_{H_0} \ker(L, H) \leq \text{ord}\, L.$$

Proof. Assume that $A_0 = 1$. Let $u_1, \ldots, u_{n+1} \in \ker(L, H)$. Then for $s = 1, \ldots, n+1$ we have

$$-D^n u_s = A_1 D^{n-1} u_s + A_2 D^{n-2} u_s + \cdots + A_n u_s$$

so that $w(u_1, \ldots, u_{n+1}) = 0$, whence $\{u_1, \ldots, u_{n+1}\}$ are linearly dependent over H_0. Thus we conclude that $\dim_{H_0} \ker(L, H) \leq n$. **Q.E.D.**

Example 1.3.

a) Let $H = \mathbf{F}_p(X)$ where X is an indeterminate, let $D = \dfrac{d}{dX}$, and assume $p \geq 3$.

If $L = D^2$ then $1, X \in \ker(L, H)$ so $\dim_{H_0} \ker(L, H) = 2 = \text{ord}\, L$.

If $L = D^{p+1}$ then obviously $1, X, \ldots, X^{p-1} \in \ker(L, H)$; it follows that $\dim_{H_0} \ker(L, H) = p < \text{ord}\, L$, since $H_0 = \mathbf{F}_p(X^p)$ and the elements $1, X, \ldots, X^{p-1}$ constitute a basis of H over H_0.

b) H as above and $\delta = X \dfrac{d}{dX}$. Then δ^2 is an operator of order 2 whose kernel has dimension 1.

We say that $L \in H[D]$ is *trivial* on H if $\dim_{H_0} \ker(L, H) = \text{ord}\, L$.

Lemma 1.4. If $L = L_1 \circ \cdots \circ L_m$ with $L_i \in H[D]$ and L is trivial on H, then $L_1, \ldots, L_m$ are also trivial on H.

Proof. It suffices to handle the case $m = 2$. Then

$$\mathrm{ord}\, L = \dim_{H_0} \ker(L, H) \leq \dim_{H_0} \ker(L_1, H) + \dim_{H_0} \ker(L_2, H)$$
$$\leq \mathrm{ord}\, L_1 + \mathrm{ord}\, L_2 = \mathrm{ord}\, L .$$

Thus all the inequalities must be equalities. **Q.E.D.**

The converse of this Lemma is not true in general, as Example 1.3 (b) shows.

Lemma 1.5. Let $L = D^n + A_1 D^{n-1} + \cdots + A_n \in H[D]$ with $n \geq 1$ and let $B \in H$. Then there exists a differential extension Ω of H in which the equation

$$Ly = B$$

has a non-zero solution.

Proof. Let $\Omega = H(z_0, z_1, \ldots, z_{n-1})$ be a transcendental extension of H of degree of transcendence n over H. We can extend the derivation D to Ω as follows

$$Dz_i = z_{i+1} \qquad (0 \leq i \leq n-2)$$
$$Dz_{n-1} = -A_1 z_{n-1} - A_2 z_{n-2} - \cdots - A_n z_0 + B .$$

Then $Lz_0 = B$.   **Q.E.D.**

Corollary 1.6. Let $L \in H[D]$. Then there exists an extension E of H such that L is trivial on E.

Proof. We proceed by induction on $n = \mathrm{ord}\, L$. If $n = 1$ the assertion is evident. Assume $n > 1$. Let Ω be an extension of H with $0 \neq u \in \Omega$ such that $Lu = 0$. If we divide L by $D - \dfrac{Du}{u}$ in $\Omega[D]$ (on the right) we obtain

$$L = L_1 \circ \left(D - \frac{Du}{u} \right) + L_2$$

where $L_1, L_2 \in \Omega[D]$ and L_2 is an element of Ω since its order must be 0. Applying both sides of this identity to u we obtain that $L_2 = 0$. Hence in $\Omega[D]$ we have

$$L = L_1 \circ \left(D - \frac{Du}{u} \right) .$$

Since $\operatorname{ord} L_1 = n - 1$ there exists an extension $\Omega' \supseteq \Omega$ such that L_1 is trivial on Ω'; thus there exist $v_1, \ldots, v_{n-1} \in \Omega'$ such that $v_1, \ldots, v_{n-1}$ lie in $\ker(L_1, \Omega')$ and $w(v_1, \ldots, v_{n-1}) \neq 0$.

Now, by Lemma 1.5, we can find an extension E of Ω' and non zero elements $u_1, \ldots, u_{n-1}$ in Ω' such that

$$\left(D - \frac{Du}{u} \right) u_i = v_i \qquad \text{for} \quad i = 1, \ldots, n - 1 \, . \tag{1.6}$$

Since $Lu_i = L_1 \circ \left(D - \dfrac{Du}{u} \right) u_i = L_1 v_i = 0$ we have that $u, u_1, \ldots, u_{n-1}$ are elements of $\ker(L, E)$. We claim that $u, u_1, \ldots, u_{n-1}$ are linearly independent over $E_0 = \ker(D, E)$. Let $\tau_i = u_i/u$; then from (1.6) it follows that

$$D\tau_i = \frac{Du_i}{u} + u_i D \left(\frac{1}{u} \right) = \frac{1}{u} \left(v_i + \frac{Du}{u} u_i \right) + u_i D \left(\frac{1}{u} \right) = \frac{v_i}{u} \, .$$

It suffices to prove that $1, \tau_1, \ldots, \tau_{n-1}$ are independent over E_0. Now we have

$$w(1, \tau_1, \ldots, \tau_{n-1}) = w(D\tau_1, \ldots, D\tau_{n-1}) = w \left(\frac{v_1}{u}, \ldots, \frac{v_{n-1}}{u} \right) \, .$$

But $v_1, \ldots, v_{n-1}$ are linearly independent over $(\Omega')_0$ so that they are also independent over E_0 and from this it follows that $\dfrac{v_1}{u}, \ldots, \dfrac{v_{n-1}}{u}$ are independent over E_0.   **Q.E.D.**

2. Nilpotent Differential Operators. Katz-Honda Theorem.

Let $\mathcal{F} = K(X)$ be the field of rational functions in the indeterminate X over the field K of characteristic $p \neq 0$; let $D = d/dX$; we write $\mathcal{R} = \mathcal{F}[D]$. Let $L \in \mathcal{R}$. We say that L is *nilpotent* if L *splits* in $\mathcal{R}$ into trivial factors, that is if there exist $L_1, \ldots, L_m \in \mathcal{R}$, trivial over $\mathcal{F}$, such that $L = L_1 \circ \cdots \circ L_m$.

Theorem 2.1. *Let $L \in \mathcal{R}$ be of order n. Then*
  i) if L is nilpotent then $D^{p^n} \in \mathcal{R}L$, that is there exists $A \in \mathcal{R}$ such that $D^{pn} = AL$;
  ii) if $D^{p\mu} \in \mathcal{R}L$ for some μ, then L is nilpotent.

Proof. i) Let us assume L to be trivial on $\mathcal{F}$. Then

$$D^p = AL + B \, ,$$

with $A, B \in \mathcal{R}$, ord $B <$ ord $L = n$. Since $\dim_{\mathcal{F}_0} \ker(L, \mathcal{F}) = n$ and $Bv = 0$ for every $v \in \ker(L, \mathcal{F})$, we must have, by Lemma 1.2, $B = 0$, so that $D^p \in \mathcal{R}L$.

Let L be non-trivial. We proceed by induction on $n = \operatorname{ord} L$. Since L is nilpotent, $L = L_1 \circ L_2$ with $n_i = \operatorname{ord} L_i < n$ for $i = 1, 2$ and L_1 and L_2 are nilpotent so by induction there exist $B_1, B_2 \in \mathcal{R}$ such that

$$D^{pn_1} = B_1 L_1 \quad \text{and} \quad D^{pn_2} = B_2 L_2 .$$

Therefore

$$B_2 B_1 L = B_2 (B_1 L_1) L_2 = B_2 D^{pn_1} L_2 = D^{pn_1} B_2 L_2$$
$$= D^{pn_1} D^{pn_2} = D^{p(n_1 + n_2)} = D^{pn},$$

since, as is easily verified,

$$D^p \circ L = L \circ D^p \qquad \text{for every} \quad L \in \mathcal{R} .$$

ii) Assume $D^{p\mu} \in \mathcal{R}L$, ord $L > 0$, so that $\mu \geq 0$; moreover, assume μ to be minimal with this property. Since in $\mathcal{R}$ every ideal is principal, there exists $L_1 \in \mathcal{R}$ such that

$$\mathcal{R}D^p + \mathcal{R}L = \mathcal{R}L_1 .$$

If $1 \in \mathcal{R}D^p + \mathcal{R}L$ then $D^{p(\mu-1)} \in \mathcal{R}D^{p\mu} + \mathcal{R}L \subseteq \mathcal{R}L$ and this contradicts the minimality of μ. Therefore ord $L_1 > 0$ and $L = A_1 L_1$ for a suitable $A_1 \in \mathcal{R}$. Since $D^p \in \mathcal{R}L_1$ and D^p is trivial on $\mathcal{F}$, we have, by Lemma 1.4, that L_1 is trivial on $\mathcal{F}$. Let E be an extension of $\mathcal{F}$ on which L is trivial and let $u \in \ker(L, E)$. Then we have $D^{p\mu}u = 0$ and $0 = L_1 D^{p\mu}u = D^{p\mu}(L_1 u)$, so that $D^{p\mu}$ annihilates $L_1(\ker(L, E))$ which by the proof of Lemma 1.4 is precisely $\ker(A_1, E)$. The dimension of this last vector space over E_0 is ord A_1. Let

$$D^{p\mu} = A_2 A_1 + B$$

with $A_1, A_2, B \in \mathcal{R}$ and ord $B <$ ord A_1. Then B annihilates $\ker(A_1, E)$ and thus we have $B = 0$ and $D^{p\mu} \in \mathcal{R}A_1$. Since ord $L_1 > 0$, we have ord $A_1 < n$, so that, by induction, A_1 is nilpotent. Therefore, we have $L = A_1 L_1$ with A_1 nilpotent and L_1 trivial on $\mathcal{F}$ and we can conclude that L is nilpotent, as desired. **Q.E.D.**

Theorem 2.1 establishes the equivalence of the following conditions for the operator L to be nilpotent:

— L is the product of trivial factors,

— $D^\mu \in \mathcal{R}L$ for a suitable positive integer μ.

There is yet another characterization of nilpotence. To establish it, let Ω be an extension of $\mathcal{F} = K(X)$ such that D^{p^s} is trivial over Ω and let

$$\Omega_s = \ker(D^{p^s}, \Omega) .$$

We observe that Ω_s is not merely a vector space over Ω_0, but is in fact a differential field. Indeed by Leibniz's formula, bearing in mind that, for $0 < i < p^s$, the binomial coefficients $\binom{p^s}{i}$ are all divisible by p, we have

$$D^{p^s}(yz) = yD^{p^s}z + zD^{p^s}y .$$

Hence if $yz \neq 0$ and if two of the three elements y, z, and yz lie in Ω_s then so does the third. Let E be any extension of Ω and put $E_s = \ker(D^{p^s}, E)$, which is again a field. Since D^{p^s} is trivial on Ω it is also trivial on Ω_s and on $E_s \supseteq \Omega_s$. Moreover we have $[E_s : E_0] = [\Omega_s : \Omega_0] = p^s$ and so

$$E_s = \Omega_s E_0 .$$

This shows that the choice of the extension on which D^{p^s} is trivial is not important.

Proposition 2.2. *Let $L \in \mathcal{R}$ be of order n and assume $p^s \geq pn$. Then L is nilpotent if and only if L is trivial over Ω_s.*

Proof. If L is nilpotent then $D^{pn} \in \mathcal{R}L$ by Theorem 2.1, so that $D^{p^s} \in \mathcal{R}L$ and this implies that L is trivial over Ω_s.

Conversely, let

$$D^{p^s} = AL + B$$

with A, $B \in \mathcal{R}$, and ord $B < n$. Since D^{p^s} and L both annihilate $\ker(L, \Omega_s)$ we have that $\ker(B, \Omega_s) \supseteq \ker(L, \Omega_s)$. On the other hand, since L is trivial on Ω_s we have $\dim_{\Omega_0} \ker(L, \Omega_s) = n$, so that $B = 0$ and L is nilpotent. **Q.E.D.**

Let $\delta = X\, d/dX$ so that $\mathcal{F}[D] = \mathcal{F}[\delta] = \mathcal{R}$ and let

$$L = \delta^n + A_1\delta^{n-1} + \cdots + A_n \in \mathcal{R} .$$

(Observe that after multiplying by a suitable power of X any element of order n in $\mathcal{R}$, monic with respect to D, can be written in this form.)

If we take the Laurent expansion of the A_j at 0, that is if we consider the A_j as elements of $K((X))$, and observe that

$$\delta^q X^s = s^q X^s$$

for every non-negative integer q and all $s \in \mathbf{Z}$, then it is easily seen that there exists an integer m such that for each $s \in \mathbf{Z}$

$$LX^s = X^m \left[\phi_0(s)X^s + \phi_1(s+1)X^{s+1} + \cdots\right] \qquad (2.1)$$

where $\phi_0, \phi_1, \ldots \in K[s]$. The degree of each ϕ_i is at most the order of L. The polynomial ϕ_0 is called the *indicial polynomial* of L at 0 and its roots are called the *exponents* of L at 0. In general the exponents, which are algebraic over K, do not belong to K. Similarly, by submitting X to a change of coordinates of the type $X \to X - a$ or $X \to 1/X$, we can define the indicial polynomial and the exponents of L for every point of the projective line over K.

We say that zero is a *regular point* (or a *regular singularity*) for L if no A_j has a pole at $X = 0$, that is if each A_j is an element of $K[[X]]$, and similarly for the other points of the projective line. Observe that if L is regular at zero, the integer m appearing in (2.1) is 0.

Theorem 2.3 (Katz-Honda). Let $L = \delta^n + A_1\delta^{n-1} + \cdots + A_n \in \mathcal{R}$. If L is nilpotent then
    i) zero is a regular point of L;
    ii) all exponents of L at zero lie in the prime field $\mathbf{F}_p$.

Proof.   We proceed by induction on $n = \operatorname{ord} L$.
    If $n = 1$, then L is trivial over $K(X)$ so that there exists $u \in K(X)$, $u \neq 0$, such that $Lu = 0$ whence

$$L = \delta - \frac{\delta u}{u}.$$

We can write $u = X^l v$ with v a unit in $K[[X]]$, so that

$$\frac{\delta u}{u} = l + \frac{\delta v}{v}$$

whence

$$L = \delta - l - \frac{\delta v}{v}.$$

Since $\delta v/v \in K[[X]]$ it follows that 0 is a regular point. Moreover

$$LX^s = (\delta - l)X^s + \frac{\delta v}{v}X^s \in (s - l)X^s + (X^{s+1})$$

where (X^{s+1}) denotes the ideal of $K[[X]]$ generated by X^{s+1}. Thus, $\phi_0(s) = s - l$ and the exponent at 0 is $l \in \mathbf{F}_p$.

In general, for $n > 1$, we can write $L = L_1 \circ L_2$ with L_1 trivial over $K(X)$ of order one and L_2 nilpotent of order $n - 1$ (this follows immediately from the fact that a trivial operator is a product of trivial operators of order 1). Then, by induction, we can write

$$L_1 = \delta + B_1$$
$$L_2 = \delta^{n-1} + C_1\delta^{n-2} + \cdots + C_{n-1}$$

where $B_1, C_1, C_2, \ldots, C_{n-1} \in K(X) \cap K[[X]]$. Then $A_1 = B_1 + C_1$, $A_2 = B_1C_1 + \delta C_1 + C_2, \ldots$, so that the A_i are again elements of $K[[X]]$.

To prove (ii) we need only observe that if (with the same conventions as above)

$$L_1X^s \in \tilde{\phi}_0(s)X^s + (X^{s+1})$$
$$L_2X^s \in \hat{\phi}_0(s)X^s + (X^{s+1})$$

then

$$(L_1 \circ L_2)X^s \in \tilde{\phi}_0(s)\hat{\phi}_0(s)X^s + (X^{s+1}) \ .$$

By the induction hypothesis all roots of $\tilde{\phi}_0(s)$ and $\hat{\phi}_0(s)$ lie in $\mathbf{F}_p$. Hence, the same is true for their product, the indicial polynomial of L. **Q.E.D.**

The same result also holds for every point of the projective line over K. Assuming, conversely, that all such points are regular and with exponents in the prime field does not, however, guarantee the nilpotence of L.

As an illustration of the theorem we shall show that the operator

$$L = d/dX - 1$$

is not nilpotent. In fact, the operator $XL = \delta - X$ is regular at zero, but, if we put $Y = 1/X$, we have

$$XL = -Y\,d/dY - 1/Y$$

so that XL is not regular at infinity. As a consequence, we find that there is no "exponential function" in $K(X)$.

3.   Differential Systems.

We now wish to discuss systems of differential equations, or equivalently, *differential modules*. Let M be a vector space of dimension n over a differential field H with non-trivial derivation D and let Δ be an extension of D to M, that is Δ is a map

$$\Delta : M \longrightarrow M$$

such that

$$\Delta(u + v) = \Delta u + \Delta v$$
$$\Delta(zu) = z\Delta u + (Dz)u$$

for all $u, v \in M$ and $z \in H$. If $\{e_1, e_2, \dots, e_n\}$ is an H-basis for M we can write

$$\Delta \begin{pmatrix} e_1 \\ e_2 \\ \vdots \\ e_n \end{pmatrix} = G \begin{pmatrix} e_1 \\ e_2 \\ \vdots \\ e_n \end{pmatrix} \tag{3.1}$$

with $G \in \mathcal{M}_n(H)$. If s is a positive integer the matrix G_s is defined by

$$\Delta^s \begin{pmatrix} e_1 \\ e_2 \\ \vdots \\ e_n \end{pmatrix} = G_s \begin{pmatrix} e_1 \\ e_2 \\ \vdots \\ e_n \end{pmatrix} ;$$

the matrices G_s satisfy the recursion relation

$$G_{s+1} = DG_s + G_s G.$$

Knowledge of the matrices G_s is important for the study of differential modules. If we change the basis by setting $e' = Pe$ with $P \in G\ell(n, H)$ then the new matrix G' is given by

$$G' = PGP^{-1} + (DP)P^{-1}.$$

To the differential module M we associate the differential system

$$D \begin{pmatrix} y_1 \\ y_2 \\ \vdots \\ y_n \end{pmatrix} = G \begin{pmatrix} y_1 \\ y_2 \\ \vdots \\ y_n \end{pmatrix} ; \tag{3.2}$$

as before we write

$$D^s \begin{pmatrix} y_1 \\ y_2 \\ \vdots \\ y_n \end{pmatrix} = G_s \begin{pmatrix} y_1 \\ y_2 \\ \vdots \\ y_n \end{pmatrix} .$$

By a *matrix solution* for the system (3.2) over a differential extension field Ω of H we mean a matrix $\mathcal{U} \in G\ell(n, \Omega)$ such that

$$D\mathcal{U} = G\mathcal{U} .$$

We say that the system (3.2), or, simply, that $D - G$ is *trivial* over the extension Ω of H if it has a matrix solution over Ω.

We say that M is a *cyclic (differential) module* if there exists a basis $e_1, \ldots, e_n$ for M such that $e_l = \Delta e_{l-1}$ for $l = 2, \ldots, n$. In this case the matrix G has the form

$$G = \begin{pmatrix} 0 & 1 & 0 & \ldots & 0 \\ 0 & 0 & 1 & \ldots & 0 \\ & & \ldots\ldots\ldots\ldots & & \\ 0 & 0 & \ldots\ldots & & 1 \\ A_n & A_{n-1} & \ldots\ldots\ldots & & A_1 \end{pmatrix}$$

and the system can be written

$$y_i = D^{i-1}y_1 \qquad \text{for} \quad i = 2, \ldots, n$$
$$Ly_1 = 0$$

where $L = D^n - A_1 D^{n-1} - \cdots - A_n$ is the *associated scalar operator*.

In this case the matrix solution of $D - G$ is precisely the Wronskian matrix for L, and so by Lemma 1.5 there exists an extension Ω in which $D - G$ is trivial.

Using the symbol D to represent both D and Δ we may say that as an $\mathcal{R}$-module, we have $M \cong \mathcal{R}/\mathcal{R}L$.

Proposition 3.1. *In the preceding notation, for $G \in \mathcal{M}_n(H)$ there exists a differential extension Ω of H over which $D - G$ is trivial.*

Proof. Let M be the differential module associated to the system $D\vec{y} = G\vec{y}$ and let $0 \neq m \in M$. Assume that $m, \Delta m, \ldots, \Delta^{l-1}m$ are linearly independent over H with $l < n$, but that the element $\Delta^l m$ lies in $\langle m, \Delta m, \ldots, \Delta^{l-1}m \rangle$. Let $f_{l+1}, \ldots, f_n \in M$ be such that

$\{m, \Delta m, \ldots, \Delta^{l-1} m, f_{l+1}, \ldots, f_n\}$ is a basis for M. Then the matrix of Δ with respect to this basis takes the form

$$\begin{pmatrix} G_1 & 0 \\ G_3 & G_4 \end{pmatrix}$$

where $G_1 \in \mathcal{M}_l(H)$, $G_3 \in \mathcal{M}_{(n-l) \times l}(H)$ and $G_4 \in \mathcal{M}_{n-l}(H)$. We look for a solution of the type

$$D \begin{pmatrix} \mathcal{U}_1 & 0 \\ \mathcal{U}_3 & \mathcal{U}_4 \end{pmatrix} = \begin{pmatrix} G_1 & 0 \\ G_3 & G_4 \end{pmatrix} \begin{pmatrix} \mathcal{U}_1 & 0 \\ \mathcal{U}_3 & \mathcal{U}_4 \end{pmatrix},$$

that is, we look for matrices $\mathcal{U}_1$, $\mathcal{U}_3$, $\mathcal{U}_4$, with $\det \mathcal{U}_1 \neq 0$, $\det \mathcal{U}_4 \neq 0$, such that

$$D\mathcal{U}_1 = G_1 \mathcal{U}_1$$
$$D\mathcal{U}_4 = G_4 \mathcal{U}_4$$
$$D\mathcal{U}_3 = G_3 \mathcal{U}_1 + G_4 \mathcal{U}_3 .$$

For the first equation we find a solution $\mathcal{U}_1$ since $D - G_1$ is equivalent to a scalar equation; for the second equation we find a solution $\mathcal{U}_4$ by induction. To solve the third equation we write

$$\mathcal{U}_3 = \mathcal{U}_4 T$$

where T is an $(n-l) \times l$ matrix. From $(D\mathcal{U}_4)T + \mathcal{U}_4 DT = G_3 \mathcal{U}_1 + G_4 \mathcal{U}_4 T$, we obtain $DT = \mathcal{U}_4^{-1} G_3 \mathcal{U}_1$ and then again we use Lemma 1.5 to find T.
Q.E.D.

We now extend the notion of nilpotence. Let $\mathcal{F} = K(X)$, $D = d/dX$; since $D^p \mathcal{F} = 0$, from Leibniz's formula we obtain

$$\Delta^p(hm) = h\Delta^p m$$

for every $h \in \mathcal{F}$, $m \in M$, so that Δ^p is an H-linear map. Then, from (3.1) and $\Delta^{p\mu} = (\Delta^p)^\mu$ it follows that

$$G_{p\mu} = (G_p)^\mu .$$

Therefore $\Delta^{p\mu}$ annihilates M if and only if $G_{p\mu} = (G_p)^\mu = 0$, that is if and only if the matrix G_p is nilpotent. We say that the differential module M (or its associated differential system $(D-G)\vec{y} = 0$) is *nilpotent* if the matrix G_p is nilpotent.

We recall that, if Ω is an extension of $\mathcal{F}$ over which D^{p^s} is trivial, we set $\Omega_s = \ker(D^{p^s}, \Omega)$.

Proposition 3.2. Let $G \in \mathcal{M}_n(\mathcal{F})$ and assume $p^s \geq pn$. Then $D - G$ is nilpotent if and only if $D - G$ is trivial over Ω_s for some extension Ω of $\mathcal{F}$ on which D^{p^s} is trivial.

Proof. Let $\mathcal{U} \in G\ell(n, \Omega)$ be a solution matrix of $D - G$, where Ω is an extension of $\mathcal{F}$ over which both $D - G$ and D^{p^s} are trivial. If $D - G$ is nilpotent then $D^{pn}\mathcal{U} = G_p^n\mathcal{U} = 0$; since $p^s \geq pn$ we also have that $D^{p^s}\mathcal{U} = 0$ so that $\mathcal{U} \in G\ell(n, \Omega_s)$.

Conversely, if $\mathcal{U} \in G\ell(n, \Omega_s)$ then $G_{p^s}\mathcal{U} = D^{p^s}\mathcal{U} = 0$ so that $G_p^{p^{s-1}} = G_{p^s} = 0$ and G_p is nilpotent. **Q.E.D.**

4. The Theorem of the Cyclic Vector.

We will now establish the relation between differential systems and scalar equations. To achieve this end we will exploit the theory of (non-commutative) elementary divisors. Recall that $\mathcal{F} = K(X)$ and $\mathcal{R} = \mathcal{F}[D]$ with K of characteristic $p \neq 0$ and $D = d/dX$. Then, given a matrix $G \in \mathcal{M}_n(\mathcal{F})$, by the theory of elementary divisors in euclidean rings there exist invertible matrices Λ_1 and $\Lambda_2 \in \mathcal{M}_n(\mathcal{R})$ such that

$$\Lambda_1(D - G)\Lambda_2 = \begin{pmatrix} L_1 & 0 & \ldots & 0 \\ 0 & L_2 & \ldots & 0 \\ \vdots & \vdots & \ddots & \vdots \\ 0 & 0 & \ldots & L_n \end{pmatrix} \tag{4.1}$$

where $0 \neq L_n \in \mathcal{R}$, $L_{i+1} \in \mathcal{R}L_i \cap L_i\mathcal{R}$.

Lemma 4.1. Let (M, Δ) be a differential module over $\mathcal{F} = K(X)$. Then M is a direct sum of cyclic $\mathcal{R}$–modules

$$M = M_1 \oplus \cdots \oplus M_n .$$

Let L_j be the scalar operator associated with M_j (hence L_j cannot be zero and $n = \sum_j \text{ord } L_j$). Then M is nilpotent if and only if each L_j is nilpotent.

Proof. If $M = M_n$, that is if M is cyclic, then the second assertion follows from Propositions 2.2 and 3.2 using the fact that the solution matrix of M coincides with the Wronskian matrix of the associated scalar operator L_n.

If M is a direct sum $M_1 \oplus M_2$ of submodules then Δ^p is represented as the direct sum of the matrices representing the restrictions $\Delta^p|M_1$ and $\Delta^p|M_2$. Thus $(\Delta^p)^\mu = 0$ if and only if the same holds for the restrictions to both M_1 and M_2. Thus the second assertion follows from the first.

For the proof of the first assertion, let $e_1, \ldots, e_n$ be a basis for M over $\mathcal{F}$ and let $G = (G_{ij}) \in \mathcal{M}_n(\mathcal{F})$ be such that (3.1) holds; let $\overline{M}$ be the free left $\mathcal{R}$-module on n symbols $\overline{e}_1, \ldots, \overline{e}_n$ and let $\overline{M}_0$ be the $\mathcal{R}$-submodule generated by the n elements

$$\left\{ D\overline{e}_i - \sum_{j=1}^{n} G_{ij}\overline{e}_j \right\}_{1 \le i \le n} .$$

Then using the symbol D for both D and Δ, we may view M as the $\mathcal{R}$-module $\overline{M}/\overline{M}_0$. Recalling (4.1) and putting

$$\begin{pmatrix} \overline{e}_1 \\ \vdots \\ \overline{e}_n \end{pmatrix} = \Lambda_2 \begin{pmatrix} \overline{f}_1 \\ \vdots \\ \overline{f}_n \end{pmatrix}$$

we find a new set of $\mathcal{R}$-generators for $\overline{M}$ and

$$\overline{M}_0 = \bigoplus_{i=1}^{n} \mathcal{R}L_i\overline{f}_i .$$

Thus

$$M \cong \bigoplus_{i=1}^{n} \mathcal{R}\overline{f}_i/\mathcal{R}L_i\overline{f}_i .$$

Q.E.D.

Remark. The proof of the first assertion of Lemma 4.1 holds with $(\mathcal{F}, \mathcal{R})$ replaced by $(\Omega, \Omega[D])$ where (Ω, D) is an arbitrary differential field.

Let H be a differential field of characteristic zero. We fix the derivation D and an element $X \in H$ in such a way that $DX = 1$. Let M be a differential module over H with operator Δ extending $\delta = XD$.

Theorem 4.2 (Theorem of the cyclic vector). *In the preceding notation there exists an element* $m \in M$ *such that*

$$m \wedge \Delta m \wedge \cdots \wedge \Delta^{n-1} m \neq 0$$

where n *is the dimension of* M *over* H.

Proof. Assume not. Then there exists a minimal integer $\mu \leq n$ such that

$$m \wedge \Delta m \wedge \cdots \wedge \Delta^{\mu-1} m = 0 \tag{4.2}$$

for all $m \in M$. Now let $t \in H_0$ and $z \in M$. Then substituting $m + tz$ for m in (4.2) we obtain an element of the form $z_0 + tz_1 + t^2 z_2 + \cdots + t^\mu z_\mu$, with $z_i \in \bigwedge^\mu M$, which must be zero for all $t \in H_0$. Since H_0 is an infinite field we obtain that each z_i must be zero. In particular,

$$z_1 = \sum_{i=0}^{\mu-1} m \wedge \Delta m \wedge \cdots \wedge \Delta^i z \wedge \cdots \wedge \Delta^{\mu-1} m = 0 \tag{4.3}$$

for all $m, z \in M$. If we now substitute $X^s z$ for z in (4.3), divide the result by X^s, and take, for $h = 0, 1, \ldots, \mu - 1$,

$$\Lambda_h(m, z) =$$
$$\sum_{k=0}^{\mu-h-1} \binom{h+k}{h} m \wedge \cdots \wedge \Delta^{h+k-1} m \wedge \Delta^k z \wedge \Delta^{h+k+1} m \wedge \cdots \wedge \Delta^{\mu-1} m \, ,$$

we find that

$$\sum_{h=0}^{\mu-1} s^h \Lambda_h(m, z) = 0 \tag{4.4}$$

for every $m, z \in M$ and every $s \in \mathbf{Z}$. (Observe that, up to this point, all we have said is valid in any characteristic; in fact we only need the field H_0 to be infinite and this is the case even if the field H has characteristic $p \neq 0$ since the element X is then necessarily transcendental over $\mathbf{F}_p$ and $H_0 \supseteq \mathbf{F}_p(X^p)$.)

If we now set $s = 1, 2, \ldots, \mu - 1$ in (4.4) and observe that the Vandermonde matrix

$$\begin{pmatrix} 1 & 2 & \cdots & \mu - 1 \\ 1^2 & 2^2 & \cdots & (\mu - 1)^2 \\ \vdots & \vdots & \cdots & \vdots \\ 1^{\mu-1} & 2^{\mu-1} & \cdots & (\mu - 1)^{\mu-1} \end{pmatrix}$$

is invertible (this remains true in characteristic p if $n < p$), we obtain that $\Lambda_h(m, z) = 0$ for every $m, z \in M$ and each $h = 0, 1, \ldots, \mu - 1$. In particular, taking $h = \mu - 1$, we have

$$\left(m \wedge \Delta m \wedge \Delta^2 m \wedge \cdots \wedge \Delta^{\mu-2} m\right) \wedge z = 0$$

for every $m, z \in M$ which implies

$$m \wedge \Delta m \wedge \Delta^2 m \wedge \cdots \wedge \Delta^{\mu-2} m = 0$$

for every $m \in M$ and this contradicts the minimality of μ.   **Q.E.D.**

5.   The Generic Disk. Radius of Convergence.

We will now study the relation between the nilpotence of differential operators over fields of positive characteristic and the radius of convergence of solutions of differential systems over a valued field of characteristic zero.

   Let K be a field of characteristic zero with a valuation which we denote by $|-|$; we assume that the residue field $\overline{K}$ has characteristic $p > 0$.

   Let X be an indeterminate over K; we consider a differential system

$$\frac{d}{dX}\vec{y} = G(X)\vec{y}, \qquad y = \begin{pmatrix} y_1 \\ \vdots \\ y_n \end{pmatrix}, \qquad G \in \mathcal{M}_n\left(K(X)\right). \tag{5.1}$$

We will study the formal solutions in the neighborhood of a point α in a complete, algebraically closed extension Ω of K. We will see that these solutions

$$\sum_{j=0}^{\infty} \vec{A}_j (X - \alpha)^j$$

where $\vec{A}_j \in \Omega^n$ have a positive radius of convergence if α is not a pole of G. We will be interested in the case of α generic, since then $G(X)$ has no pole in the residue class of α and the calculations are easier; later we specialize α.

   Let t be a transcendental element over K (independent of the X above). The valuation of K can be extended to $K(t)$ in many different ways. We choose the Gauss valuation on $K(t)$, that is we set

$$\left| \frac{\sum_i a_i t^i}{\sum_j b_j t^j} \right| = \frac{\sup_i |a_i|}{\sup_j |b_j|} .$$

We now take the complete, algebraically closed extension Ω of K in such a way that $t \in \Omega$ and that the valuation of Ω is an extension of the Gauss valuation on $K(t)$. From these conditions it follows that the residue class $\bar{t}$ of t (in the residue field $\overline{\Omega}$ of Ω) is transcendental over $\overline{K}$. The open disk $D(t, 1^-) = \{x \in \Omega \mid |x - t| < 1\}$ will be called the *generic disk* and t the *generic point* with respect to K.

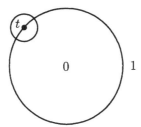

The Generic Disk

If, for $s \geq 0$, we write

$$\frac{d^s \vec{y}}{dX^s} = G_s(X)\vec{y}$$

then the solution matrix of (5.1) at the generic point t is given by

$$\mathcal{U} = \sum_{s=0}^{\infty} \frac{G_s(t)}{s!}(X - t)^s . \tag{5.2}$$

As we observed in Section 3, the matrices $G_s \in \mathcal{M}_n(K(X))$ satisfy the following recursion relation

$$G_s = \frac{d}{dX}G_{s-1} + G_{s-1}G . \tag{5.3}$$

If $f(X) \in K(X)$, we denote by

$$|f|_{\text{gauss}} = |f(X)|_{\text{gauss}} = |f(t)|$$

the Gauss valuation of $f(X)$ with respect to X, and if $G = (g_{ij}) \in \mathcal{M}_n(K(X))$ we set

$$|G|_{\text{gauss}} = \sup_{ij}\{|g_{ij}|_{\text{gauss}}\} .$$

Note that the Gauss norm of the operator d/dX, i.e. the norm of d/dX as operator on $K(X)$ with Gauss norm, satisfies

$$\left|\frac{d}{dX}\right|_{\text{gauss}} \le 1 \qquad (5.4)$$

since obviously if $|f|_{\text{gauss}} \le 1$ then $|df/dX|_{\text{gauss}} \le 1$. By (5.3) and (5.4) we have

$$|G_s|_{\text{gauss}} \le \sup(1, |G|_{\text{gauss}})^s . \qquad (5.5)$$

We fix an element $\pi \in \Omega$ such that $\pi^{p-1} = -p$. Then as immediately follows from (II.4.6)

$$|s!| \ge |\pi|^s .$$

Thus we have

$$\left|\frac{G_s(t)}{s!}\right| |x - t|^s \le \frac{\sup(1, |G|_{\text{gauss}})^s}{|s!|} |x - t|^s \le \frac{\sup(1, |G|_{\text{gauss}})^s}{|\pi|^s} |x - t|^s$$

and so we have convergence of (5.2) for

$$|x - t| \frac{\sup(1, |G|_{\text{gauss}})}{|\pi|} < 1 .$$

We define the *radius of convergence of the system* $d/dX - G$ to be the radius of convergence of $\mathcal{U}$. Therefore, if R is the radius of convergence of the system $d/dX - G$, we have

$$R \ge \frac{|\pi|}{\sup(1, |G|_{\text{gauss}})} .$$

In particular if

$$|G|_{\text{gauss}} \le 1$$

then $R \ge |\pi|$.

If $|G|_{\text{gauss}} \le 1$ then we can reduce the elements of G modulo $\mathfrak{p}_K[X]$. If we denote the reduced matrix by $\overline{G}$ we obtain the following differential system in characteristic p

$$\frac{d}{dX} \vec{y} = \overline{G}(X)\vec{y}, \qquad y = \begin{pmatrix} y_1 \\ \vdots \\ y_n \end{pmatrix}, \qquad \overline{G} \in \mathcal{M}_n\left(\overline{K}(X)\right) . \qquad (5.6)$$

From (5.5) it follows that if $|G|_{\text{gauss}} \leq 1$ then $|G_s|_{\text{gauss}} \leq 1$, so that on taking the reductions $\overline{G}_s$ we find

$$\frac{d^s}{dX^s}\vec{y} = \overline{G}_s(X)\vec{y}\,.$$

We can now state the relation between nilpotence of the differential system (5.6) and the radius of convergence R of the system (5.1), that is of the matrix solution in (5.2).

Proposition 5.1. *Let* $G \in \mathcal{M}_n(K(X))$ *and assume* $|G|_{\text{gauss}} \leq 1$. *Let* R *be the radius of convergence of the system (5.1). Then the system (5.6) is nilpotent if and only if* $R > |\pi| = |p|^{1/(1-p)}$.

Proof. By the definition of nilpotence, our task is equivalent to proving that $\overline{G}_{pn} = 0$ if and only if $R > |\pi|$. We set $D = d/dX$.

Assume $R > |\pi|$ and let $R > R_1 > |\pi|$. Then $|G_s/s!|\, R_1^s \to 0$ for $s \to \infty$. There exist subsequences s_j such that $R_1^{s_j}/|s_j!| \to \infty$ as $j \to \infty$ (take, for example, the subsequence $s_j = p^j$ for $j = 1, 2, \dots$); therefore, for any such subsequence we have $|G_{s_j}| \to 0$ as $j \to \infty$. Since from (5.3) and (5.4) it easily follows that for $s' > s$ we have $|G_{s'}| \leq |G_s|$, we obtain $|G_s| \to 0$ as $s \to \infty$ whence $|G_s| < 1$ for $s > s_0$. Therefore we have $|G_{p\mu}| < 1$ for $p\mu > s_0$ which implies that $\overline{G}_{p\mu} = 0$ and $D - \overline{G}$ is nilpotent.

Conversely, if $\overline{G}_{pn} = 0$, then $|G_{pn}| < 1$; this is equivalent to saying $\operatorname{ord} G_{pn} \geq 1/e$ for some $e \geq 1$ (in general e depends upon G but if the valuation of K is discrete then we may take e to be $e(K/\mathbf{Q})$, the absolute ramification of K). We claim that

$$\operatorname{ord} G_{pns} \geq \frac{s}{e}\,. \tag{5.7}$$

We proceed by induction on s. Let $\mathcal{U}$ be a matrix solution at t of the system (5.1). Then

$$G_{pn(s+1)}\mathcal{U} = D^{pn}(G_{pns}\mathcal{U}) = \sum_{i=0}^{pn} \binom{pn}{i} (D^i G_{pns}) D^{pn-i}\mathcal{U}$$

so that

$$G_{pn(s+1)} = \sum_{i=0}^{n} \binom{pn}{pi} (D^{pi}G_{pns}) G_{pn-pi} + \sum_{\substack{i=1\\(p,i)=1}}^{pn-1} \binom{pn}{i} (D^i G_{pns}) G_{pn-i}\,.$$

If we now observe that by the inductive hypothesis we have

$$\operatorname{ord}(G_{pns}G_{pn}) \geq \frac{s+1}{e} \,,$$

that for $i \neq 0$ we have

$$\operatorname{ord} D^{pi}G_{pns} \geq 1 + \operatorname{ord} G_{pns} \geq 1 + \frac{s}{e} \geq \frac{s+1}{e} \,,$$

that for $p \!\!\not|\, i$ we have

$$\operatorname{ord}\binom{pn}{i} \geq 1$$

and that $\operatorname{ord} G_{s'} \geq \operatorname{ord} G_s$ if $s' \geq s$ (as follows from (5.3) and (5.4)), we obtain the desired inequality (5.7).

From (5.7), using once again that $\operatorname{ord} G_{s'} \geq \operatorname{ord} G_s$ if $s' \geq s$, we obtain

$$\operatorname{ord} G_s \geq \left[\frac{s}{pn}\right]\frac{1}{e}$$

where $[s/pn]$ is the integer part of s/pn. The matrix solution $\mathcal{U}$ of (5.2) will certainly converge at $X = x$ if

$$s \operatorname{ord}(x - t) + \frac{1}{e}\left[\frac{s}{pn}\right] - \frac{s}{p-1} \longrightarrow +\infty$$

as $s \to \infty$. For this to occur it suffices that

$$\operatorname{ord}(x - t) > \frac{1}{p-1} - \frac{1}{epn} \,.$$

This gives $R > |\pi|p^{1/epn}$, as desired.   **Q.E.D.**

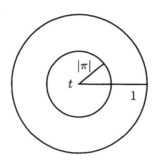

Remark 5.2.   From the second part of the proof of Proposition 5.1, if we add the hypothesis that the valuation of K is discrete, it follows that if $D - \overline{G}$ is nilpotent then the radius of convergence of $D - G$ is at least $|\pi| C_{n,e}$ where $C_{n,e} > 1$ is a constant which depends only on the order of the system and the absolute ramification of K. This shows that, when the valuation of K is discrete, there can not be intermediate radii of convergence: if we have any system $D - G$, with $G \in \mathcal{M}_n(K(X))$ and $|G|_{\text{gauss}} \leq 1$, having radius of convergence $R > |\pi|$, then in fact $R \geq |\pi| C_{n,e}$ since, by Proposition 5.1, $D - \overline{G}$ is nilpotent.

Remark 5.3.   For the differential systems satisfied by the periods of families of algebraic varieties defined over $\mathbf{Q}$ it is known that $R_p = 1$ for almost all primes p.

The operator $d/dX - 1$ over $\mathbf{Q}$ is an example for which $R_p = |\pi_p|_p$ for every prime p. As we shall see any equation with irregular singularities will also be an example of this type. Harder cases in which $R_p = |\pi_p|_p$ for almost all p have been examined by Deligne as well as in Katz [1] and Chudnovsky [2].

It has been conjectured that if $R_p > |\pi_p|_p$ for almost all p, then in fact $R_p = 1$ for almost all p.

Example 5.4.   We illustrate the role played by the absolute ramification index in determining the radius of convergence of a power series. We consider the binomial series

$$(1 - X)^{-\alpha} = \sum_{s=0}^{\infty} \frac{\alpha(\alpha + 1) \cdots (\alpha + s - 1)}{s!} X^s .$$

We assume that α belongs to a ramified extension of $\mathbf{Q}_p$ and $\text{ord}\, \alpha = 1/e$; then this series converges when

$$\text{ord}\, x > \frac{1}{p-1} - \frac{1}{ep} .$$

To establish this fact, following the general lines of the proof of Proposition 5.1, it suffices to observe that

$$\text{ord}\, (\alpha(\alpha + 1) \cdots (\alpha + s - 1)) = \left(1 + \left[\frac{s-1}{p}\right]\right) \frac{1}{e} \geq \frac{s}{pe} .$$

See also Proposition IV.7.3.

6. Global Nilpotence. Katz's Theorem.

Let K be a number field, that is a finite field extension of $\mathbf{Q}$, let X an indeterminate over K and let $L \in K(X)[D]$ be of the form

$$L = \delta^n + A_1 \delta^{n-1} + \cdots + A_n \qquad (6.1)$$

with $\delta = XD$, $D = d/dX$. For every non-archimedean valuation (i.e. finite prime) $\mathfrak{p}$ of K we have the corresponding Gauss norm on $K(X)$, which we will denote by $|-|_{\mathfrak{p},\text{gauss}}$. Consider the finite set of finite primes of K defined by

$$\mathcal{S} = \{ \mathfrak{p} \mid |A_j|_{\mathfrak{p},\text{gauss}} > 1 \quad \text{for at least one } j, \quad 1 \leq j \leq n \} \,.$$

For a finite prime $\mathfrak{p}$ of K, with $\mathfrak{p} \notin \mathcal{S}$, we denote by $L_{\mathfrak{p}} \in \overline{K}_{\mathfrak{p}}(X)[D]$ the reduction of L modulo $\mathfrak{p}$ where, as usual, $\overline{K}_{\mathfrak{p}}$ is the residue field of K. We say that the operator L is *globally nilpotent* if $L_{\mathfrak{p}}$ is nilpotent for almost every $\mathfrak{p}$.

The definitions of *regular singularity, indicial polynomial* and *exponents* for an operator L in characteristic 0 are the same as those given in Section 2 for a field of characteristic $p > 0$.

Theorem 6.1 (Katz). *Let K be a number field and let L be a differential operator over $K(X)$.*
i) If $L_{\mathfrak{p}}$ is nilpotent for an infinite set of finite primes $\mathfrak{p}$ of K then L has only regular singularities.
ii) If L is globally nilpotent, then all the exponents of L are rational numbers.

Proof. Take L as in (6.1).
 (i) It clearly suffices to prove the assertion for the point 0. Let

$$A_j = \frac{A_{j,m}}{X^m} + \cdots + \frac{A_{j,1}}{X} + B_{j,0} + B_{j,1} X + \cdots$$

with the $A_{j,s}$, $B_{j,l} \in K$. Then for almost all $\mathfrak{p}$ the function A_j has no pole in the punctured disk $D(0, 1^-) \backslash \{0\}$ and so by Remark 6.2 for $\mathfrak{p} \notin \mathcal{S}$ and excluding exceptional primes,

$$|A_j|_{\mathfrak{p},\text{gauss}} = \sup_{s,l}(|A_{j,s}|_{\mathfrak{p}}, |B_{j,l}|_{\mathfrak{p}}) \leq 1 \,.$$

If $L_{\mathfrak{p}}$ is nilpotent, then by Theorem 2.3 the rational function $(\overline{A}_j)_{\mathfrak{p}}$, the reduction modulo $\mathfrak{p}$ of A_j, does not have a pole at zero, so that $(\overline{A}_{j,s})_{\mathfrak{p}} = 0$ for $s = 1, 2 \ldots, m$. If this holds for infinitely many $\mathfrak{p}$ then $A_{j,s} = 0$ for all s, whence A_j is analytic at zero and zero is a regular singularity for L (we recall that if $\alpha \in K$ and $|\alpha|_{\mathfrak{p}} < 1$ for infinitely many $\mathfrak{p}$, then $\alpha = 0$).

(ii) To prove the second assertion we will use the following somewhat deeper result from algebraic number theory which is a consequence of the Čebotarev Density Theorem (cf. Lang [1]; Theorem 10 of Chapter VIII): if α is an element of the number field K and satisfies $\alpha \pmod{\mathfrak{p}} \in \mathbf{F}_p$ (where $p = \mathfrak{p} \cap \mathbf{Q}$) for almost all primes $\mathfrak{p}$ of K then $\alpha \in \mathbf{Q}$.

We recall that the indicial polynomial $\phi_0(s)$ of L is defined by the identity

$$LX^s = \phi_0(s)X^s + \phi_1(s+1)X^{s+1} + \cdots$$

and that $\phi_i \in K[s]$ and $\deg \phi_0 = n$. If $|A_j|_{\mathfrak{p},\text{gauss}} \leq 1$ for $j = 1, 2, \ldots, n$ then the coefficients of ϕ_0 are bounded by 1 in the valuation $\mathfrak{p}$, and so the indicial polynomial of $L_{\mathfrak{p}}$ is the reduction mod $\mathfrak{p}$ of $\phi_0(s)$. Since by Theorem 2.3 all roots of this reduction are in $\mathbf{F}_p$, and this holds for almost all $\mathfrak{p}$, the exponents of L are rational. **Q.E.D.**

Remark 6.2. If $\xi \in K(X)$ has a representation in $K((X))$

$$\xi = \sum_{j=-m}^{\infty} c_j X^j$$

then $|\xi|_{\text{gauss}} = \sup_j |c_j|$ provided that ξ has no pole in the punctured disk $D(0, 1^-) \backslash \{0\}$. The proof may be reduced to the case in which $\xi = f(X)/g(X)$ where $g(X) = 1 + a_1 X + \cdots + a_l X^l$ with $|a_j| \leq 1$ for $j = 1, \ldots, l$, so that $|g|_{\text{gauss}} = 1$ and $|\xi|_{\text{gauss}} = |f|_{\text{gauss}}$. From

$$\xi = f(X) \sum_{s=0}^{\infty} (-a_1 X - \cdots - a_l X^l)^s$$

we see that $|c_j| \leq |f|_{\text{gauss}}$. Conversely, from

$$f(X) = g(X) \sum_{j=-m}^{\infty} c_j X^j$$

we deduce that $|f|_{\text{gauss}} \leq \sup_j |c_j|$, which completes the proof.

Remark 6.3. In statement (ii) of Theorem 6.1 it would suffice to have nilpotence of $L_{\mathfrak{p}}$ for all primes of K dividing a set of primes of $\mathbf{Q}$ of density strictly greater than $1/2$ since Čebotarev Theorem is valid for such sets of primes.

By way of illustration, let $K = \mathbf{Q}(\sqrt{5})$; then $\sqrt{5} \pmod{\mathfrak{p}} \in \mathbf{F}_p$ if and only if 5 is a quadratic residue modulo p for $p \neq 5$; such primes have density $1/2$ and in fact $\sqrt{5}$ is not rational. This explains the necessity of a stronger condition on the set of primes in the second assertion.

7. Regular Singularities. Fuchs' Theorem.

Consider the differential system

$$\frac{d}{dX}\,\vec{y} = G(X)\,\vec{y}\,, \tag{7.1}$$

where $G(X) \in \mathcal{M}_n(\mathcal{F})$ and $\mathcal{F}$ is the field of complex meromorphic functions in an open disk $\mathcal{D}$ centered at the origin. We can assume that the radius of $\mathcal{D}$ is so small that zero is the only singularity of $G(X)$. Let $z_0 \in \mathcal{D}' = \mathcal{D}\backslash\{0\}$. Then the vector space of solutions of (7.1) in a neighborhood of z_0 has dimension n, by Cauchy's theorem (cf. Appendix III, Poole [1], p. 5). Let $\mathcal{F}_{z_0}$ be the ring of complex analytic functions at z_0 and let $\mathcal{U} \in G\ell(n, \mathcal{F}_{z_0})$ be a matrix solution to (7.1) at z_0, that is let

$$\frac{d\mathcal{U}}{dX} = G(X)\mathcal{U}\,.$$

We recall that, having fixed a closed circuit $\gamma \subset \mathcal{D}'$ starting at z_0 and wrapping once counterclockwise around zero, analytic continuation of an element $f \in \mathcal{F}_{z_0}$ along γ produces in general a new element $Tf \in \mathcal{F}_{z_0}$. We call the map T the *monodromy map* (relative to γ). As an example consider any determination of $\log X$; then $T \log X = \log X + 2\pi i$. One easily verifies that analytic continuation commutes with differentiation.

Now the matrix $T\mathcal{U}$ is a new matrix solution of (7.1) since $TG = G$. Indeed, that $\det(T\mathcal{U}) \neq 0$ follows from the fact that $T \det \mathcal{U} = \det(T\mathcal{U})$ and the fact that $\det \mathcal{U}$ satisfies the following differential equation

$$\frac{d}{dX}\det \mathcal{U} = (\mathrm{Tr}G)(\det \mathcal{U})\,.$$

Thus there exists a matrix $C \in G\ell(n, \mathbf{C})$ such that

$$T\mathcal{U} = \mathcal{U}\,C\,.$$

If, for $A \in \mathcal{M}_n(\mathbf{C})$ we define X^A by

$$X^A = \exp(A \log X) = \sum_{s=0}^{\infty} \frac{(A \log X)^s}{s!}$$

then we have

$$TX^A = \exp(A \log X + 2\pi i A) = X^A \exp(2\pi i A).$$

If we take A such that $\exp(2\pi i A) = C$ then

$$T(\mathcal{U} X^{-A}) = \mathcal{U} CC^{-1} X^{-A} = \mathcal{U} X^{-A}$$

so that $W = \mathcal{U} X^{-A}$ is "uniform", that is, W is analytic in $\mathcal{D}'$. There are now two possibilities: either zero is an essential singularity of W or W has at worst a pole at zero. Classically if the latter occurs the singularity is called *regular* and otherwise *irregular*.

By the theorem of the cyclic vector, system (7.1) is equivalent to a scalar equation $Ly = 0$ where

$$L = \delta^n + A_1 \delta^{n-1} + \cdots + A_n$$

with the $A_i \in \mathcal{F}$. In analogy with the definition given in Section 2 we say that L has a *regular singularity* at zero if the A_i are analytic functions at zero. The following Theorem shows that there is no inconsistency in the two uses of the word "regular".

Theorem 7.1 (Fuchs). *The differential system (7.1) has a regular singularity at zero if and only if one (and so any) associated scalar equation is regular at zero.*

If the singularity is irregular it is very difficult to calculate the matrices A and W since it is difficult to compute the map T. On the contrary, if the singularity is regular there is an algebraic method for computing A which avoids analytic continuation. This is particularly important for our aims since we shall be interested in the p-adic case where analytic continuation is not available (if two disks overlap then one is contained in the other by the strong triangle inequality).

8.   Formal Fuchsian Theory.

In order to give an algebraic treatment of the Fuchsian theory we first require a definition of the matrix X^A encountered in its classical incarnation in the preceding Section. Let K be an algebraically closed field of characteristic zero, and let $\mathcal{F}$ be a differential extension field of $K((X))$ containing the algebraic closure of $K((X))$ and containing a solution, which we will call $\log X$, of the equation $\delta y = 1$, where as usual $\delta = X d/dX$. For every positive integer n we fix a root, which we call $X^{1/n}$, of the equation $Z^n = X$ in such a way that $(X^{1/mn})^n = X^{1/m}$, for all m, $n > 0$. Moreover, for all m, $n > 0$, we define

$$X^{m/n} = (X^{1/n})^m , \quad X^{-m/n} = (X^{m/n})^{-1} .$$

Clearly, for $\alpha \in \mathbf{Q}$ we have $\delta X^\alpha = \alpha X^\alpha$.

We now wish to extend the definition of X^α to all $\alpha \in K$. To this end, for each non-zero class $\overline{\alpha}$ of $(K\backslash\mathbf{Q})/\mathbf{Z}$ we choose a representative α and we enlarge $\mathcal{F}$ so as to contain a non-zero solution in $\mathcal{F}$, which we call X^α, for the equation $\delta y = \alpha y$. Clearly any other non-zero solution in $\mathcal{F}$ of this equation is obtained by multiplying X^α by a non-zero constant in $\mathcal{F}$, that is an element of the subfield $\mathcal{F}_0 = \ker(\delta, \mathcal{F})$. We also assume that

$$X^{-\alpha} = (X^\alpha)^{-1} ,$$

that is we take $-\alpha$ to be the representative of $-\overline{\alpha}$, and choose $X^{-\alpha}$ as just shown. Finally, we define

$$X^{\alpha+m} = X^\alpha X^m$$

for all $m \in \mathbf{Z}$. In this way X^α is defined for every $\alpha \in K$. Moreover the two preceding equalities are easily verified for any $\alpha \in K$ and each $m \in \mathbf{Z}$.

Warning.   Let α and $\beta \in K$. Then there exists a non-zero constant $c_{\alpha,\beta} \in K$ such that $X^\alpha X^\beta = c_{\alpha,\beta} X^{\alpha+\beta}$. However, the constant $c_{\alpha,\beta}$ need not be equal to 1, if α, $\beta \notin \mathbf{Q}$. In particular for each $\alpha \in K$ and $n \in \mathbf{Z}$ there exists a non-zero $c \in \mathcal{F}_0$, which depends on α and n, such that

$$(X^\alpha)^n = c X^{n\alpha} .$$

Observe that the symbol $(X^n)^\alpha$ for $n > 1$ has not been defined.

Now let $\Delta \in \mathcal{M}_n(K)$ be the diagonal matrix

$$\Delta = \begin{pmatrix} \Delta_1 & & & \\ & \Delta_2 & & \\ & & \ddots & \\ & & & \Delta_n \end{pmatrix} .$$

Then we define the matrix $X^\Delta \in \mathcal{M}_n(\mathcal{F})$ to be the diagonal matrix

$$X^\Delta = \begin{pmatrix} X^{\Delta_1} & & & \\ & X^{\Delta_2} & & \\ & & \ddots & \\ & & & X^{\Delta_n} \end{pmatrix} . \qquad (8.1)$$

On the other hand, if $N \in \mathcal{M}_n(K)$ is a nilpotent matrix then we define

$$X^N = \sum_{j=0}^\infty \frac{N^j}{j!} (\log X)^j = \exp(N \log X) . \qquad (8.2)$$

Let Δ, N as above and assume $\Delta N = N \Delta$. Then, from $(\Delta_i - \Delta_j)N_{ij} = 0$, it follows immediately that $X^\Delta N = N X^\Delta$ so that

$$X^\Delta X^N = X^N X^\Delta .$$

Moreover it is clear from the definitions that

$$(X^\Delta)^{-1} = X^{-\Delta}$$
$$(X^N)^{-1} = X^{-N} .$$

For a general $A \in \mathcal{M}_n(K)$ let $P \in G\ell(n, K)$ be such that $A = P(\Delta + N)P^{-1}$ with Δ diagonal and N nilpotent satisfying $\Delta N = N\Delta$. Then we define

$$X^A = P X^\Delta X^N P^{-1} . \qquad (8.3)$$

We now check that this gives a well defined matrix in the sense that if we choose some other representation, say $A = P'(\Delta' + N')P'^{-1}$ then we obtain the same formula for X^A as an element of $\mathcal{M}_n(\mathcal{F})$. To this end, it suffices to remark that PNP^{-1} is uniquely defined by A so that $PNP^{-1} = P'N'P'^{-1}$. On the other hand, by definition,

$$X^{PNP^{-1}} = \sum_{j=0}^\infty \frac{(PNP^{-1})^j}{j!} (\log X)^j$$

$$= P \left(\sum_{j=0}^\infty \frac{N^j}{j!} (\log X)^j \right) P^{-1}$$

$$= P X^N P^{-1} .$$

Likewise we obtain, $X^{P'N'P'^{-1}} = P'X^{N'}P'^{-1}$ and therefore $PX^N P^{-1} = P'X^{N'}P'^{-1}$. So there remains only to check that if $P\Delta P^{-1} = P'\Delta'P'^{-1}$ then $PX^\Delta P^{-1} = P'X^{\Delta'}P'^{-1}$. Replacing P by $(P')^{-1}P$ we have reduced to proving that if $P\Delta = \Delta'P$ then $PX^\Delta = X^{\Delta'}P$. To see this, let $P = (P_{ij})$. Then $P\Delta = \Delta'P$ means $P_{ij}\Delta_j = \Delta'_i P_{ij}$ for all i, j so that $P_{ij}X^{\Delta_j} = X^{\Delta'_i}P_{ij}$ as desired.

It is not difficult to verify that for every $A \in \mathcal{M}_n(K)$ the matrix X^A is invertible in $\mathcal{M}_n(\mathcal{F})$; more precisely

$$(X^A)^{-1} = X^{-A} .$$

Moreover, X^A is a matrix solution of the system $\delta - A$, that is

$$\delta X^A = A X^A .$$

Remark 8.1. In the sequel whenever we refer to X^A in a neighborhood of zero we will mean the object introduced above. For any non-zero a in an extension field of K we shall also use the symbol $(X/a)^A$ (but when $a = 1$ we again write simply X^A rather that $(X/1)^A$) to denote the normalized matrix solution of $\delta Y = AY$ at $x = a$. In this case

$$\left(\frac{X}{a}\right)^A = g_A\left(\frac{X}{a} - 1\right) ,$$

where

$$g_A(X) = \sum_{s=0}^{\infty} \binom{A}{s} X^s ,$$

$\binom{A}{0} = I_n$ and for any positive integer s

$$\binom{A}{s} = \frac{1}{s!}A(A - I_n)\cdots(A - (s-1)I_n)$$

(cf. Section IV.7).

Before proceeding we introduce some useful notation. Let (F, δ) be any differential field and consider the system

$$\delta y = Gy \qquad \text{with} \quad G \in \mathcal{M}_n(F) .$$

If we set $z = Hy$ with $H \in G\ell(n, F)$ and we put

$$G_{[H], \delta} = G_{[H]} = (\delta H)H^{-1} + HGH^{-1} , \qquad (8.4)$$

then we obtain

$$\delta z = G_{[H]}z .$$

Note that if $H_1, H_2 \in G\ell(n, F)$ then

$$G_{[H_1][H_2]} = (G_{[H_1]})_{[H_2]} = G_{[H_2 H_1]} . \qquad (8.5)$$

In the following lemma we describe what are classically referred to as *shearing transformations*.

Lemma 8.2. *Let $G \in \mathcal{M}_n(K[[X]])$ and let α be an eigenvalue of $G(0)$. Then there exists $H \in G\ell(n, K[X, X^{-1}])$ such that $G_{[H]} \in \mathcal{M}_n(K[[X]])$ and such that the eigenvalues of $G_{[H]}(0)$ coincide including multiplicities with those of $G(0)$, except that the eigenvalue α is replaced by $\alpha + 1$. A similar statement holds with α replaced by $\alpha - 1$.*

Proof. Assume $G(0)$ to be upper triangular, and assume further

$$G(0) = \begin{pmatrix} J & B \\ 0 & D \end{pmatrix}$$

where $B \in \mathcal{M}_{(n-m) \times m}(K)$, $D \in \mathcal{M}_m(K)$, $J \in \mathcal{M}_{n-m}(K)$ and both J and D are upper triangular matrices

$$J = \begin{pmatrix} \alpha & & * \\ & \ddots & \\ 0 & & \alpha \end{pmatrix}, \quad D = \begin{pmatrix} \alpha_1 & & * \\ & \ddots & \\ 0 & & \alpha_m \end{pmatrix}$$

with $\alpha_i \neq \alpha$ for all i. Let

$$H = \begin{pmatrix} X I_{n-m} & 0 \\ 0 & I_m \end{pmatrix} .$$

If we write

$$G = \begin{pmatrix} G_1 & G_2 \\ G_3 & G_4 \end{pmatrix}$$

then

$$G_{[H]} = \begin{pmatrix} X I_{n-m} & 0 \\ 0 & 0 \end{pmatrix} \begin{pmatrix} \frac{1}{X} I_{n-m} & 0 \\ 0 & I_m \end{pmatrix}$$
$$+ \begin{pmatrix} X I_{n-m} & 0 \\ 0 & I_m \end{pmatrix} \begin{pmatrix} G_1 & G_2 \\ G_3 & G_4 \end{pmatrix} \begin{pmatrix} \frac{1}{X} I_{n-m} & 0 \\ 0 & I_m \end{pmatrix}$$
$$= \begin{pmatrix} I_{n-m} & 0 \\ 0 & 0 \end{pmatrix} + \begin{pmatrix} G_1 & X G_2 \\ \frac{1}{X} G_3 & G_4 \end{pmatrix} .$$

Since $G_3(0) = 0$ we have $X^{-1} G_3 \in K[[X]]$ and

$$G_{[H]}(0) = \begin{pmatrix} J + I_{n-m} & 0 \\ * & D \end{pmatrix} .$$

Thus the matrix $G_{[H]}(0)$ has $\alpha + 1$ as an eigenvalue with the same multiplicity as that of α for $G(0)$, while the eigenvalues $\alpha_1, \ldots, \alpha_m$ remain unchanged. If $G(0)$ is not upper triangular, we find a matrix $C \in G\ell(n, K)$ such that $C G(0) C^{-1}$ is upper triangular and then apply the preceding technique to $G_{[C]}$.

The transformation under which $\alpha \to \alpha - 1$ and $\alpha_i \to \alpha_i$ may be obtained with a preliminary reduction of $G(0)$ to lower triangular form, followed by an application of the transformation with matrix

$$H = \begin{pmatrix} \frac{1}{X} I_{n-m} & 0 \\ 0 & I_m \end{pmatrix} .$$

Q.E.D.

Let $G \in \mathcal{M}_n(K[[X]])$ and let $\alpha_1, \ldots, \alpha_n$ be the eigenvalues of $G(0)$. We say that $G(0)$ has *prepared eigenvalues* if the following two conditions are satisfied

$$\begin{aligned} &\text{if} \quad \alpha_i \in \mathbf{Z} \quad \text{then} \quad \alpha_i = 0 , \\ &\text{if} \quad \alpha_i - \alpha_j \in \mathbf{Z} \quad \text{then} \quad \alpha_i = \alpha_j . \end{aligned} \tag{8.6}$$

Corollary 8.3. *Let G be as in Lemma 8.2. There exists a matrix H in $G\ell(n, K[X, X^{-1}])$ such that the eigenvalues of $G_{[H]}(0)$ are prepared.*

Proof. This follows immediately on applying Lemma 8.2 a finite number of times. **Q.E.D.**

Lemma 8.4. *Let K be any field of characteristic zero. Let $A, B \in \mathcal{M}_n(K)$ and consider the linear map*

$$\psi_{A,B} : \mathcal{M}_n(K) \longrightarrow \mathcal{M}_n(K)$$
$$X \longmapsto AX - XB .$$

Then

$$\det(tI - \psi_{A,B}) = \prod_{\mu, \Delta_B(\mu)=0} \Delta_A(t + \mu) \qquad (8.7)$$

where $\Delta_A(t)$, $\Delta_B(t)$ denote the characteristic polynomials of A and B respectively, and I denotes the identity map of $\mathcal{M}_n(K)$. In particular, if $\lambda_1, \dots, \lambda_n$ are the eigenvalues of A and $\mu_1, \dots, \mu_n$ are the eigenvalues of B then $\lambda_i - \mu_j$, for $i, j = 1, 2, \dots, n$, are the eigenvalues of $\psi_{A,B}$.

Proof. Let e_{ij} be the matrix with 1 in the (i, j) position and 0 elsewhere; if $A = (A_{ij})$, $B = (B_{ij})$ then

$$Ae_{ij} - e_{ij}B = \sum_{s,k} A_{sk} e_{sk} e_{ij} - \sum_{s,k} B_{sk} e_{ij} e_{sk} = \sum_s A_{si} e_{sj} - \sum_k B_{jk} e_{ik} .$$

If B is diagonal, that is $B_{ij} = \mu_i \delta_{ij}$ then

$$\psi_{A,B} e_{ij} = \sum_{s=1}^n A_{si} e_{sj} - \mu_j e_{ij} .$$

Letting V_j be the span of $\{e_{1j}, \dots, e_{nj}\}$, we see that

$$\mathcal{M}_n(K) = V_1 \oplus \cdots \oplus V_n$$

and that $\psi_{A,B}$ is stable on each V_j with matrix $A - \mu_j I_n$ relative to the basis indicated. Thus

$$\det\left(tI_n - (\psi_{A,B}|V_j)\right) = \det\left((t + \mu_j)I_n - A\right) = \Delta_A(t + \mu_j) .$$

Thus (8.7) holds if B is diagonal. Now observe that if the B_{ij} are independent indeterminates then the matrix B is diagonalizable and so (8.7) holds (after extending the base field). Since the right hand side of (8.7) is a symmetric polynomial in the roots of Δ_B it follows that both sides are polynomials over **Z** in t and the coefficients of A and B. Thus it is possible to specialize to any matrix B.   **Q.E.D.**

In what follows we will use ψ_A to denote $\psi_{A,A}$.

Proposition 8.5. *Let $G \in \mathcal{M}_n(K[[X]])$ and assume that the eigenvalues of $G(0)$ are prepared. Then the system $\delta - G$ has a matrix solution of the type $W X^{G(0)}$ where $W \in G\ell(n, K[[X]])$ and $W(0) = I_n$.*

Proof. We seek a

$$W = I_n + W_1 X + W_2 X^2 + \cdots$$

with $W_i \in \mathcal{M}_n(K)$ such that

$$\delta(W X^{G(0)}) = G(X) W X^{G(0)} ,$$

that is

$$\delta W + W G(0) = GW . \tag{8.8}$$

If we write

$$G = G_0 + G_1 X + G_2 X^2 + \cdots$$

with $G_i \in \mathcal{M}_n(K)$, and $G_0 = G(0)$, then (8.8) gives an infinite sequence of equations:

$$
\begin{aligned}
0 =& G_0 - G_0 \\
W_1 =& G_0 W_1 - W_1 G_0 + G_1 \\
2W_2 =& G_0 W_2 - W_2 G_0 + (G_1 W_1 + G_2) \\
& \cdots \\
mW_m =& G_0 W_m - W_m G_0 + (G_1 W_{m-1} + G_2 W_{m-2} + \cdots + G_m) \\
& \cdots .
\end{aligned}
$$

We set $W_0 = I_n$ and construct the sequence W_m by induction. Suppose then that $W_1, W_2, \ldots, W_{m-1} \in \mathcal{M}_n(K)$ have been determined. To find $W_m \in \mathcal{M}_n(K)$ we will apply Lemma 8.4. Indeed we observe that $\det(mI - \psi_{G_0}) \neq 0$ for $m \in \mathbf{N}$, $m \geq 1$, by hypothesis, whence the linear map $mI - \psi_{G_0}$ is invertible so that

$$W_m = (mI - \psi_{G_0})^{-1}(G_1 W_{m-1} + \cdots + G_m) .$$

Q.E.D.

If G satisfies the hypotheses of Proposition 8.5, the matrix W described therein will be called the *uniform part* of the matrix solution at zero of the system $\delta - G$ and will be denoted by Y_G.

Theorem 8.6. *Let K be an algebraically closed field of characteristic zero and let $G \in \mathcal{M}_n(K((X)))$. If the system $\delta - G$ has a solution matrix of the type WX^A with $A \in \mathcal{M}_n(K)$ and $W \in G\ell(n, K((X)))$, then there exists $H \in G\ell(n, K(X))$ such that $G_{[H]} \in \mathcal{M}_n(K[[X]])$. Conversely, if $G \in \mathcal{M}_n(K[[X]])$, there exists a matrix solution WX^A of $\delta - G$ with $W \in G\ell(n, K((X)))$ and $A \in \mathcal{M}_n(K)$.*

Proof. If $W \in G\ell(n, K[[X]])$ then from $\delta(WX^A) = GWX^A$ we deduce $G = (\delta W + WA)W^{-1} \in G\ell(n, K[[X]])$. So for the first part of the theorem it is enough to show that given $W \in G\ell(n, K((X)))$ there exists $H \in G\ell(n, K(X))$ such that $HW \in G\ell(n, K[[X]])$. By the theorem of elementary divisors we have $W = \Lambda X^\Delta \Lambda'$, where Λ and Λ' lie in $G\ell(n, K[[X]])$ and Δ is a diagonal element of $\mathcal{M}_n(\mathbf{Z})$. We choose $m \in \mathbf{N}$ such that both $X^m X^\Delta$ and $X^m X^{-\Delta}$ lie in $\mathcal{M}_n(K[X])$. We choose $H_1 \in \mathcal{M}_n(K[X])$ such that $\Lambda \equiv H_1 \mod X^{1+2m}$. Hence $H_1(0) = \Lambda(0)$ is invertible and so $H_1 \in G\ell(n, K(X)) \cap G\ell(n, K[[X]])$. We put $H = X^{-\Delta} H_1^{-1}$ and assert that $HW \in G\ell(n, K[[X]])$. It is enough to prove that $H\Lambda X^\Delta \in G\ell(n, K[[X]])$. Indeed there exists a matrix $B \in \mathcal{M}_n(K[[X]])$ such that we have

$$
\begin{aligned}
H\Lambda X^\Delta &= X^{-\Delta}(H_1^{-1}\Lambda)X^\Delta \\
&= X^{-\Delta}\left(I_n + X^{1+2m}B\right)X^\Delta \\
&= I_n + (X^m X^{-\Delta})XB(X^m X^\Delta).
\end{aligned}
$$

This last term lies in $I_n + X\mathcal{M}_n(K[[X]]) \subset G\ell(n, K[[X]])$ as desired.

Conversely, assume $G \in \mathcal{M}_n(K[[X]])$. Then Corollary 8.3 gives a matrix $H \in G\ell\left(n, K[X, X^{-1}]\right)$ such that $G_{[H]}(0)$ has prepared eigenvalues, so by Proposition 8.5 the required solution is $HY_{G_{[H]}}X^{G_{[H]}(0)}$. **Q.E.D.**

Remark 8.7. Let $\mathcal{F}$ be any differential field with non-trivial derivation δ. Let $y_1, \ldots, y_n \in \mathcal{F}$ have Wronskian determinant (see (1.3)) $w(y_1, \ldots, y_n) \neq 0$. Then we have

$$
\frac{w(y, y_1, \ldots, y_n)}{w(y_1, \ldots, y_n)} = \delta^n y + A_1 \delta^{n-1} y + \cdots + A_n y
$$

with all $A_i \in \mathcal{F}$. If we write $L = \delta^n + A_1 \delta^{n-1} + \cdots + A_n$ then $Ly = 0$ is the only monic equation of order n having $y_1, \ldots, y_n$ as solutions.

Now assume that

$$
(y_1, \ldots, y_n) = (u_1, \ldots, u_n)X^B
$$

with $u_i \in K((X))$ and $B \in \mathcal{M}_n(K)$. Then
$$\delta(y_1, \dots, y_n) = \big(\delta(u_1, \dots, u_n) + (u_1, \dots, u_n)B\big)X^B \;,$$
and $\delta(u_1, \dots, u_n) + (u_1, \dots, u_n)B$ has elements in $K((X))$. Iterating, we obtain
$$\begin{pmatrix} y_1 & \cdots & y_n \\ \delta y_1 & \cdots & \delta y_n \\ \cdots\cdots\cdots\cdots \\ \delta^n y_1 & \cdots & \delta^n y_n \end{pmatrix} = \begin{pmatrix} v_{01} & \cdots & v_{0n} \\ v_{11} & \cdots & v_{1n} \\ \cdots\cdots\cdots \\ v_{n1} & \cdots & v_{nn} \end{pmatrix} X^B = V X^B$$
with $v_{ij} \in K((X))$. From this it easily follows that for $j = 1, \dots, n$ we have
$$A_j = \pm \frac{\det(V_{n-j}X^B)}{\det(V_n X^B)}$$
where V_i is the matrix V with the row of index i removed ($i = 0, \dots, n$). We can conclude that $A_j \in K((X))$ and $L \in K((X))[\delta]$.

Lemma 8.8. *Let*
$$L = \delta^n + A_1 \delta^{n-1} + \cdots + A_n$$
with $A_i \in K((X))$. For $\alpha \in K$, let
$$L' = \frac{1}{X^\alpha} \circ L \circ X^\alpha \;.$$
Then L is regular at zero if and only if L' is regular at zero. The same statement holds for
$$L' = \frac{1}{u} \circ L \circ u$$
with $u \in K((X))$, $u \neq 0$.

Proof.   Obviously it suffices to prove the statement only in one direction.

We have
$$\frac{1}{X^\alpha} \circ \delta^j \circ X^\alpha = (\delta + \alpha)^j$$
so that
$$L' = (\delta + \alpha)^n + A_1(\delta + \alpha)^{n-1} + \cdots + A_n$$
$$= \delta^n + \left(\binom{n}{1}\alpha + A_1 \right) \delta^{n-1} + \cdots + (\alpha^n + A_1\alpha^{n-1} + \cdots + A_n)$$

and the coefficients of L' are in $K[[X]]$ if $A_1, \dots, A_n \in K[[X]]$.

If $0 \neq u \in K((X))$ and $u = X^\alpha v$, where $\alpha \in \mathbf{Z}$ and v is a unit in $K[[X]]$, then the preceding case allows one to take $\alpha = 0$. If $L \in K[[X]][\delta]$ then clearly $v^{-1} \circ L \circ v \in K[[X]][\delta]$.   **Q.E.D.**

Theorem 8.9. Let $u_1, \ldots, u_n \in K((X))$, and $A \in \mathcal{M}_n(K)$. Let

$$(y_1, \ldots, y_n) = (u_1, \ldots, u_n) X^A$$

with $w(y_1, \ldots, y_n) \neq 0$. Let $L \in K((X))[\delta]$ be defined as above. Then zero is a regular singularity of L.

Proof. Assume that A is similar to the matrix

$$\begin{pmatrix} \alpha I_l + N & 0 \\ 0 & A' \end{pmatrix}$$

where N is a nilpotent matrix of order l of the form

$$\begin{pmatrix} 0 & 1 & \cdots & 0 \\ 0 & 0 & \ddots & 0 \\ \vdots & \vdots & \ddots & 1 \\ 0 & 0 & \cdots & 0 \end{pmatrix}.$$

Then there exists $C \in G\ell(n, K)$ such that

$$A = C \begin{pmatrix} \alpha I_l + N & 0 \\ 0 & A' \end{pmatrix} C^{-1}.$$

Now $\vec{y} = \vec{u} X^A = \vec{u} C C^{-1} X^A C C^{-1}$; but $\vec{y} C = (z_1, \ldots, z_n)$ is again a basis for the solutions of L and $\vec{u} C = (v_1, \ldots, v_n)$ is an element of $K((X))^n$. Moreover,

$$C^{-1} X^A C = X^{\begin{pmatrix} \alpha I_l + N & 0 \\ 0 & A' \end{pmatrix}}.$$

Therefore we can assume from the beginning that

$$A = \begin{pmatrix} \alpha I_l + N & 0 \\ 0 & A' \end{pmatrix}.$$

Now letting L act on components we have

$$0 = L(\vec{u} X^A) = (L \circ X^\alpha)(\vec{u} X^{A - \alpha I_n})$$

so that $L' = X^{-\alpha} \circ L \circ X^\alpha$ has $\vec{u} X^{A - \alpha I_n}$ as a basis of the solution space. Therefore Lemma 8.8 allows us to suppose that

$$A = \begin{pmatrix} N & 0 \\ 0 & A' \end{pmatrix}.$$

Then

$$X^A = \begin{pmatrix} X^N & 0 \\ 0 & X^{A'} \end{pmatrix}$$

and

$$X^N = \begin{pmatrix} 1 & \log X & \frac{1}{2}\log^2 X & \ldots & \cdot \\ 0 & 1 & \log X & \ldots & \cdot \\ 0 & 0 & 1 & \ldots & \cdot \\ \multicolumn{5}{c}{\dotfill} \\ 0 & 0 & 0 & \ldots & \log X \\ 0 & 0 & 0 & \ldots & 1 \end{pmatrix}$$

which shows that

$$\vec{y} = (u_1, y_2, \ldots, y_n)$$

so that $u_1 \neq 0$, $u_1 \in K((X))$. Now

$$0 = L\vec{y} = (L \circ u_1)\left(\frac{1}{u_1}\vec{y}\right)$$

so that

$$\left(\frac{1}{u_1} \circ L \circ u_1\right)\left(\frac{1}{u_1}\vec{y}\right) = 0 .$$

By Lemma 8.8 we can assume

$$\vec{y} = (1, y_2, \ldots, y_n) = (1, u_1, \ldots, u_{n-1})X^{\begin{pmatrix} N & 0 \\ 0 & A' \end{pmatrix}}$$

with $w(1, y_2, \ldots, y_n) \neq 0$ which implies that $w(\delta y_2, \ldots, \delta y_n) \neq 0$. On the other hand,

$$(0, \delta y_2, \ldots, \delta y_n) = \delta(1, y_2, \ldots, y_n)$$

$$= \left((0, \delta u_1, \ldots, \delta u_{n-1}) + (1, u_1, \ldots, u_{n-1})\begin{pmatrix} N & 0 \\ 0 & A' \end{pmatrix}\right)X^A ,$$

$$= (0, v_1, \ldots, v_{n-1})\begin{pmatrix} X^N & 0 \\ 0 & X^{A'} \end{pmatrix}$$

so that projecting on the last $n - 1$ components we obtain

$$(\delta y_2, \ldots, \delta y_n) = (v_1, \ldots, v_{n-1})\begin{pmatrix} X^{N'} & 0 \\ 0 & X^{A'} \end{pmatrix}$$

where N' is the canonical nilpotent matrix of order $l - 1$ (obtained by removing the first row and first column of N).

We can now use induction on n. Let L' be the equation for δy_2, $\ldots, \delta y_n$; then

$$L' \circ \delta = L .$$

By induction $L' = \delta^{n-1} + B_1 \delta^{n-2} + \cdots + B_{n-1}$ with the B_j in $K[[X]]$. Therefore also L has coefficients in $K[[X]]$. **Q.E.D.**

Remark 8.10. Let

$$G = \begin{pmatrix} 0 & 1 & \cdots & 0 \\ \vdots & \vdots & \ddots & \vdots \\ 0 & 0 & \cdots & 1 \\ -A_n & -A_{n-1} & \cdots & -A_1 \end{pmatrix}$$

with $A_i \in K((X))$ and let $\mathcal{U} = W X^A$, with $W \in G\ell(n, K((X)))$, $A \in \mathcal{M}_n(K)$, be a solution matrix of $\delta - G$. If $(u_1, \ldots, u_n)$ is the first row of the matrix W, then $(y_1, \ldots, y_n) = (u_1, \ldots, u_n) X^A$ is a set of n linearly independent solutions of the scalar equation $Ly = 0$, where $L = \delta^n + A_1 \delta^{n-1} + \cdots + A_n$. In fact $w(y_1, \ldots, y_n) = \det \mathcal{U} \neq 0$.

We summarize the results of this section. Let $G \in \mathcal{M}_n(K((X)))$. Theorems 8.6 and 8.9 and the Theorem of the Cyclic Vector give the equivalence of the following conditions:

- the system $\delta - G$ has a solution matrix of the type $W X^A$ where $W \in G\ell(n, K((X)))$, and $A \in \mathcal{M}_n(K)$;
- $G_{[H]} \in \mathcal{M}_n(K[[X]])$ for some $H \in G\ell(n, K((X)))$;
- one (and then any) scalar equation associated to $\delta - G$ has a regular singularity at zero.

We say that the system $\delta - G$ has a *regular singularity* at zero if one of the preceding conditions is satisfied.

Observe finally that if we take $K = \mathbf{C}$ the results of this section give an algebraic proof for all of Fuchs' Theorem except the point regarding the convergence of solutions. For this point see Appendix II at the end of the book.

The eigenvalues of the matrix A are uniquely determined modulo $\mathbf{Z}$ (for an explicit proof see Lemma V.2.4). They are called the *exponents* of $\delta - G$ at the origin.

CHAPTER IV

EFFECTIVE BOUNDS. ORDINARY DISKS

1. p-adic Analytic Functions.

We will need the p-adic analogue of the classical Cauchy estimates on the coefficients of a Laurent series.

Proposition 1.1. *Let Ω be a field with non-archimedean valuation $|-|$ and with infinite residue field $\overline{\Omega}$. Let*

$$f(X) = \sum_{s=-\infty}^{\infty} A_s X^s$$

with $A_s \in \Omega$ and let $R \in G_\Omega$, the valuation group of Ω. If $f(x)$ converges for $|x| = R$, then

$$\sup_{|x|=R} |f(x)| = \sup_s(|A_s|R^s).$$

Proof. Let $M = \sup_s(|A_s|R^s)$ (which is certainly finite in view of $\lim_{s \to \pm\infty} |A_s|R^s = 0$). Furthermore, if $|x| = R$ then

$$|f(x)| \le \sup(|A_s|R^s) = M ,$$

so it will suffice to prove the other inequality.

We may suppose without loss of generality that $R = 1$. Moreover, there are only finitely many indices such that $|A_s| = M$ so that we may write

$$f(X) = g(X) + h(X)$$

where

$$g(X) = \sum_{|A_s|<M} A_s X^s$$

114

and

$$h(X) = \sum_{|A_s|=M} A_s X^s$$

is a Laurent polynomial. Since the valuation is non-archimedean, if $|x| = 1$ then $|g(x)| < M$. Finally, for the polynomial $h(X)$ we have

$$\sup_{|x|=1} |h(x)| = |h|_{\text{gauss}} = M$$

since $\overline{\Omega}$ is infinite. In fact, let $y \in \Omega$ be such that $|y| = M$ and consider $y^{-1}h(x) = h_1(x) \in \mathcal{O}_\Omega$; its reduction $\overline{h}_1$ modulo $\mathfrak{p}_\Omega$ is not the zero polynomial so that there exists $t \in \mathcal{O}_\Omega$ with $|t| = 1$ such that $\overline{h}_1(\overline{t}) \neq 0$, that is, $|h_1(t)| = 1$ as desired. An alternative approach to this last point is the following. The Newton polygon of $h(X)$ has only one finite side, and that side is horizontal with height $m = \text{ord}\,A_s$ where A_s is any coefficient of $h(X)$. Therefore, we can write

$$h(X) = A \prod_i (X - \alpha_i)$$

where $A \in \Omega$ has order m and all the α_i belong to an extension of Ω and have order zero. Since there are infinitely many distinct residue classes we can surely find $x \in \Omega$ with $\text{ord}\,x = 0$ and such that $\text{ord}\,(x - \alpha_i) = 0$ for all i. Therefore $\text{ord}\,h(x) = \text{ord}\,A = m$.   **Q.E.D.**

We now assume that the field Ω is "sufficiently large", and let

$$f(X) = \sum_{s=-\infty}^{\infty} A_s X^s$$

be convergent for $|x| = R$. We define

$$|f|_0(R) = \sup_{\substack{x \in \Omega \\ |x|=R}} |f(x)| . \tag{1.1}$$

Now let

$$g(X) = \sum_{s=-\infty}^{\infty} B_s X^s$$

be another Laurent series convergent for $|x| = R$. Then it is easily seen that the series $C_n = \sum_s A_s B_{n-s}$ converges in Ω for all $n \in \mathbf{Z}$, so that we can define the series fg by

$$(fg)(X) = \sum_{s=-\infty}^{\infty} C_s X^s$$

which also converges for $|x| = R$. We have

$$|fg|_0(R) = |f|_0(R)|g|_0(R) \; ;$$

in fact, let $\alpha, A, B, C \in \Omega$ be such that $|\alpha| = R$, $|A| = |f|_0(R)$, $|B| = |g|_0(R)$ and $|C| = |fg|_0(R)$. Then each of the three series $A^{-1}f(\alpha X)$, $B^{-1}g(\alpha X)$, and $C^{-1}(fg)(\alpha X)$ has coefficients in $\mathcal{O}_\Omega$, and their reductions mod $\mathfrak{p}_\Omega$ are non-zero Laurent polynomials. We can then find $x \in \Omega$ with $|x| = 1$ and $|f(\alpha x)| = |A|$, $|g(\alpha x)| = |B|$, and $|(fg)(\alpha x)| = |C|$. Clearly we also have

$$|f + g|_0(R) \leq \sup\left(|f|_0(R), |g|_0(R)\right) \; ,$$

and also

$$|f|_0(R) = 0 \qquad \text{if and only if} \quad f = 0 \; .$$

Thus, the Laurent series converging for $|x| = R$ form an integral domain. We can extend the definition of $|-|_0(R)$ to the quotient field by putting

$$\left|\frac{f}{g}\right|_0(R) = \frac{|f|_0(R)}{|g|_0(R)} \; .$$

Clearly this definition is effective. Note, however, that $|f/g|_0(R)$ is always finite, even if $f(X)/g(X)$ has a pole on $|x| = R$. For example, $|X/(X-1)|_0(1) = 1$. For any $a \in \Omega$ we may define $|f|_a(R)$ in a completely analogous fashion for $f = \sum_s A_s(X - a)^s$ converging for $|x - a| = R$.

Theorem 1.2 (Maximum Modulus Principle). *Let Γ be the annulus $R_1 \leq |x| \leq R_2$ and let $\partial\Gamma$ be the boundary $\left\{ x \,\middle|\, |x| \in \{R_1, R_2\} \right\}$. Let*

$$f = \sum_{s=-\infty}^{\infty} A_s X^s$$

be convergent on Γ. Then

$$\sup_{x \in \Gamma} |f(x)| = \sup_{x \in \partial\Gamma} |f(x)| \; .$$

Proof. For $R \in [R_1, R_2]$ let $\lambda_R = -\log R$. By Proposition 1.1 the number $-\log|f|_0(R)$ is the y-intercept of the line of support of slope $-\lambda_R$ to the Newton polygon of f. This intercept is a continuous, locally linear function of λ_R which decreases (resp. increases) with λ_R if the abscissa of the point of contact is positive (resp. negative). Thus the intercept assumes its most negative value at either R_1 or R_2. **Q.E.D.**

Let $\mathcal{A}_0$ be the ring of *analytic functions* in $D(0, 1^-)$, that is the subring of $\Omega[[X]]$ consisting of series converging for $|x| < 1$, and let $\mathcal{A}_0'$ be the quotient field of $\mathcal{A}_0$. We call $\mathcal{A}_0'$ the field of *meromorphic functions* on $D(0, 1^-)$. For $f \in \mathcal{A}_0'$, the function $R \longmapsto |f|_0(R)$ is continuous as one verifies using the theory of the Newton polygon, but not necessarily bounded for $0 \le R < 1$, as is shown by the example $f(X) = \log(1 - X)$. We define the *boundary seminorm* $\|-\|_0$ on $\mathcal{A}_0'$ by setting

$$\| f \|_0 = \limsup_{R \to 1} |f|_0(R) .$$

It is clear that

$$\| f + g \|_0 \le \sup (\| f \|_0, \| g \|_0)$$

for all $f, g \in \mathcal{A}_0'$ and that

$$\| fg \|_0 \le \| f \|_0 \| g \|_0$$

whenever the right side is defined (the only case excluded is $\| f \|_0 = 0$ $\| g \|_0 = \infty$). Note also that it is quite possible to have $\| f \|_0 = 0$ with $f \ne 0$, as the function $1/\log(1 - X)$ illustrates.

If $f = \sum_s A_s X^s \in \mathcal{A}_0$, then it is easily verified that

$$\| f \|_0 = \sup_s |A_s| .$$

Let $\mathcal{B}_0 = \{f \in \mathcal{A}_0 \,|\, \| f \|_0 < \infty\}$ and let $\mathcal{B}_0'$ be its field of quotients. We call $\mathcal{B}_0$ the ring of *bounded analytic functions* on $D(0, 1^-)$. Then for f, g in $\mathcal{B}_0'$ we have

$$\| fg \|_0 = \| f \|_0 \| g \|_0$$

so that $\| f \|_0$ is a norm on $\mathcal{B}_0'$. Clearly we have

$$\mathcal{B}_0 = \left\{ \sum_{s=0}^{\infty} A_s X^s \,\Big|\, \sup_s |A_s| < \infty \right\}$$

and the ring $\mathcal{B}_0$ is complete with respect to the norm $\|-\|_0$ as one can immediately check.

If $h(X) = f(X)/g(X)$ with f, $g \in \Omega[X]$ is a rational function, then clearly $h \in \mathcal{B}'_0$; moreover we have

$$\| h \|_0 = |h|_{\text{gauss}} \cdot$$

Indeed $\| h \|_0 = \| f \|_0 / \| g \|_0 = |f|_0(1)/|g|_0(1) = |f|_{\text{gauss}}/|g|_{\text{gauss}} = |h|_{\text{gauss}}$.

Analogous definitions may be given for the rings $\mathcal{A}_{a,\rho}$, $\mathcal{A}'_{a,\rho}$, $\mathcal{B}_{a,\rho}$ and $\mathcal{B}'_{a,\rho}$ for every $a \in \Omega$ and every $\rho > 0$ ($\rho \in G_\Omega$) and their relative boundary semi-norms; for example

$$\mathcal{A}_{a,\rho} = \left\{ \sum_s A_s(X - a)^s \text{converging in } D(a, \rho^-) \right\}$$

and, for $f \in \mathcal{A}_{a,\rho}$,

$$\| f \|_{a,\rho} = \limsup_{R \to \rho} |f|_a(R) \cdot$$

Clearly the map

$$\sum_s A_s X^s \longmapsto \sum_s \alpha^{-s} A_s(X - a)^s$$

where $\alpha \in \Omega$ has $|\alpha| = \rho$, is an isometry of $\mathcal{A}_0$ (respectively $\mathcal{A}'_0$, $\mathcal{B}_0$, $\mathcal{B}'_0$) onto $\mathcal{A}_{a,\rho}$ (respectively $\mathcal{A}'_{a,\rho}$, $\mathcal{B}_{a,\rho}$, $\mathcal{B}'_{a,\rho}$).

Now let $f \in \mathcal{A}_0$ with f, $g \in \mathcal{A}_0$ and let $D = d/dX$. Then, if $f(X) = \sum_s A_s X^s$ we have $Df(x) = \sum_s sA_s X^{s-1}$, so that $Df(X) \in \mathcal{A}_0$ and D is a non-trivial derivation of $\mathcal{A}_0$ which can be extended in the usual way to $\mathcal{A}'_0$.

For $\xi \in \mathcal{A}'_0$ we have

$$\left\| \frac{D^s \xi}{s! \xi} \right\|_0 \leq 1 . \tag{1.2}$$

Indeed, if $g \in \mathcal{A}_0$ and $g\xi \in \mathcal{A}_0$ then by Leibniz's formula we have

$$\frac{D^s \xi}{s! \xi} = \frac{D^s(g\xi)}{s! g\xi} - \sum_{n=0}^{s-1} \frac{D^{s-n}g}{(s-n)!g} \frac{D^n \xi}{n! \xi} ,$$

so that it suffices to prove (1.2) for $\xi \in \mathcal{A}_0$ and then apply induction. Let $\xi = \sum_s A_s X^s \in \mathcal{A}_0$ and let $0 < R < 1$. We have

$$\frac{1}{n!} D^n \xi = \sum_s A_s \binom{s}{n} X^{s-n}$$

so that

$$\left| \frac{1}{n!} D^n \xi \right|_0 (R) \leq \sup_s(|A_s| R^{s-n}) = \frac{1}{R^n} |\xi|_0(R) \,.$$

Therefore

$$\left| \frac{D^n \xi}{n! \xi} \right|_0 (R) \leq \frac{1}{R^n} \,,$$

from which (1.2) follows immediately. We observe that if $\xi \in \mathcal{B}'_0$ then we also have

$$\left\| \frac{D^n \xi}{n!} \right\|_0 \leq \| \xi \|_0 \,. \tag{1.3}$$

Alternative proof. Let $f, g \in \mathcal{A}_0$ and assume that g has no zero z with $|z| = R < 1$. Then $\xi = f/g$ can be represented as a Laurent series

$$\xi = \sum_{s=-\infty}^{\infty} A_s X^s$$

converging in the annulus $R_1 < |x| \leq R$ where $R_1 = \max\{|\alpha| \,|\, g(\alpha) = 0, |\alpha| < R\}$ (this can be shown as follows: the theory of the Newton polygon allows us to write $g(X) = P(X)h(X)$ where

$$P(X) = A \prod_{|\alpha| < R_1} (X - \alpha)$$

and $h(X) \in \mathcal{A}_0$ has no zeros in the disk $|x| \leq R$. Therefore $1/h(X)$ is a power series in X converging for $|x| \leq R$ and $1/P(X)$ is a finite product of terms of the type

$$\frac{1}{X - \alpha} = \frac{1}{X \left(1 - \frac{\alpha}{X}\right)} = \frac{1}{X} \sum_{s=0}^{\infty} \left(\frac{\alpha}{X}\right)^s$$

which converges for $R_1 < |x|$). Now it suffices to repeat the calculation given above and observe that it is possible to find R arbitrarily close to 1 such that $g(X)$ has no zero on the circumference of radius R.

2.   Effective Bounds. The Dwork-Robba Theorem.

Let $u_1, \dots, u_n \in \mathcal{A}'_0$ be such that $w(u_1, \dots, u_n) \neq 0$. For any non-negative integer s we define $G_{s,0}, G_{s,1}, \dots, G_{s,n-1} \in \mathcal{A}'_0$ by means of

$$\frac{1}{s!} D^s (u_1 \ \dots \ u_n) = (G_{s,0} \quad G_{s,1} \quad \dots \quad G_{s,n-1}) W(u_1, \dots, u_n)$$

where

$$W(u_1, \dots, u_n) = \begin{pmatrix} u_1 & \cdots & u_n \\ u_1' & \cdots & u_n' \\ \cdots\cdots\cdots\cdots\cdots\cdots \\ u_1^{(n-1)} & \cdots & u_n^{(n-1)} \end{pmatrix}$$

is the Wronskian matrix.

The following theorem is due to Dwork and Robba.

Theorem 2.1 (Theorem of Effective Growth). *In the preceding notations we have, for all $s \geq 0$ and $0 \leq j \leq n - 1$,*

$$\| G_{s,j} \|_0 \leq \{s, n - 1\}_p \overset{\text{def}}{=} \sup \frac{1}{|\lambda_1 \lambda_2 \cdots \lambda_{n-1}|_p}$$

where the sup is extended over all sequences $1 \leq \lambda_1 < \lambda_2 < \cdots < \lambda_{n-1} \leq s$.

Proof. We proceed by induction on n. For $n = 1$ it suffices to observe that $\{s, 0\}_p = 1$ and recall (1.2). As an illustration we consider the case $n = 2$ as well. We then have

$$\frac{1}{s!} \begin{pmatrix} u_1^{(s)} & u_2^{(s)} \end{pmatrix} = \begin{pmatrix} G_{s,0} & G_{s,1} \end{pmatrix} \begin{pmatrix} u_1 & u_2 \\ u_1' & u_2' \end{pmatrix}.$$

If we write $u_2 = \tau u_1$ we have

$$\frac{1}{s!} u_1^{(s)} = G_{s,0} u_1 + G_{s,1} u_1'$$
$$\frac{1}{s!} (\tau u_1)^{(s)} = G_{s,0} \tau u_1 + G_{s,1} (\tau u_1)'.$$

If we multiply the first equation by τ and subtract the result from the second to which we have applied Leibniz's rule we find

$$G_{s,1} = \sum_{i=1}^{s} \frac{\tau^{(i)}}{i! \tau'} \frac{u_1^{(j)}}{j! u_1} = \sum_{i=1}^{s} \frac{1}{i} \frac{(\tau')^{(i-1)}}{(i-1)! \tau'} \frac{u_1^{(j)}}{j! u_1}.$$

Hence we obtain

$$\| G_{s,1} \|_0 \leq \sup_{1 \leq i \leq s} \frac{1}{|i|_p} = \{s, 1\}_p.$$

Moreover

$$G_{s,0} = \frac{1}{s!} \frac{u_1^{(s)}}{u_1} - G_{s,1} \frac{u_1'}{u_1},$$

so that

$$\| G_{s,0} \|_0 \leq \sup(1, \| G_{s,1} \|_0) \leq \sup_{1 \leq i \leq s} \frac{1}{|i|_p}.$$

General case: we suppose the theorem known for n and we write

$$(u_1 \cdots u_{n+1}) = u(1\ \tau_1 \cdots \tau_n),$$

so that

$$\frac{1}{s!} D^s (u_1 \cdots u_{n+1}) = \sum_{i+j=s} \frac{u^{(i)}}{i!} \frac{D^j}{j!} (1\ \tau_1 \cdots \tau_n).$$

We now consider the $\infty \times (n+1)$-matrix

$$\mathcal{U} = \begin{pmatrix} u_1 & \cdots & u_{n+1} \\ u_1' & \cdots & u_{n+1}' \\ \cdots\cdots\cdots\cdots\cdots \\ \frac{1}{s!}u_1^{(s)} & \cdots & \frac{1}{s!}u_{n+1}^{(s)} \\ \cdots\cdots\cdots\cdots\cdots \end{pmatrix} = u\, P \begin{pmatrix} 1 & \tau_1 & \cdots & \tau_n \\ 0 & \tau_1' & \cdots & \tau_n' \\ 0 & \frac{1}{2!}\tau_1'' & \cdots & \frac{1}{2!}\tau_n'' \\ \cdots\cdots\cdots\cdots\cdots \end{pmatrix}$$

where P is the lower triangular $\infty \times \infty$ matrix

$$P = \frac{1}{u} \begin{pmatrix} u & 0 & 0 & \cdots \\ u' & u & 0 & \cdots \\ \frac{1}{2!}u'' & u' & u & \cdots \\ \vdots & \vdots & \vdots & \ddots \end{pmatrix} = \begin{pmatrix} 1 & 0 & 0 & \cdots \\ u'/u & 1 & 0 & \cdots \\ \frac{1}{2!}u''/u & u'/u & 1 & \cdots \\ \vdots & \vdots & \vdots & \ddots \end{pmatrix}.$$

Now we can write

$$\begin{pmatrix} \tau_1' & \cdots & \tau_n' \\ (\tau_1')' & \cdots & (\tau_n')' \\ \frac{1}{2!}(\tau_1')'' & \cdots & \frac{1}{2!}(\tau_n')'' \\ \cdots\cdots\cdots\cdots\cdots \end{pmatrix} = HW(\tau_1', \ldots, \tau_n')$$

where

$$W(\tau_1', \ldots, \tau_n') = \begin{pmatrix} \tau_1' & \cdots & \tau_n' \\ (\tau_1')' & \cdots & (\tau_n')' \\ \cdots\cdots\cdots\cdots\cdots\cdots\cdots \\ (\tau_1')^{(n-1)} & \cdots & (\tau_n')^{(n-1)} \end{pmatrix}$$

and H is an $\infty \times n$-matrix whose $(s+1)$-th row has boundary-norm bounded by $\{s, n-1\}_p$ in view of the inductive hypothesis. If we define the diagonal $\infty \times \infty$-matrix Δ by

$$\Delta = \begin{pmatrix} 1 & 0 & 0 & \cdots \\ 0 & 1/2 & 0 & \cdots \\ 0 & 0 & 1/3 & \cdots \\ \vdots & \vdots & \vdots & \ddots \end{pmatrix}$$

we can then write

$$\mathcal{U} = uP \begin{pmatrix} 1 & \tau_1 & \cdots & \tau_n \\ 0 & \Delta H W(\tau_1', \ldots, \tau_n') \end{pmatrix}. \tag{2.1}$$

By definition we have

$$\mathcal{U} = GW \tag{2.2}$$

where G is the $\infty \times (n+1)$-matrix whose $(s+1)$-th row is the vector $(\, G_{s,0} \quad \cdots \quad G_{s,n} \,)$, and

$$W = \begin{pmatrix} u & u\tau_1 & \cdots & u\tau_n \\ u' & (u\tau_1)' & \cdots & (u\tau_n)' \\ \cdots\cdots\cdots\cdots\cdots\cdots\cdots\cdots\cdots \\ u^{(n)} & (u\tau_1)^{(n)} & \cdots & (u\tau_n)^{(n)} \end{pmatrix}$$

$$= \begin{pmatrix} u & 0 & \cdots & & 0 \\ u' & u & 0 & \cdots & 0 \\ u'' & \binom{2}{1}u' & u & 0 & \cdots & 0 \\ \cdots\cdots\cdots\cdots\cdots\cdots\cdots\cdots\cdots \\ u^{(n)} & \binom{n}{n-1}u^{(n-1)} & \cdots & & u \end{pmatrix} \begin{pmatrix} 1 & \tau_1 & \cdots & \tau_n \\ 0 & \tau_1' & \cdots & \tau_n' \\ \cdots\cdots\cdots\cdots\cdots\cdots \\ \cdots\cdots\cdots\cdots\cdots\cdots \\ 0 & \tau_1^{(n)} & \cdots & \tau_n^{(n)} \end{pmatrix}$$

$$= uP_n \begin{pmatrix} 1 & \tau_1 & \cdots & \tau_n \\ 0 & W(\tau_1', \ldots, \tau_n') \end{pmatrix}$$

where

$$P_n = \begin{pmatrix} 1 & 0 & 0 & \ldots & 0 \\ u'/u & 1 & 0 & \ldots & 0 \\ \multicolumn{5}{c}{\cdots\cdots\cdots\cdots\cdots\cdots\cdots\cdots\cdots} \\ u^{(n)}/u & \binom{n}{n-1}u^{(n-1)}/u & & \ldots & 1 \end{pmatrix} .$$

Comparing (2.1) and (2.2), putting $\tau = (\tau_1 \quad \ldots \quad \tau_n)$ and multiplying on the right by

$$u^{-1}\begin{pmatrix} 1 & \tau \\ 0 & W(\tau_1', \ldots, \tau_n') \end{pmatrix}^{-1}$$

we obtain

$$GP_n = P\begin{pmatrix} 1 & \tau \\ 0 & \Delta\, H\, W(\tau_1', \ldots, \tau_n') \end{pmatrix}\begin{pmatrix} 1 & -\tau W(\tau_1', \ldots, \tau_n')^{-1} \\ 0 & W(\tau_1', \ldots, \tau_n')^{-1} \end{pmatrix}$$

$$= P\begin{pmatrix} 1 & 0 \\ 0 & \Delta H \end{pmatrix} .$$

Therefore

$$G = P\begin{pmatrix} 1 & 0 \\ 0 & \Delta\, H \end{pmatrix} P_n^{-1} .$$

We observe that $\| P \|_0 \leq 1$, $\| P_n \|_0 \leq 1$ (where, as usual, we set $\| (\xi_{ij}) \|_0 \overset{\text{def}}{=} \sup_{i,j} \| \xi_{ij} \|_0$). Since $\det P_n = 1$ we have $\| P_n^{-1} \|_0 \leq 1$. Since the h-th row of ΔH is bounded by $(1/|h|_p)\{h - 1, n - 1\}_p$ the $(h + 1)$-th row of

$$\begin{pmatrix} 1 & 0 \\ 0 & \Delta\, H \end{pmatrix} P_n^{-1} \tag{2.3}$$

has the same bound. Since the $(s+1)$-th row of G involves the $(s+1)$-th row of P and elements of rows of the matrix (2.3) up to and including the $(s + 1)$-th, we can conclude that the $(s + 1)$-th row of G is bounded by

$$\sup_{1 \leq h \leq s} \frac{1}{|h|_p} \{h - 1, n - 1\}_p \leq \{s, n\}_p$$

as desired.   **Q.E.D.**

Remark 2.2. If we assume that $u_1, \ldots, u_n \in \mathcal{A}'_{a,\rho}$ with $a \in \Omega$, $\rho > 0$ and that $w(u_1, \ldots, u_n) \neq 0$ the estimate in Theorem 2.1 becomes

$$\| G_{s,j} \|_{a,\rho} \leq \rho^{-s+j} \{s, n-1\}_p .$$

Application 2.3. Let $u_1, \ldots, u_n \in \mathcal{A}_0$ and assume that the Wronskian

$$w(u_1, \ldots, u_n)$$

is never zero in $D(0, 1^-)$. If we write

$$Lu = \frac{w(u, u_1, \ldots, u_n)}{w(u_1, \ldots, u_n)},$$

then

$$L = D^n + A_1 D^{n-1} + \cdots + A_n$$

where the $A_i \in \mathcal{A}'_0$. Let

$$\mathcal{U} = \begin{pmatrix} u_1 & \cdots & u_n \\ u'_1 & \cdots & u'_n \\ \frac{1}{2!}u''_1 & \cdots & \frac{1}{2!}u''_n \\ \cdots\cdots\cdots\cdots\cdots \end{pmatrix} .$$

Then

$$\mathcal{U} = G\, W(u_1, \ldots, u_n)$$

where $G = (G_{s,j})$ and $W(u_1, \ldots, u_n)$ is the Wronskian matrix. If we denote the adjoint matrix of $W(u_1, \ldots, u_n)$ by $\tilde{W}$ then

$$\mathcal{U} \tilde{W} = wG ,$$

so that $wG_{s,j} \in \mathcal{A}_0$ for all $s \geq 0$ and $0 \leq j \leq n-1$. Then, by the maximum modulus principle we have

$$|wG_{s,j}(x)| \leq \| wG_{s,j} \|_0 \leq \| w \|_0 \| G_{s,j} \|_0 \leq \| w \|_0 \{s, n-1\}_p .$$

Now we observe that since w has no zeros in $D(0, 1^-)$ the Newton polygon of w has no side of negative slope, so that $|w(x)| = |w(0)|$ for all $x \in D(0, 1^-)$ and so $\| w \|_0 = |w(0)|$. Therefore we have

$$|G_{s,j}(0)| \leq \{s, n-1\}_p.$$

Now let

$$u(X) = \sum_s C_s X^s$$

be a solution of the equation $L = 0$, with the initial conditions

$$u^{(i)}(0) = b_i \qquad \text{for} \quad 0 \le i \le n - 1 \,.$$

From Taylor's formula we have

$$u(X) = \sum_s \frac{1}{s!} u^{(s)}(0) X^s \,.$$

Since by definition

$$\frac{1}{s!} u^{(s)} = G_{s,0} u + G_{s,1} u' + \cdots + G_{s,n-1} u^{(n-1)}$$

we obtain

$$\frac{1}{s!} u^{(s)}(0) = \sum_{i=0}^{n-1} G_{s,i}(0) b_i \,,$$

whence

$$|C_s| \le \{s, n-1\}_p \sup_{0 \le i \le n-1} |b_i| \,.$$

Example 2.4. The series defined by

$$\ell_2(X) = \sum_{s=1}^{\infty} \frac{X^s}{s^2}$$

is called the *dilogarithm function*. Together with 1 and $\log x$ it satisfies the third order differential equation

$$\frac{1}{X(1-X)} D \circ (1 - X) \circ D \circ X \circ Dy = 0 \,.$$

Observe that $\ell_2(0) = 0$, $\ell_2{}'(0) = 1$ and $\ell_2{}''(0) = 1/2$ and if $p \ne 2$

$$\{s, 2\}_p \sup_{0 \le i \le 2} \ell_2^{(i)}(0) = \{s, 2\}_p \,.$$

If the preceding estimates held we would have $|1/s|_p^2 \le \{s, 2\}_p$ which is easily seen to be false (take $s = p$). In fact the equation satisfied by the dilogarithm has a singularity at zero, so there are only two solutions in $\Omega[[X]]$. Thus the preceding estimate cannot be applied to the solutions at the origin (but see Theorem V.2.1).

3. Effective Bounds for Systems.

We wish to extend the results of the preceding section to the case of systems. We use the same notation as above.

Theorem 3.1. *Let* $Y \in G\ell(n, \mathcal{A}'_0)$ *and let*

$$A_s = \frac{1}{s!}(D^s Y)Y^{-1} \in \mathcal{M}_n(\mathcal{A}'_0) . \tag{3.1}$$

Then

$$\| A_s \|_0 \leq \{s, n-1\}_p \sup_{0 \leq i \leq n-1} (\| i! A_i \|_0) .$$

Proof. It suffices to prove the result for the first row of A_s.

Let $\vec{u} = (u_1, \dots, u_n)$ be the first row of Y. Assume K to be a field of definition for $u_1, \dots, u_n$; that is, assume that $u_1, \dots, u_n \in K((X))$ and also that $K \supseteq \ker(D, K(u_1, \dots, u_n))$. Let $q = \dim_K \langle u_1, \dots, u_n \rangle$ and let $z_1, \dots, z_q$ be a basis of $\langle u_1, \dots, u_n \rangle$. If we put $\vec{z} = (z_1, \dots, z_q)$ then there exists an element $E \in G\ell(n, K)$ such that

$$\vec{u} E = (\vec{z}, \vec{0}_{n-q}) .$$

Let Z be the (invertible $q \times q$) Wronskian matrix of $\vec{z}$, that is

$$Z = \begin{pmatrix} \vec{z} \\ D\vec{z} \\ \vdots \\ D^{q-1}\vec{z} \end{pmatrix} ,$$

and let $G_s = (G_{s,0}, \dots, G_{s,q-1})$ be the q-vector defined by

$$\frac{1}{s!}D^s \vec{z} = G_s Z .$$

Now consider the $q \times n$-matrix

$$T = \begin{pmatrix} \vec{u} \\ D\vec{u} \\ \vdots \\ D^{q-1}\vec{u} \end{pmatrix} Y^{-1} .$$

We then have

$$\left(\frac{1}{s!}D^s\vec{u}\right)E = \frac{1}{s!}D^s(\vec{u}E) = \frac{1}{s!}D^s(\vec{z},\vec{0}) = G_s(Z\,|\,0)$$

where the last 0 indicates the $q \times (n-q)$ zero matrix. On the other hand,

$$TYE = \begin{pmatrix} \vec{u} \\ D\vec{u} \\ \vdots \\ D^{q-1}\vec{u} \end{pmatrix} E = \begin{pmatrix} \vec{u}\,E \\ D(\vec{u}\,E) \\ \vdots \\ D^{q-1}(\vec{u}\,E) \end{pmatrix} = \begin{pmatrix} \vec{z},\vec{0}_{n-q} \\ D\vec{z},\vec{0}_{n-q} \\ \vdots \\ D^{q-1}\vec{z},\vec{0}_{n-q} \end{pmatrix} = (Z\,|\,0)\,.$$

Therefore

$$\left(\frac{1}{s!}D^s\vec{u}\right)E = G_s TYE$$

and, since E is invertible,

$$\left(\frac{1}{s!}D^s\vec{u}\right) = G_s TY\,.$$

From (3.1) it follows that the first row of A_s is the vector

$$\left(\frac{1}{s!}D^s\vec{u}\right)Y^{-1} = G_s T$$

and therefore

$$\|\,\text{first row of } A_s\,\|_0 \le \|\,G_s\,\|_0\,\|\,T\,\|_0\,.$$

But the $(i+1)$-th row of T is equal to

$$(D^i\vec{u})Y^{-1} = (i!)(\text{first row of } A_i)\,,$$

so that

$$\|\,T\,\|_0 \le \sup_{0\le i\le n-1}\,(\|\,i!A_i\,\|_0)\,.$$

On the other hand, by Theorem 2.1,

$$\|\,G_s\,\|_0\|\,T\,\|_0 \le \{s,q-1\}_p\|\,T\,\|_0 \le \{s,n-1\}_p\|\,T\,\|_0\,,$$

as desired.   **Q.E.D.**

Remark 3.2. If $Y \in G\ell\left(n, \mathcal{A}'_{a,\rho}\right)$ then the estimate becomes

$$\| A_s \|_{a,\rho} \leq \rho^{-s} \{s, n-1\}_p \sup_{0 \leq i \leq n-1} \left(\rho^i \| i! A_i \|_{a,\rho}\right) .$$

4. Analytic Elements.

In this section K will be an algebraically closed field of characteristic zero complete with respect to a non-archimedean valuation extending the p-adic valuation of $\mathbf{Q}$.

Let $K(X)$ be the field of rational functions over K. We recall that the Gauss norm is defined on $K(X)$. Let E denote the completion of $K(X)$ with respect to the Gauss norm. The elements of E are called *analytic elements (on the generic disk) defined over K*. If $\xi = (\xi_n)$ is a Cauchy sequence with $\xi_n \in K(X)$, then

$$|\xi|_{\text{gauss}} = \lim_{n \to \infty} |\xi_n|_{\text{gauss}} .$$

For $a \in K$ we define E_a to be the closure in E of the subring of $K(X)$ consisting of the rational functions with no pole in the disk $D(a, 1^-)$. The ring E_a can be identified with a subring of the ring of functions from $D(a, 1^-) \cap K$ to K endowed with the topology of uniform convergence on $D(a, 1^-)$. If $\xi = (\xi_n)$, where $\xi_n \in K(X)$ has no pole in $D(a, 1^-)$, then for $x \in D(a, 1^-)$ we define

$$\xi(x) = \lim_{n \to \infty} \xi_n(x) .$$

Thus the elements of E_a can be viewed as uniform limits of rational functions with no poles in $D(a, 1^-)$, and clearly

$$|\xi|_{\text{gauss}} = \sup_{|x-a|<1} |\xi(x)| . \tag{4.1}$$

We also have an imbedding of E_a in $\mathcal{B}_a$. This is easily obtained as follows: to each $\xi \in K(X)$ with no poles in $D(a, 1^-)$ we associate its Taylor expansion at a; the map we obtain is an isometry which can be extended to an isometry of all of E_a into $\mathcal{B}_a$ by continuity (recall that $\mathcal{B}_a$ is complete). This isometry will be assumed implicitly, so that we will identify the elements of E_a with their Taylor expansions. If $\xi = \sum_s A_s (X - a)^s \in E_a$ we have

$$|\xi|_{\text{gauss}} = \| \xi \|_0 = \sup_s |A_s| . \tag{4.2}$$

We call E_a the ring of *analytic elements on the disk* $D(a, 1^-)$ *defined over* K, and denote its field of quotients by E'_a.

In general the elements of E can not be interpreted as functions defined on arbitrary disks.

Example 4.1.   The series

$$\frac{1}{X} + \sum_{\nu=1}^{\infty} \frac{p^\nu}{X^{p^{\nu-1}} - 1}$$

is an element of E, since

$$\lim_{\nu \to \infty} \left| \frac{p^\nu}{X^{p^{\nu-1}} - 1} \right|_{\text{gauss}} = \lim_{\nu \to \infty} |p^\nu| = 0 \; .$$

This analytic element has singularities in every residue class of $\mathbf{Q}_p^{\text{alg}}$.

However, if t is a generic point with respect to K each element of $K(X)$ has no pole in the generic disk $D(t, 1^-)$, and therefore we can repeat the discussion given above. If Ω is a complete algebraically closed extension of K containing t, then the elements ξ of E can be viewed both as functions on $D(t, 1^-)$ with values in Ω and as Taylor series

$$\sum_{s=0}^{\infty} \frac{\xi^{(s)}(t)}{s!} (X - t)^s \; .$$

There is another interpretation of the elements of E as Laurent series (with coefficients in K) which we now explain. We introduce the *Amice ring*

$$K\{\{X\}\} = \left\{ \sum_{s \in \mathbf{Z}} A_s X^s \; \middle| \; A_s \in K, \; \sup_s |A_s| < \infty, \; \lim_{s \to -\infty} |A_s| = 0 \right\} \; .$$

If $\xi = \sum_s A_s X^s$, $\eta = \sum_s B_s X^s \in K\{\{X\}\}$ let

$$C_s = \sum_t A_t B_{s-t} \; .$$

This series is easily seen to converge, so that we can define

$$\xi \eta = \sum_s C_s X^s$$

and again one checks easily that $\xi\eta \in K\{\{X\}\}$ which is therefore a ring (actually $K\{\{X\}\}$ is a field as is not difficult to verify). Clearly $\mathcal{B}_0 \subset K\{\{X\}\}$. If $\xi = \sum_s A_s X^s$ we define

$$\| \xi \|_0 = \sup_s |A_s| \, .$$

As we will now show, this defines a norm on the ring $K\{\{X\}\}$. It is obvious that $K\{\{X\}\}$ is complete with respect to this norm. The only point requiring proof is that

$$\| \xi\eta \|_0 = \| \xi \|_0 \, \| \eta \|_0 \, .$$

We know that this equality holds for the elements of $\mathcal{B}_0$. To simplify the typography in the proof of this equality we write $|\xi|$ for $\| \xi \|_0$. Clearly $|\xi\eta| \leq |\xi||\eta|$. Let $\xi = \sum_s A_s X^s$; for $r \in \mathbf{Z}$ we put

$$\xi_r = \sum_{s \geq -r} A_s X^s \, .$$

Let $\eta = \sum_s B_s X^s$. We choose $r > 0$ sufficiently large so that

$$|\xi - \xi_r| < |\xi| \quad \text{and} \quad |\eta - \eta_r| < |\eta| \, .$$

Then $|\xi| = |\xi_r|$ and $|\eta| = |\eta_r|$.
We observe that $X^r \xi_r$ and $X^r \eta_r \in \mathcal{B}_0$ and

$$|\xi_r \eta_r| = |X^{2r} \xi_r \eta_r| = |X^r \xi_r||X^r \eta_r| = |\xi_r||\eta_r| \, .$$

Therefore

$$|\xi\eta| = |(\xi - \xi_r)\eta + \xi_r(\eta - \eta_r) + \xi_r \eta_r| = |\xi_r \eta_r| \, .$$

In general the elements of $K\{\{X\}\}$ are Laurent series converging on no annulus, so that also in this case they can not be interpreted as functions.

We can imbed $K(X)$ in $K\{\{X\}\}$. Indeed, to the element $(X-a)^{-1}$ with $a \in K$ we associate the series

$$-\frac{1}{a} \sum_{s=0}^{\infty} \left(\frac{X}{a} \right)^s \quad \text{if } |a| \geq 1$$

and the series

$$\frac{1}{X} \sum_{s=0}^{\infty} \left(\frac{a}{X}\right)^s \quad \text{if } |a| < 1 .$$

This imbedding is extended to all of E by continuity, and if to the element $\xi \in E$ there corresponds the series $\sum_s A_s X^s$ we have

$$|\xi|_{\text{gauss}} = \sup_s |A_s| \tag{4.3}$$

(since this equality holds for polynomials by definition). In a similar way we can imbed E in $K\{\{X - a\}\}$ and we have $\mathcal{B}_a \subset K\{\{X - a\}\}$. It is often useful to identify an element $\xi \in E$ with its image in $K\{\{X - a\}\}$.

Now fix $a \in K$ and let ρ be a real number satisfying $0 < \rho < 1$. Let $\xi = \sum_s A_s (X - a)^s \in E$. We say that ξ is *analytic in the annulus* $\rho \le |x - a| < 1$ if the series $\sum_s A_s (X - a)^s$ converges there. Such an element ξ can obviously be interpreted as a function on the annulus and we have

$$|\xi|_{\text{gauss}} \le \sup_{\rho \le |x - a| < 1} |\xi(x)| < \infty .$$

Counterexample. Let $G \in \mathcal{M}_n(\mathcal{B}_0)$ and let $\vec{v} \in (K\{\{X\}\})^n$ satisfy $D\vec{v} = G\vec{v}$ where $D = d/dX$. Is it necessarily the case that $\vec{v} \in \mathcal{B}_0^n$? The answer is no. Following A. Ogus, for $m > 1$ let

$$G = \begin{pmatrix} 0 & 1 \\ -m(m-1)X^{m-2} & -mX^{m-1} \end{pmatrix} .$$

A solution of the equation $D\vec{v} = G\vec{v}$ is given by $\vec{v} = \begin{pmatrix} z \\ Dz \end{pmatrix}$ where z is solution of the scalar equation $D \circ (D + mX^{m-1})z = 0$. Now $z_1 = \exp(-X^m)$ is a solution of the scalar equation but there is also a unique solution in $\mathbf{Q}((1/X))$, the asymptotic series (where $(a)_s = a(a+1)\cdots(a+s-1)$ for $s \ge 1$ and $(a)_0 = 1$)

$$z_2 = \sum_{s=0}^{\infty} \left(\frac{m-1}{m}\right)_s \frac{1}{X^{m-1+ms}}$$

and this lies in $K\{\{X\}\}\backslash\mathcal{B}_0$ if $m > 1$, $p \nmid m$. For a more subtle related question see Berthelot–Messing [1], pages 19,20.

We now give a p-adic analogue of Rouché's Theorem. Let E_0' be the quotient field of

$$E_0 = \{\xi \in E \,|\, \xi \text{ analytic on } D(0,1^-)\} ,$$

and for $f \in E_0'$ define $\#(f)$ to be the number of zeros of f in $D(0,1^-)$.

Proposition 4.2. *If $f \in E_0$ then $\#(f)$ is finite.*

Proof. We may assume that $|f|_{\text{gauss}} = 1$. There exists a rational function g with no pole in $D(0, 1^-)$ such that $|f - g|_{\text{gauss}} < 1$; therefore we also have $|g|_{\text{gauss}} = 1$. We write $f = g + h$ so that $h \in E_0$ and $|h| < 1$. Clearly $\#(g)$ is finite. We claim that $\#(f) = \#(g)$. Since the Gauss norm of g is 1, there exists a point $(N, 0)$ on its Newton polygon with minimal N (cf. diagram).

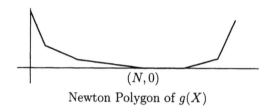

Newton Polygon of $g(X)$

Then $\#(g) = N$. On considering the Newton polygon of $h = b_0 + b_1 X + \cdots$ and bearing in mind that each b_i satisfies $|b_i| < 1$, we see that the Newton polygon of $g + h = f$ produces the same point $(N, 0)$ (although the new Newton polygon may very well be different from that of g).   **Q.E.D.**

Remark 4.3. We have seen that E_0 may be viewed as a subring of $\mathcal{B}_0$. We now show that the inclusion is strict. Consider the power series

$$S(X) = p + p^{1/2} X + p^{1/4} X^2 + p^{1/8} X^3 + \cdots ;$$

Clearly $S(X)$ has radius of convergence 1, but since $|p^{1/2^n}|$ tends to 1 it is not convergent for $|x| = 1$, although it is bounded by 1 in $D(0, 1^-)$. We observe that the Newton polygon of $S(X)$ has infinitely many finite sides, so that $S(X)$ has infinitely many zeros. Therefore $S(X) \in \mathcal{B}_0$ but $S(X) \notin E_0$.

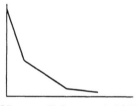

Newton Polygon of $S(X)$

5.   Some Transfer Theorems.

In this section we will assume that Ω is a complete algebraically closed p-adic field sufficiently large so as to contain all the generic points we need.

A.   Transfer from a disk with apparent singularities into a disk with no singularity.

Proposition 5.1. *Let K be a subfield of Ω and let $G \in \mathcal{M}_n(K(X))$. Assume that the equation*

$$\frac{d\vec{y}}{dx} = G\vec{y} \tag{5.1}$$

has a solution matrix in $G\ell(n, \mathcal{A}_0')$. Assume moreover that the matrix G is analytic in $D(\alpha, 1^-)$ where $\alpha \in \Omega$ has $|\alpha| = 1$. Then the solution matrix $Y_\alpha \in \mathcal{M}_n(K(\alpha)[[X - \alpha]])$ such that $Y_\alpha(\alpha) = I_n$ converges in the disk $D(\alpha, 1^-)$.

Proof.   Let $Y_0 \in G\ell(n, \mathcal{A}_0')$ be a matrix solution of (5.1). If

$$G_s = \left(\frac{1}{s!}\frac{d^s Y_0}{dX^s}\right) Y_0^{-1}$$

we know from Theorem 3.1 that there is a κ not depending on s such that

$$\|G_s\|_0 \leq \kappa\{s, n - 1\}_p .$$

We recall that the recursion relation for G_s is

$$(s + 1)G_{s+1} = DG + G_s G \qquad (G_0 = I_n, \quad G_1 = G)$$

so that, since $G \in \mathcal{M}_n(K(X))$, each $G_s \in \mathcal{M}_n(K(X))$. Therefore

$$\|G_s\|_{\text{gauss}} \leq \kappa\{s, n - 1\}_p .$$

On the other hand, for the same reason, $G_s \in \mathcal{M}_n(\mathcal{A}_\alpha)$ and therefore by the maximum modulus principle we have

$$|G_s(\alpha)| \leq \|G_s\|_{\text{gauss}} \leq \kappa\{s, n - 1\}_p .$$

Now the Taylor formula for Y_α is given by

$$Y_\alpha(X) = \sum_s G_s(\alpha)(X - \alpha)^s$$

and this converges in the disk $D(\alpha, 1^-)$ since $\{s, n-1\}_p \leq s^{n-1}$. **Q.E.D.**

B.   *Transfer from an annulus into a contiguous disk with no singularity.*

Again let K be a subfield of Ω. For $0 < R < 1$ we denote by $t_R \in \Omega$ a generic point with respect to K satisfying $|t_R| = R$ (for example, if $\log R / \log p \in \mathbf{Q}$ and t is the generic point defined in Section III.5 we can take $t_R = \beta t$ with $|\beta| = R$ and β algebraic over $\mathbf{Q}$).

Proposition 5.2. *Let $G \in \mathcal{M}_n(E)$ and assume that the elements of G are analytic in the annulus $R_0 \leq |x| < 1$ and also analytic in the disk $D(\alpha, 1^-)$ with $|\alpha| = 1$. For every R with $R_0 < R < 1$ let Y_R be the solution matrix of the system (5.1) at t_R, normalized by $Y_R(t_R) = I_n$. Assume that Y_R converges in the disk $D(t_R, \rho(R)^-)$ with $\rho(R) \leq R$, and moreover that*

$$\lim_{R \to 1} \rho(R) = 1 .$$

Then the solution matrix Y_α of the system (5.1) with $Y_\alpha(\alpha) = I_n$ converges in the disk $D(\alpha, 1^-)$.

Proof.   Obviously the matrix G_s is analytic in the annulus for every $s \geq 0$, so we can write

$$G_s(X) = \sum_{m=-\infty}^{\infty} G_{s,m} X^m$$

where $G_{s,m}$ is an $n \times n$ matrix with elements in K. We know that

$$Y_R(X) = \sum_{s=0}^{\infty} G_s(t_R)(X - t_R)^s$$

converges for $|x - t_R| < \rho(R)$. Therefore, applying Remark 3.2 and Theorem 1.2 we have

$$|G_s(t_R)| \leq \|G_s\|_{t_R, \rho(R)} \leq \rho(R)^{-s} \sup_{0 \leq i \leq n-1} \left(R^i \| i! G_i \|_{t_R, \rho(R)} \right) \{s, n-1\}_p .$$

Since $\rho(R) \leq R$ and G_i is analytic on $D(t_R, R^-)$ we have

$$\|G_i\|_{t_R, \rho(R)} = |G_i|_{t_R} (\rho(R)) \leq \sup_{R_0 \leq |x| < 1} |G_i(x)|$$

$$\leq \sup \left(|G_i|_0(R_0), |G_i|_{\text{gauss}} \right) .$$

Therefore we have

$$|G_s(t_R)| \leq \kappa\{s, n-1\}_p \rho(R)^{-s}$$

where

$$\kappa = \sup_{0 \leq i \leq n-1} \sup \left(|G_i|_0(R_0), |G_i|_{\text{gauss}} \right) .$$

On the other hand, since t_R is generic we must have

$$|G_s(t_R)| = \sup_{m \in \mathbf{Z}} |G_{s,m}| R^m .$$

Thus, we have

$$|G_{s,m}| R^m \rho(R)^s \leq \kappa\{s, n-1\}_p ,$$

and letting R tend to 1 we obtain

$$|G_{s,m}| \leq \kappa\{s, n-1\}_p .$$

Therefore

$$|G_s(\alpha)| \leq \sup_{m \in \mathbf{Z}} |G_{s,m}| \leq \kappa\{s, n-1\}_p ,$$

as desired.   **Q.E.D.**

C.   *Transfer from a disk with one regular singularity having exponents in $\mathbf{Z}_p$ to a contiguous disk without singularity.*

Corollary 5.3. *Let $A \in \mathcal{M}_n(K)$ with eigenvalues in $\mathbf{Z}_p$. If (see Section III.8) $Y(X) = U(X) \cdot X^A$ is a solution matrix of (5.1) with U converging in the disk $D(0, 1^-)$ then $\rho(R) \geq R$ for all $R \in (0, 1)$.*

Proof.   Since U is analytic at t_R, a solution matrix for (5.1) at t_R is given by $U V_A$ where $V_A(t_R) = I_n$, V_A a solution matrix of $\delta y = AY$. Since U is analytic in $D(t_R, R^-)$ it is enough to show that the same holds for V_A. There exists $P \in G\ell(n, K)$ such that $A = P^{-1}(N + \Delta)P$ where $N\Delta = \Delta N$, N is nilpotent and Δ is diagonal. Then

$$V_A = P^{-1} V_{N+\Delta} P$$

and so we may assume $A = N + \Delta$. Now

$$V_\Delta = \sum_{s=0}^{\infty} \binom{\Delta}{s} (\frac{X}{t_R} - 1)^s ,$$

and

$$V_N = \exp\left(N \log \frac{X}{t_R}\right),$$

a polynomial in

$$\log \frac{X}{t_R} = -\sum_{s=1}^{\infty} \frac{1}{s}\left(1 - \frac{X}{t_R}\right)^s .$$

Thus V_N converges in $D(t_R, R^-)$ and the same also holds for V_Δ since $\begin{pmatrix} \Delta_i \\ s \end{pmatrix} \in \mathbf{Z}_p$ for each diagonal element Δ_i of Δ (see Proposition I.4.3). The formula for V_Δ shows that $V_\Delta N = N V_\Delta$ and hence

$$\delta(V_\Delta V_N) = \Delta V_\Delta V_N + V_\Delta N V_M = (\Delta + N) V_\Delta V_N$$

and so $V_{\Delta+N} = V_\Delta V_N$.   **Q.E.D.**

Proposition 5.4. *Let $G \in \mathcal{M}_n(E_0)$ and let $G^{(k)}$ be a sequence of matrices in $\mathcal{M}_n(K(X))$ whose elements have no poles in $D(0, 1^-)$. Assume that the sequence $G^{(k)}$ converges uniformly in $D(0, 1^-)$ to the matrix G. Assume moreover that each system $D - G^{(k)}$ has a solution matrix $Y^{(k)}$ which converges on $D(0, 1^-)$ and let $Y^{(k)}(0) = I_n$ for all k. Then the sequence $Y^{(k)}$ converges on $D(0, 1^-)$ to a solution matrix Y of $D - G$. The convergence is uniform on every disk $D(0, R^-)$ with $R < 1$.*

Proof. We define as usual the matrices $G_s^{(k)}$ by the following recursion formulas

$$G_{s+1}^{(k)} = G_s^{(k)} G^{(k)} + \frac{d}{dX} G_s^{(k)} \quad \text{with} \quad G_0^{(k)} = I_n, \quad G_1^{(k)} = G^{(k)}$$

for every k. Clearly the elements of each $G_s^{(k)}$ are rational functions with no pole in $D(0, 1^-)$. With this definition we know that

$$Y^{(k)}(X) = \sum_s \frac{1}{s!} G_s^{(k)}(0) X^s .$$

As is easily verified, for each fixed s the sequence $G_s^{(k)}$ converges to $G_s \in \mathcal{M}_n(E_0)$ uniformly on $D(0, 1^-)$.

From Theorem 3.1 it follows that

$$\left| \frac{1}{s!} G_s^{(k)} \right|_{\text{gauss}} \leq \{s, n-1\}_p \sup_{0 \leq i \leq n-1} \left(|G_i^{(k)}|_{\text{gauss}} \right) .$$

Since for fixed i the sequence $|G_i^{(k)}|_{\text{gauss}}$ converges (to $|G_i|_{\text{gauss}}$) there exists a constant C independent of k such that

$$\left| \frac{1}{s!} G_s^{(k)}(0) \right| \leq C\{s, n-1\}_p \tag{5.2}$$

for every $k \geq 0$. From this it follows immediately that (5.2) also holds for $G_s(0)$. Hence

$$Y(X) = \sum_{s=0}^{\infty} \frac{1}{s!} G_s(0) X^s$$

converges in $D(0, 1^-)$.

As to the uniform convergence, given an $\varepsilon > 0$ and $R < 1$ we must find a positive integer k_0 such that for $k > k_0$ one has

$$|Y(x) - Y^{(k)}(x)| < \varepsilon$$

for all $x \in D(0, R^-)$. To this end, we note that we may choose a sufficiently large positive integer s_0 so that

$$R^s C\{s, n-1\}_p < \varepsilon, \quad \text{for all} \quad s \geq s_0$$

(since $\{s, n-1\}_p \leq s^{n-1}$). It follows that

$$\sup_{|x| \leq R} \left| \sum_{s=s_0}^{\infty} \frac{1}{s!} G_s^{(k)}(0) x^s \right| < \varepsilon$$

for all k and

$$\sup_{|x| \leq R} \left| \sum_{s=s_0}^{\infty} \frac{1}{s!} G_s(0) x^s \right| < \varepsilon, \quad \text{for} \quad s \geq s_0$$

also holds. Having fixed such an s_0, we see that there exists a positive integer N such that

$$\frac{1}{|s!|} |G_s^{(k)}(0) - G_s(0)| < \varepsilon$$

for $k \geq N$ and $s = 1, 2, \ldots, s_0$. Therefore for such k and for $|x| < R < 1$ we have

$$|Y^{(k)}(x) - Y(x)| = \left| \sum_{s=0}^{\infty} \frac{1}{s!} (G_s^{(k)}(0) - G_s(0)) x^s \right| < \varepsilon,$$

as desired. **Q.E.D.**

Corollary 5.5. *With the same hypotheses as in Proposition 5.4, let $\vec{a} \in K^n$ and let $\vec{y}^{(k)}$ be a solution of the system*

$$\frac{d}{dX}\vec{y} = G^{(k)}\vec{y}$$

such that $\vec{y}^{(k)}(0) = \vec{a}$. Then the sequence $\vec{y}^{(k)}$ converges uniformly on $D(0, R^-)$, for all $R < 1$, to a solution $\vec{y}$ of the system

$$\frac{d}{dX}\vec{y} = G\vec{y}$$

such that $\vec{y}(0) = \vec{a}$.

Proof.   With the notation of Proposition 5.4 we have $\vec{y}^{(k)} = Y^{(k)}\vec{a}$. **Q.E.D.**

6.   Logarithms.

For $k \geq 1$, let $L_k = \delta^2 - p^k\delta$ where $\delta = X d/dX$. Clearly as $k \to \infty$ we have $L_k \to L = \delta^2$. The associated differential systems are

$$\delta\begin{pmatrix} u_1 \\ u_2 \end{pmatrix} = \begin{pmatrix} 0 & 1 \\ 0 & p^k \end{pmatrix}\begin{pmatrix} u_1 \\ u_2 \end{pmatrix} .$$

Now the operator L_k has a normalized set of solutions at $x = 1$ given by the functions

$$1 \quad \text{and} \quad u_k(X) = \frac{X^{p^k} - 1}{p^k} ,$$

and the operator L has a normalized set of solutions at 1 given by

$$1 \quad \text{and} \quad u(X) = \log X = \log(1 + (X - 1)) = (X - 1) - \frac{(X - 1)^2}{2} + \cdots .$$

Thus, by Corollary 5.5, we have that $u_k \longrightarrow u$ uniformly on $D(1, R)$ for $R < 1$. (This was obtained in a different way in Leopoldt [1] by estimating the error in

$$\frac{(1 + X)^{p^k} - 1}{p^k} = \frac{\exp\left(p^k \log(1 + X)\right) - 1}{p^k}$$

$$= \log(1 + X) + (\text{error})_k \longrightarrow \log(1 + X) .)$$

We apply this result to determine the zeros of $\log(1+X)$ (cf. Proposition II.1.2).

Let ζ satisfy $\zeta^{p^l} = 1$; if $k > l$ then clearly $u_k(\zeta) = 0$ so that $\log \zeta = 0$.

Conversely assume that $\zeta \in D(1, 1^-)$ satisfies $\log \zeta = 0$. Then, as we saw in the proof of Proposition II.1.2, there exists a positive integer l such that $|\zeta^{p^l} - 1| < |\pi|$ where, as always, $\pi^{p-1} = -p$. If $k > l$ then

$$u_k(\zeta) = \frac{\zeta^{p^k} - 1}{p^k} = \frac{(\zeta^{p^l})^{p^{k-l}} - 1}{p^{k-l}} \frac{1}{p^l} \, .$$

Since $\log \zeta = 0$ we have $\lim_k u_k(\zeta) = 0$ so that if we write $w = \zeta^{p^l}$ then $\lim_k u_k(w) = 0$ with $|w - 1| < |\pi|$. If we write $w = 1 + x$ then, since

$$\left| \binom{p}{i} x^i \right| < |p| |x|$$

for $i = 2, \ldots, p$, we have $|w^p - 1| = |p| |x| = |p| |w - 1|$. Iterating we obtain

$$|w^{p^k} - 1| = |p^k| |w - 1|$$

and therefore $w = 1$.

Another application is the following. Let $z, w \in D(1, 1^-)$. Then as is well-known $\log(zw) = \log z + \log w$. A new proof of this relation can be obtained by observing that

$$\frac{1}{p^k} \left[(zw)^{p^k} - z^{p^k} - w^{p^k} + 1 \right] = p^k \frac{z^{p^k} - 1}{p^k} \cdot \frac{w^{p^k} - 1}{p^k}$$

and that on taking the limit for $k \to \infty$ the left side tends to $\log(zw) - \log z - \log w$ while the right side tends to zero.

7.   The Binomial Series.

Let K be an extension field of $\mathbf{Q}_p$, complete under a valuation extending
the p-adic valuation on $\mathbf{Q}_p$. Let $\alpha \in K$ and let $g_\alpha(X) = (1+X)^\alpha$ denote
the formal solution $y(X)$ at $x = 0$ to the equation

$$(1+X)\frac{dy}{dX} = \alpha y \qquad (7.1)$$

with the initial condition $y(0) = 1$. Clearly $g_\alpha(X)$ is the *binomial series*,
that is

$$g_\alpha(X) = (1+X)^\alpha = \sum_{s=0}^{\infty} \binom{\alpha}{s} X^s$$

where

$$\binom{\alpha}{s} = \frac{\alpha(\alpha - 1) \cdots (\alpha - s + 1)}{s!}.$$

The following identities of formal power series are easily seen to
hold

$$(1+X)^\alpha = \exp \alpha \log(1 + X),$$

$$(1+X)^\alpha (1+X)^\beta = (1+X)^{\alpha+\beta},$$

$$((1+X)^\alpha)^\beta = (1+X)^{\alpha\beta},$$

$$((1+X)(1+Y))^\alpha = (1+X)^\alpha (1+Y)^\alpha.$$

We give another application of Corollary 5.5. Assume $\alpha \in \mathbf{Z}_p$ and
let α_k be a sequence of positive integers converging to α. Then $y_k = (1 + X)^{\alpha_k}$ is a solution of

$$(1+X)\frac{dy}{dX} = \alpha_k y$$

with the initial condition $y_k(0) = 1$, and y_k converges everywhere. Moreover, the function y_k is bounded in the disk $D(0, 1^-)$ by 1 as is easily
verified. Therefore by Proposition 5.4 the series $(1 + X)^\alpha$ converges
and is bounded by 1 (in view of the uniform boundedness of the y_k
on $D(0, 1^-)$) in the disk $D(0, 1^-)$. Hence by Cauchy's inequality (cf.
Proposition 1.1) this gives another proof of Proposition I.4.3.

We shall need the following lemma concerning convergence of composed power series.

Lemma 7.1. *Let Ω be a field with a non-archimedean valuation. Let $R_1, R_2 > 0$*

$$f(X) = \sum_{r=0}^{\infty} A_r X^r, \qquad g(X) = \sum_{s=1}^{\infty} B_s X^s$$

with A_r, $B_s \in \Omega$. Assume that $f(X)$ converges in $D(0, R_1^+)$ and that $g(X)$ converges in $D(0, R_2^+)$ and maps $D(0, R_2^+)$ into $D(0, R_1^+)$. Then the formal power series $h(X) = f(g(X))$ has radius of convergence $\geq R_2$ and if $|x| \leq R_2$ then $h(x) = f(g(x))$.

Proof. Let $h(X) = \sum_{n=0}^{\infty} C_n X^n$. Then $C_0 = A_0$, while for $n \geq 1$

$$C_n = \sum_{j=1}^{n} A_j \left(\sum_{s_1 + \cdots + s_j = n} B_{s_1} \cdots B_{s_j} \right).$$

As g converges in $D(0, R_2^+)$ and maps $D(0, R_2^+)$ into $D(0, R_1^+)$, by Proposition 1.1 it follows that

$$|B_s| R_2^s \leq R_1$$

for every s. Therefore

$$|A_j B_{s_1} \cdots B_{s_j}| R_2^n = |A_j| |B_{s_1}| R_2^{s_1} \cdots |B_{s_j}| R_2^{s_j} \leq |A_j| R_1^j$$

and

$$|C_n| R_2^n \leq \sup_{1 \leq j \leq n} |A_j| R_1^j.$$

Since the right hand side of the inequality tends to zero as $n \to \infty$, the series $h(X)$ converges for $|x| \leq R_2$.

Assume $|x| \leq R_2$ and let

$$f_N(X) = \sum_{r=0}^{N} A_r X^r,$$

$$h_N(X) = f_N(g(X)) = \sum_{n=0}^{\infty} C_{n,N} X^n \in \Omega[[X]].$$

Since $f_N(X)$ is a polynomial, we have $h_N(x) = f_N(g(x))$. Therefore to prove $h(x) = f(g(x))$ it suffices to prove that

$$\lim_{N \to \infty} h_N(x) = h(x)$$

and that

$$\lim_{N \to \infty} f_N(g(x)) = f(g(x)) .$$

The second assertion is clear since $|g(x)| \leq \sup_s |B_s| R_2^s \leq R_1$ and $f(X)$ converges for $|x| \leq R_1$. As to the first assertion, we note that

$$C_n - C_{n,N} = \begin{cases} 0 & \text{if } n \leq N \\ \displaystyle\sum_{N \leq j \leq n} A_j \sum_{s_1 + \cdots + s_j = n} B_{s_1} \cdots B_{s_j} & \text{otherwise,} \end{cases}$$

so that

$$\left| \sum_{n=0}^{\infty} (C_n - C_{n,N}) x^n \right| \leq \sup_n \left(|C_n - C_{n,N}| R_2^n \right) \leq \sup_{j \geq N} \left(|A_j| R_1^j \right)$$

and this last term tends to zero as $N \to \infty$. **Q.E.D.**

Corollary 7.2. *Let* $R, R_1, \ldots, R_n$ *be positive real numbers and let* Δ *be the closed disk* $D(0, R^+)$ *and* U *the polydisk* $D(0, R_1^+) \times \ldots \times D(0, R_n^+)$.

(i) For $1 \leq i \leq n$ *let* $F_i \in X\Omega[[X]]$ *be convergent in* Δ *and map it into* $D(0, R_i^+)$. *(So* $F = (F_1, \ldots, F_n)$ *maps* Δ *into* U). *Let* $G \in \Omega[[Y_1, \ldots, Y_n]]$ *be convergent on the polydisk* U. *Let* $H(X)$ *be the formal power series* $H(X) = G(F_1(X), \ldots, F_n(X))$. *Then* H *is convergent on* Δ *and for* $x \in \Delta$ *we have*

$$H(x) = G(F_1(x), \ldots, F_n(x)) .$$

(ii) Let $F \in \Omega[[X]]$ *converge on* Δ *and let* $G \in \Omega[[Y_1, \ldots, Y_n]]$ *be convergent on the polydisk* U *and map it into* Δ. *Assume moreover that* $G(0) = 0$. *Let* H *be the formal power series in* $Y_1, \ldots, Y_n$ *defined by*

$$H(Y_1, \ldots, Y_n) = F(G(Y_1, \ldots, Y_n)) .$$

Then H *is convergent on* U *and for* $(y_1, \ldots, y_n) \in U$ *we have*

$$H(y_1, \ldots, y_n) = F(G(y_1, \ldots, y_n)) .$$

Proof.   The proof of (i) parallels that of Lemma 7.1 and may be omitted. For the proof of (ii) let Z be an indeterminate in Ω. Putting $(y_1 Z, \dots, y_n Z)$ for $(Y_1, \dots, Y_n)$ we obtain an element of $\Omega[[Z]]$

$$H(y_1 Z, \dots, y_n Z) = F(G(y_1 Z, \dots, y_n Z))$$

which by Lemma 7.1 converges in $D(0, 1^+)$; specializing at $Z = 1$ we obtain the assertion. **Q.E.D.**

We return to our discussion of the binomial series. If α, $\beta \in K$ then, as noted above, we have the formal identity

$$g_\alpha(X) g_\beta(X) = g_{\alpha+\beta}(X) \,.$$

By the discussion of products of Laurent series (Section 1) or by Corollary 7.2 (i) we know that if g_α and g_β converge on $D(0, R^+)$ then the same holds for $g_{\alpha+\beta}$ and that if $x \in D(0, R^+)$ then $g_\alpha(x) g_\beta(x) = g_{\alpha+\beta}(x)$.

Proposition 7.3. *Let $\alpha \in K$.*
  (i) Let g_α converge in $D(0, R^+)$ and x, y lie in that disk. Then

$$g_\alpha(x) g_\alpha(y) = g_\alpha(x + y + xy)$$

(and so the map $z \mapsto g_\alpha(z-1)$ is a character of the multiplicative group $D(1, R^+)$).
  (ii) If $|\alpha| > 1$ then $g_\alpha(x)$ converges if and only if $|x| < |\pi|/|\alpha|$.
  (iii) If $\alpha \in \mathbf{Z}_p$ then $g_\alpha(x)$ converges for $|x| < 1$.
  (iv) If $\sup\{\mathrm{ord}\,(\alpha - b) \,|\, b \in \mathbf{Z}_p\} = m + \varepsilon$, with $m \in \mathbf{N}$, and $\varepsilon \in [0, 1)$ then $g_\alpha(x)$ converges if and only if

$$\mathrm{ord}\, x > \frac{1}{p^m(p-1)} - \varepsilon \frac{1}{p^{m+1}} \,.$$

  (v) If $|x| < 1$ and x lies in the disk of convergence of g_α then $g_\alpha(x) = 1$ implies that $(1 + x)^{p^s} = 1$ for some s.
  (vi) Let $\mathrm{ord}\,\alpha > l$. If $\zeta^{p^{l+1}} = 1$ then $g_\alpha(\zeta - 1) = 1$. Conversely if

$$\mathrm{ord}\,(\zeta - 1) \geq \frac{1}{p^l(p-1)}$$

and $g_\alpha(\zeta - 1) = 1$ then $\zeta^{p^{l+1}} = 1$.
  (vii) If g_α converges in $D(0, 1^-)$ then $\alpha \in \mathbf{Z}_p$.

Proof.   (i) We have already noted that as formal power series in X, Y

$$g_\alpha(X) g_\alpha(Y) = g_\alpha(X + Y + XY) \,.$$

Letting $G(X, Y) = X + Y + XY$, the assertion follows from Corollary 7.2 (ii).

(ii) For $|\alpha| > 1$ we have

$$\left| \binom{\alpha}{s} \right| = |\alpha|^s / |s!|$$

and so convergence is precisely the same as for $\exp \alpha X$.

(iii) This follows immediately from Proposition I.4.3.

(iv) **Case 1.** $\operatorname{ord} \alpha = m + \varepsilon$. Then since $\operatorname{ord}(\alpha - s) \leq \operatorname{ord} \alpha$ for all $s \in \mathbf{N}$, we have

$$\operatorname{ord}(\alpha(\alpha - 1) \ldots (\alpha - s + 1)) = \sum_{j=1}^{m} (1 + [\frac{s-1}{p^j}]) + \varepsilon(1 + [\frac{s-1}{p^{m+1}}])$$

while

$$\operatorname{ord} s! = \sum_{j=1}^{\infty} [\frac{s}{p^j}] .$$

Thus by a routine calculation

$$\lim_{s \to \infty} \frac{1}{s} \operatorname{ord} \binom{\alpha}{s} = \frac{\varepsilon}{p^{m+1}} - \frac{1}{p^m(p-1)} .$$

This confirms the asserted radius of convergence. To show that there can be no convergence on the boundary we use the above formula to compute for $s = p^{1+m+l}$, $l \in \mathbf{N}$

$$\operatorname{ord} \binom{\alpha}{s} + s\left(\frac{1}{p^m(p-1)} - \frac{\varepsilon}{p^{m+1}} \right) = \frac{1}{p-1}$$

which shows that $g_\alpha(x)$ cannot converge for

$$\operatorname{ord} x = \frac{1}{p^m(p-1)} - \frac{\varepsilon}{p^{m+1}} .$$

Case 2. By hypothesis $\alpha \notin \mathbf{Z}_p$. By compactness of $\mathbf{Z}_p$ there exists $b \in \mathbf{Z}_p$ such that $\operatorname{ord}(\alpha - b) = m + \varepsilon$. By case 1 we know the radius of convergence of $g_{\alpha-b}$ and by (iii) this is strictly smaller than the radii of convergence of g_b and of g_{-b}. Since $g_\alpha = g_{\alpha-b} g_b$, $g_{\alpha-b} = g_\alpha g_{-b}$ we conclude that g_α has precisely the same radius of convergence as $g_{\alpha-b}$.

(v) Let g_α converge in $D(0, R^+)$ and $\lambda \in D(0, R^+)$. Then

$$\lim_{s \to \infty} (1 + \lambda)^{p^s} = 1 .$$

Letting $1 + \lambda_s = (1 + \lambda)^{p^s}$, it follows from (i) that if $g_\alpha(\lambda) = 1$ then $g_\alpha(\lambda_s) = 1$ and $\lambda_s \to 0$. It follows (e.g. by Proposition 4.2) that $\{\lambda_s\}$ is a finite set with zero limit and hence $\lambda_s = 0$ for s sufficiently large.

(vi) We first explain the converse point. If $g_\alpha(\zeta - 1) = 1$ then from (v) it follows that $\zeta^{p^s} = 1$ for some s. Let s be minimal; then by Example I.9.3

$$\text{ord}(\zeta - 1) = \frac{1}{(p-1)p^{s-1}}$$

if $s \geq 1$ (the case $s = 0$ is trivial). Hence $s - 1 \leq l$ as asserted.

For the direct part we consider two cases.

Case 1. $l = 0$.

In this case $\text{ord}\,\alpha > 0$, $\zeta^p = 1$. Hence $\text{ord}(\zeta - 1) \geq \frac{1}{p-1}$ and by (iv) the series g_α converges on $D(0, |\pi|^+)$. Thus by (i),

$$g_\alpha(\zeta - 1)^p = g_\alpha(\zeta^p - 1) = g_\alpha(0) = 1 .$$

Thus $g_\alpha(\zeta - 1)$ is a pth root of unity but for $s \geq 1$

$$\text{ord}\binom{\alpha}{s}(\zeta - 1)^s \geq \text{ord}\,\alpha + \frac{s}{p-1} - \text{ord}\,s! \geq \text{ord}\,\alpha + \frac{1}{p-1} > \frac{1}{p-1} .$$

Thus $\text{ord}(g_\alpha(\zeta - 1) - 1) > \frac{1}{p-1}$ and so $g_\alpha(\zeta - 1)$ cannot be a primitive p^{th} root of unity, i.e. $g_\alpha(\zeta - 1) = 1$ as desired.

Case 2. $l \geq 1$

By hypothesis $\alpha = p^l \beta$, $\text{ord}\beta > 0$. Let $h_l(X) = (1 + X)^{p^l} - 1$, $R = |\pi|^{1/p^l}$. Then h_l maps $D(0, R^+)$ into $D(0, |\pi|^+)$ while as formal power series $g_\alpha(X) = g_\beta(h_l(X))$. Thus by Lemma 7.1 (since g_β converges on $D(0, |\pi|^+)$ and $\zeta - 1 \in D(0, R^+)$) and by Case 1, since $1 + h_l(\zeta - 1)$ is a pth root of unity, we have

$$g_\alpha(\zeta - 1) = g_\beta(h_l(\zeta - 1)) = 1 .$$

(vii) This follows from (ii) and (iv).   **Q.E.D.**

Remark 7.4.

(1) Subject to the hypothesis of Case 1 in the proof of (iv) we may represent f_α as a composition of functions for

$$\text{ord } x > \frac{1}{p^m(p-1)} - \frac{1}{p^m} ,$$

in two ways: as $\exp(\alpha \log(1+X))$ and as $g_\beta(h_m(X))$ where $\alpha = \beta p^m$ and $h_m(X) = (1+X)^{p^m} - 1$. By these compositions we could obtain alternate demonstrations that g_α converges on the indicated set. The proof used shows that we have obtained the precise radius of convergence. An alternate procedure would use the formula

$$g_\beta(x) = \frac{1}{p^m} \sum_{h_m(y)=x} g_\alpha(y)$$

(for $\alpha = p^m\beta$, $|\beta| \le 1$) to estimate the radius of convergence of g_β from that of g_α.

(2) For fixed x with $|\pi| < |x| < 1$ consider $F(\alpha) = (1+x)^\alpha$ as a function of α, and suppose that $|\alpha| << 1$ so that we have convergence of the sum in

$$(1+x)^\alpha = \sum_{s=0}^\infty \binom{\alpha}{s} x^s .$$

Then, at least formally, we have

$$\frac{dF}{d\alpha}(\alpha) = F(\alpha)\log(1+x) .$$

Thus, if $\log(1+x) = 0$ for our fixed x, we also have

$$\frac{dF}{d\alpha}(\alpha) = 0$$

so that for small α we see that $F(\alpha)$ is independent of α. But $F(\alpha) = 1$ for $\alpha = 0$, hence $F(\alpha) = 1$ for all sufficiently small α.

We can generalize our discussion of equation (7.1) to the case of matrices (see also Section III.8). For $A \in \mathcal{M}_n(K)$ we set

$$g_A(X) = \sum_{s=0}^\infty \binom{A}{s} X^s \in \mathcal{M}_n(K[[X]])$$

where

$$\binom{A}{0} = I_n$$

and

$$\binom{A}{s} = \frac{1}{s!} A(A - I_n) \cdots (A - (s-1)I_n) \quad \text{for} \quad s \geq 1 .$$

Then one can easily verify that $g_A(X)$ is the formal matrix solution to the differential system

$$(1 + X)\frac{dY}{dX} = AY$$

such that $g_A(0) = I_n$. Assume that $A = P^{-1}(\Delta + N)P$ where Δ is a diagonal matrix, N is a nilpotent matrix commuting with Δ and $P \in G\ell(n, K)$. Then we consider formal matrix solutions Y_Δ and Y_N to

$$(1 + X)\frac{dY_\Delta}{dX} = \Delta Y_\Delta \quad \text{and} \quad (1 + X)\frac{dY_N}{dX} = NY_N ,$$

normalized by the condition $Y_\Delta(0) = I_n = Y_N(0)$, and we have

$$Y_\Delta = g_\Delta(X) = \sum_{s=0}^{\infty} \binom{\Delta}{s} X^s .$$

If we observe that

$$Y_\Delta N = \sum_s \binom{\Delta}{s} X^s N = N \sum_s \binom{\Delta}{s} X^s = NY_\Delta$$

we can then conclude that

$$\begin{aligned}
(1 + X)\frac{d(Y_\Delta Y_N)}{dX} &= (1 + X)\frac{dY_\Delta}{dX} Y_N + Y_\Delta(1 + X)\frac{dY_N}{dX} \\
&= (\Delta Y_\Delta Y_N + Y_\Delta N Y_N) \\
&= (N + \Delta)Y_N Y_\Delta .
\end{aligned}$$

On the other hand for Y_N we may take the formal power series

$$Y_N = \sum_{s=0}^{n-1} \frac{N^s}{s!} \log^s(1 + X)$$

which converges in the disk $D(0, 1^-)$. Therefore,

$$g_N(X) = \sum_{s=0}^{n-1} \frac{N^s}{s!} \log^s(1 + X)$$

and

$$g_A(X) = P^{-1}g_\Delta(X)g_N(X)P = P^{-1}g_\Delta(X) \left(\sum_{s=0}^{n-1} \frac{N^s}{s!} \log^s(1 + X) \right) P \, .$$

Now if

$$\Delta = \begin{pmatrix} \Delta_1 & & & 0 \\ & \Delta_2 & & \\ & & \ddots & \\ 0 & & & \Delta_n \end{pmatrix},$$

then, by Proposition 7.3, $g_\Delta(X)$ converges on $D(0, R^+)$, $R = |\pi|^{1/p^l}$, $l \geq 0$, if and only if each $\Delta_i \in \mathbf{Z} + p^{l+\varepsilon}\mathcal{O}_K$, $\varepsilon > 0$ (the point is that the inequality

$$\frac{1}{p^m(p-1)} - \frac{\varepsilon'}{p^{m+1}} < \frac{1}{(p-1)p^l} \, ,$$

where $\varepsilon' \in [0, 1)$, implies $m \geq l$ and if $m = l$ then $\varepsilon' > 0$).
We can conclude:

Proposition 7.5. *Let $A \in \mathcal{M}_n(K)$ and let $\Delta_1, \ldots, \Delta_n$ be the eigenvalues of the matrix A. Then $g_A(X)$ converges in $D(0, R^+)$, $R = |\pi|^{1/p^l}$, $l \geq 0$, if and only if*

$$\Delta_1, \ldots, \Delta_n \in \mathbf{Z} + p^{l+\varepsilon}\mathcal{O}_K$$

for some $\varepsilon > 0$. In particular if each $\Delta_i \in p^{l+\varepsilon}\mathcal{O}_K$ then

$$g_A(\zeta - 1) = I_n \tag{7.2}$$

for all ζ such that $\zeta^{p^{l+1}} = 1$. Moreover, if A is nilpotent equality (7.2) holds for all p-th power roots of unity.

Corollary 7.6. *Let A be as in Proposition 7.5. Then $g_A(X)$ converges in $D(0, 1^-)$ if and only if the eigenvalues of A belong to $\mathbf{Z}_p$.*

The following proposition due to G. Christol gives an example of a new type of transfer theorem. Here we obtain information on the singular disk $D(0, 1^-)$ from information on the generic disk.

Proposition 7.7. *Let $G \in \mathcal{M}_n(E_0)$. Let $\mathcal{U}_{G,t}$ be the solution matrix at the generic point t of the system*

$$\frac{d\vec{y}}{dX} = \frac{1}{X} G \vec{y}$$

such that $\mathcal{U}_{G,t}(t) = I_n$. Assume that $\mathcal{U}_{G,t}$ has radius of convergence 1. Then the eigenvalues of the matrix $G(0)$ belong to $\mathbf{Z}_p$.

Proof. We define

$$G_1 = \frac{1}{X} G$$

and as usual, for $s = 2, 3, \ldots$, we define the matrices G_s by

$$G_{s+1} = G_s G_1 + \frac{d}{dX} G_s$$

so that

$$\frac{d^s}{dX^s} \vec{y} = G_s \vec{y}$$

for all solutions $\vec{y}$ of the system. We know that $X G_1 \in \mathcal{M}_n(E_0)$ and we now show that indeed $X^s G_s \in \mathcal{M}_n(E_0)$ for all s. In fact, we prove by induction that

$$G_s(X) = s! \binom{G(0)}{s} \frac{1}{X^s} + \left(\frac{1}{X^{s-1}} \right)$$

where

$$s! \binom{G(0)}{s} = G(0)(G(0) - I_n) \cdots (G(0) - (s-1)I_n) .$$

The recursion formula reduces the verification to the following obvious identity

$$s! \binom{G(0)}{s} \binom{G(0)}{1} + s!(-s) \binom{G(0)}{s} = (s+1)! \binom{G(0)}{s+1} .$$

By hypothesis

$$u_{G,t}(X) = \sum_{s=0}^{\infty} \frac{G_s(t)}{s!}(X-t)^s$$

converges in $D(t, 1^-)$ and so by Theorem 2.1 we have

$$\left| \frac{G_s(t)}{s!} \right| \leq C\{s, n-1\}_p$$

for a suitable constant C. Therefore, we also have

$$\left| \frac{G_s}{s!} \right|_{\text{gauss}} \leq C\{s, n-1\}_p .$$

On the other hand

$$\left| \frac{G_s}{s!} \right|_{\text{gauss}} \geq \left| \binom{G(0)}{s} \right|$$

so that the series

$$(1+X)^{G(0)} = \sum_{s=0}^{\infty} \binom{G(0)}{s} X^s$$

converges in $D(0, 1^-)$, and therefore by Corollary 7.6 we obtain the desired result. **Q.E.D.**

8.   The Hypergeometric Function of Euler and Gauss.

Let K be as in Section 7. We consider the *hypergeometric function* defined by

$$(8.1) \qquad\qquad F(a, b, c; X) = \sum_{s=0}^{\infty} \frac{(a)_s (b)_s}{s!(c)_s} X^s$$

where $a, b, c \in K$, $c \notin \mathbf{N}$, and where for any element $\alpha \in K$ the symbol $(\alpha)_s$ is defined by

$$(\alpha)_0 = 1, \qquad (\alpha)_s = \alpha(\alpha+1)\cdots(\alpha+s-1) \quad \text{for} \quad s \geq 1 .$$

The power series (8.1) is a formal solution of the *hypergeometric differential equation*

$$(8.2) \qquad L_{a,b,c} = D^2 + \frac{c - (a+b+1)X}{X(1-X)} D - \frac{ab}{X(1-X)} .$$

It is well known that $L_{a,b,c}$ has only three singular points, namely 0,1 and ∞ with exponents given by the following table.

Exponents of $L_{a,b,c}$		
0	1	∞
0	0	a
$1-c$	$c-a-b$	b

Another formal solution to $L_{a,b,c}$ at zero has the form

$$X^{1-c}F(a+1-c, b+1-c, 2-c; X)$$

and if $c \notin \mathbf{Z}$ then this is independent of the first solution. Kummer gave a list of 24 such formal series solutions; here are four more of them:

$$F(a, b, a+b+1-c; 1-X),$$
$$(1-X)^{c-a-b}F(c-a, c-b, c+1-a-b; 1-X),$$
$$X^{-a}F(a, a+1-c, a+1-b; 1/X),$$
$$X^{-b}F(b+1-c, b, b+1-a; 1/X).$$

We have the following result regarding the domain of convergence of the formal solutions of $L_{a,b,c}$ at the generic point.

Proposition 8.1. *If a, b, and $c \in \mathbf{Z}_p$ then the solutions of $L_{a,b,c}$ at the generic point t are convergent in the disk $D(t, 1^-)$.*

Proof. Let $(a_k, b_k, c_k) \in (-\mathbf{N})^3$ be a sequence converging to (a, b, c) chosen so that

a) $-a_k < -c_k$,
b) $b_k + 1 - c_k < 0$.
   Let

$$u_k(X) = \sum_{s < 1-c_k} \frac{(a_k)_s (b_k)_s}{(c_k)_s s!} X^s.$$

By hypothesis (a) this polynomial really has degree at most $-a_k$ and, as is easily verified, is a solution of L_{a_k, b_k, c_k}. Let

$$v_k(X) = X^{1-c_k}F(a_k+1-c_k, b_k+1-c_k, 2-c_k; X),$$

which is well defined since $2 - c_k > 0$. In view of hypothesis (b) we see that v_k is also a polynomial. Now we have $u_k(0) = 1$ and $v_k(0) = 0$ so that u_k and v_k are linearly independent. Let

$$W_k(X) = W(u_k(X), v_k(X))$$

where W denotes the Wronskian matrix. Moreover, let $\tilde{u}_k(X)$, $\tilde{v}_k(X)$ be defined by

$$(\tilde{u}_k(X), \tilde{v}_k(X)) = (u_k(X), v_k(X))W_k(t)^{-1} .$$

Then $\tilde{u}_k$, $\tilde{v}_k$ are linearly independent solutions of L_{a_k,b_k,c_k} normalized by $\tilde{W}_k(t) = I_2$ (here $\tilde{W}_k = W(\tilde{u}_k, \tilde{v}_k)$) and are clearly analytic in the disk $D(t, 1^-)$. Therefore, by Proposition 5.4, the same is true for the two solutions $u(X)$ and $v(X)$ of $L_{a,b,c}$ defined by

$$u(X) = \lim_{k \to \infty} \tilde{u}_k(X) , \qquad v(X) = \lim_{k \to \infty} \tilde{v}_k(X) ,$$

where the limits are taken in the topology of uniform convergence on $D(t, R^-)$ for each $R < 1$. **Q.E.D.**

We note that the convergence of the two solutions in the generic disk does not permit one to draw conclusions regarding the convergence of solutions at the points 0, 1 and ∞. As an example of this, consider the case $a = 1$ and $c = b + 1 \notin \mathbf{Z}$ with $b \in \mathbf{Z}_p$. In this situation the exponents are given by the following table.

Exponents of $L_{1,b,b+1}$		
0	1	∞
0	0	1
$-b$	0	b

Formal Solutions at zero:

$$F(1, b, b+1; X) = \sum_{s=0}^{\infty} \frac{(1)_s (b)_s}{s!(b+1)_s} X^s = b \sum_{s \geq 0} \frac{1}{b+s} X^s$$

$$X^{-b} F(1 - b, 0, 1 - b; X) = X^{-b} .$$

The radius of convergence of the first series at zero can be zero or very small if $|b + s|$ is very small for arbitrarily large values of s, that is if b is what is called a p-adic Liouville number, see Section VI.1.

Formal Solutions at ∞:

$$\frac{1}{X}F(1, 1-b, 2-b; \frac{1}{X}) = \frac{1}{X}\sum_{s=0}^{\infty}\left(\frac{1}{X}\right)^s \frac{(1-b)_s}{(2-b)_s}$$

$$= \frac{1}{X}\sum_{s=0}^{\infty}\left(\frac{1}{X}\right)^s \frac{1}{1-b+s}$$

$$\left(\frac{1}{X}\right)^b F(0, b, b; \frac{1}{X}) = \left(\frac{1}{X}\right)^b$$

and so this case is similar to the preceding one.

Formal Solutions at 1:

$$F(1, b, 1; 1-X) = \sum_{s=0}^{\infty}\frac{(b)_s}{s!}(1-X)^s = \sum_{s=0}^{\infty}\binom{-b}{s}(X-1)^s = X^{-b} .$$

Since the two exponents at 1 are both zero, we must expect a solution with a logarithmic term. To find it we first observe that $L_{1,b,b+1}y = 0$ is equivalent to

$$\delta \circ (X-1) \circ (\delta + b)y = 0 , \qquad \text{with} \quad \delta = X\frac{d}{dX} .$$

We further note that if $P(\delta) \in K(X)[\delta]$ then from Leibniz's formula it follows that

$$P(\delta)(X^\alpha y) = X^\alpha P(\delta + \alpha)y .$$

Therefore, if $y = X^{-b}z$, from $\delta \circ (X-1) \circ (\delta + b)X^{-b}z = 0$ we find

$$(\delta - b) \circ (X-1) \circ \delta z = 0$$

so that

$$(X-1) \circ \delta z = X^b .$$

If we now put $X = 1 + U$ then we must have

$$(1+U)U\frac{d}{dU}z = (1+U)^b$$

so that

$$\frac{d}{dU}z = \frac{1}{U} + \sum_{s=1}^{\infty}\binom{b-1}{s}U^{s-1}$$

whence the other solution is

$$X^{-b}\left[\log(X-1)+\sum_{s=1}^{\infty}\binom{b-1}{s}\frac{(X-1)^s}{s}\right].$$

Aside from the factor $\log(1-X)$ both solutions converge in the disk $D(1,1^-)$, and convergence of all solutions in $D(t,1^-)$ could again be established in this case by Proposition 5.2 and Corollary 5.3.

Remark 8.2. The reader should check that if L_1 and L_2 are elements of $K(X)[D]$ each with the property that all its solutions at t converge in $D(t,1^-)$ then the same holds for $L = L_1 \circ L_2$. By applying this remark to

$$L_{1,b,b+1} = \frac{1}{X-1}\circ\delta\circ(X-1)\circ(\delta+b)$$

we deduce a third proof that all solutions of $L_{1,b,b+1}$ at t converge in the disk $D(t,1^-)$.

CHAPTER V

EFFECTIVE BOUNDS. SINGULAR DISKS

In this Chapter K will be a fixed complete algebraically closed extension of $\mathbf{Q}_p$ and Ω will be a sufficiently large extension of K so as to contain all our generic points over K.

1. The Dwork-Frobenius Theorem.

We recall the E is the field of analytic elements over K, cf. Section IV.4. We denote by $\mathfrak{p}_E$ the valuation ideal of E that is

$$\mathfrak{p}_E = \{\xi \in E \,|\, |\xi|_{\text{gauss}} < 1\} \,.$$

We observe that the residue field of E is $\overline{K}(X)$, the same as that of $K(X)$ with respect to the Gauss norm.

Lemma 1.1. *Let $G \in \mathcal{M}_n(E_0)$ and let $\mathcal{U}_{G,t}$ be the matrix solution at the generic point t of the system*

$$\frac{d}{dX}\vec{y} = G\vec{y}$$

such that $\mathcal{U}_{G,t}(t) = I_n$. Assume that $\mathcal{U}_{G,t}$ converges in the disk $D(t, |\pi|^-)$ where as usual $|\pi|^{p-1} = |p|$. Assume moreover that

$$|G|_{\text{gauss}} > 1 \,.$$

If $\alpha \in K$ satisfies $|\alpha| = |G|_{\text{gauss}}$ then the matrix $\alpha^{-1}G$ is nilpotent modulo $\mathfrak{p}_E$.

Proof. With G_s defined as in (III.5.3) we shall prove by induction that

$$\frac{G_s}{\alpha^s} \equiv \left(\frac{G}{\alpha}\right)^s \quad (\text{mod } \frac{1}{\alpha}) \,.$$

Indeed,

$$\frac{G_{s+1}}{\alpha^{s+1}} = \frac{G_s}{\alpha^s}\frac{G}{\alpha} + \frac{1}{\alpha}\frac{d}{dX}\frac{G_s}{\alpha^s}$$

and it suffices to observe that

$$\left|\frac{d}{dX}\frac{G_s}{\alpha^s}\right|_{\text{gauss}} \leq 1$$

since by the induction hypothesis $|G_s/\alpha^s|_{\text{gauss}} \leq 1$ and (III.5.4) extends to E.

Let λ be an eigenvalue of $G(t)/\alpha$ so that

$$\frac{G(t)}{\alpha}\vec{v} = \lambda\vec{v}$$

where $\vec{v}$ is an eigenvector such that $|\vec{v}| = 1$. Then

$$\frac{G_s(t)}{\alpha^s}\vec{v} = \left[\left(\frac{G(t)}{\alpha}\right)^s + \left(\frac{1}{\alpha}\right)\right]\vec{v} = \lambda^s\vec{v} + \left(\frac{1}{\alpha}\right) .$$

Assume now that $|\lambda| \geq 1$. Then $|G_s(t)\vec{v}| \geq |\alpha|^s$. Therefore the solution

$$\sum_{s=0}^{\infty}\frac{G_s(t)}{s!}\vec{v}(X - \alpha)^s$$

is not convergent if $|x - t| > |\pi|/|\alpha|$. In this way we contradict the hypothesis that all solutions are convergent in $D(t, \pi^-)$. We can conclude that $|\lambda| < 1$ as desired.   **Q.E.D.**

We proved in Section III.5 that, in the above notation, if $|G|_{\text{gauss}} \leq 1$ then $\mathcal{U}_{G,t}$ converges in the disk $D(t, |\pi|^-)$. The following is a partial converse to that result.

Theorem 1.2 (Dwork-Frobenius). *Let*

$$G = \begin{pmatrix} 0 & 1 & & \cdots & 0 \\ 0 & 0 & 1 & & \\ \vdots & & & \ddots & \vdots \\ 0 & & & & 1 \\ C_n & C_{n-1} & & \cdots & C_1 \end{pmatrix}$$

with $C_1, \ldots, C_n \in E_0$. *Assume that the matrix solution of the system*

$$\frac{d}{dX}\vec{y} = G\vec{y}$$

is convergent on $D(t, |\pi|^-)$. *Then* $|C_j|_{\text{gauss}} \leq 1$ *for* $j = 1, \ldots, n$.

Proof. Let $\lambda \in K$ be such that $|\lambda| = \sup_{1 \leq j \leq n}(|C_j|_{\text{gauss}}^{1/j})$ so that

$$\sup_{1 \leq j \leq n} (|C_j/\lambda^j|_{\text{gauss}}) = 1 .$$

Take

$$H = \begin{pmatrix} 1 & & & 0 \\ & \lambda & & \\ & & \ddots & \\ 0 & & & \lambda^{n-1} \end{pmatrix}$$

and consider $G_{[H]} = H^{-1}GH$. It follows that

$$G_{[H]} = \lambda \begin{pmatrix} 0 & 1 & \cdots & 0 \\ 0 & 0 & 1 & \\ \vdots & & \ddots & \vdots \\ 0 & & & 1 \\ \lambda^{-n}C_n & \lambda^{-n+1}C_{n-1} & \cdots & \lambda^{-1}C_1 \end{pmatrix} .$$

Hence

$$|G_{[H]}|_{\text{gauss}} = |\lambda| .$$

If $|C_j|_{\text{gauss}} > 1$ for some j then $|\lambda| > 1$ so that by Lemma 1.1 the matrix $\lambda^{-1}G_{[H]}$ is nilpotent modulo $\mathfrak{p}_E$. On the other hand, the characteristic polynomial of $\lambda^{-1}G_{[H]}$ is $X^n - \lambda^{-1}C_1X^{n-1} + \cdots + (-1)^n\lambda^{-n}C_n$, which has a root α with $|\alpha| = 1$ and this gives a contradiction. **Q.E.D.**

Remark 1.3. For a scalar equation $L = D^n + C_1D^{n-1} + \cdots + C_n$ with $C_j \in E$ and $|C_j|_{\text{gauss}} \leq 1$ for all j we have shown in Section III.5 that all solutions at the generic point converge in the disk $D(t, |\pi|^-)$, and for special points one can not generally do better. For example, the equation $d/dX - 1$ has the exponential function as a solution at zero, and this has radius of convergence $|\pi|$.

Example 1.4. By a *Heun equation* we will mean a second order differential equation with four singular points, all regular; we will suppose that all exponents belong to $\mathbf{Q} \cap \mathbf{Z}_p$. If we assume the singularities to be at the points 0, 1, λ and ∞ we may take the equation to be in the form

$$D^2 + \left(\frac{a}{X} + \frac{b}{X-1} + \frac{c}{X-\lambda} \right) D + \frac{eX + f}{X(X-1)(X-\lambda)} \, .$$

It is possible to compute the indicial polynomials at the singularities and then one sees that the exponents determine the coefficients a, b, c, e but not f which is called, in Klein's terminology, the *accessory parameter*. If $|f| \leq \sup(1, |\lambda|)$ then all the solutions at the generic point converge in the disk $D(t, |\pi|^-)$, but there may be special values of f for which the radius of convergence is larger. Not very much is known about this question. Actually this equation does not "come from geometry" in the sense that there is no integral formula for the solutions, unlike the case of the hypergeometric equation (three singularities) which we have discussed in Section IV.8.

The case $a = b = c = 1/2$, $e = -n(n+1)/4$ is known as the *Lamé equation* and is the subject of an extensive literature (cf. Dwork [2]). If λ is algebraic over $\mathbf{Q}$ and $n \in \mathbf{N}$ then there exist $2n + 1$ special values of f for which there is a solution which is the square root of an element of $\mathbf{Q}[\lambda, f, X]$ and in this case the radius of convergence at the generic point is 1 for almost all valuations of $\mathbf{Q}(\lambda, f)$. If f is algebraic over $\mathbf{Q}$ but not equal to one of these $2n + 1$ special values then the solutions of the equation are given by the exponential of an integral of the third kind on the elliptic curve $Y^2 = X(X-1)(X-\lambda)$. In this case it has been shown in Chudnovsky [2] that global nilpotence implies finiteness of the monodromy group and so under that hypothesis once again $R_v = 1$ for almost all valuations v of $\mathbf{Q}(\lambda, f)$.

For the case $n = 0$, the special value of f is $f = 0$. For $f \neq 0$ the solutions involve the exponential of an integral of the first kind and hence by the result of Chudnovsky global nilpotence is impossible. An earlier proof based on the Riemann hypothesis for elliptic curves defined over a finite field may be found in Katz [1]. We note that for f not special, for almost all primes, nilpotence implies zero p–curvature, i.e. $G_p \equiv 0$ mod v if v is an extension of p. Thus the result of the Chudnovskys is a special case of Grothendieck's conjecture that $G_p \equiv 0$ mod v for almost all p implies finiteness of monodromy. The main treatment of this conjecture has been given by Katz [2].

2.   Effective Bounds for Solutions in a Singular Disk: the Case of Nilpotent Monodromy. The Christol–Dwork Theorem: Outline of the Proof.

Consider a system

$$\delta - G \quad \text{with} \quad G \in E_0 , \qquad (2.1)$$

where $\delta = X d/dX$. The origin is a regular singularity of (2.1) (cf. the end of Section III.8). Let $\lambda_1, \ldots, \lambda_n$ be the eigenvalues of $G(0)$ and assume that they are prepared (cf. (III.8.6)). From Proposition III.8.5 we know that there exists a solution matrix $Y_G X^{G(0)}$ of (2.1) with $Y_G \in G\ell(n, K[[X]])$ and $Y_G(0) = I_n$.

Two questions now arise naturally:
- What is the radius of convergence of Y_G?
- If $Y_G = I_n + Y_1 X + Y_2 X^2 + \cdots$, where the $Y_j \in \mathcal{M}_n(K)$, what sort of estimate does one have for the growth of the coefficients Y_j?

We begin with the second question, and shall dedicate some time to providing an answer. In order to do so, it is convenient to introduce the class $\mathcal{R}$ of $n \times n$ matrices G satisfying the following four conditions:

$\mathcal{R}1$ $G \in \mathcal{M}_n(E_0)$.
$\mathcal{R}2$ There is a matrix solution $\mathcal{U}_{G,t}$ at the generic point t (normalized by $\mathcal{U}_{G,t}(t) = I_n$) which converges on $D(t, 1^-)$.
$\mathcal{R}3$ $G(0)$ is nilpotent.
$\mathcal{R}4$ $|G|_{\text{gauss}} \leq 1$.

Note that in view of $\mathcal{R}2$, Proposition IV.7.7 forces all eigenvalues of $G(0)$ to lie in $\mathbf{Z}_p$, but with condition $\mathcal{R}3$ we now impose the stronger condition that 0 be the only eigenvalue.

For systems (2.1) with $G \in \mathcal{R}$ the following theorem gives an estimate on coefficient growth.

Theorem 2.1 (Christol-Dwork).   *Let $G \in \mathcal{R}$, and let $Y_G X^{G(0)}$ be a matrix solution of (2.1) with $Y_G(0) = I_n$. Let*

$$Y_G = \sum_{s=0}^{\infty} Y_s X^s$$

where $Y_0 = I_n$ and $Y_s \in \mathcal{M}_n(K)$. Then

$$|Y_s| \leq T(s)^{n-1+\text{ord}\,(n-1)!+\beta_n} ,$$

where for each positive integer s

$$T(s) = \sup_{1 \le k \le s} |k|^{-1}$$

and

$$\beta_n = \inf \left(n - 1, \operatorname{ord} \left(\prod_{j=1}^n \binom{n}{j} \right) \right) .$$

We begin with some preliminary observations. From this theorem it will follow that Y_G converges in the disk $D(0, 1^-)$, since, roughly speaking, the estimate gives $|Y_s| < s^{l(n,p)}$ where $l(n,p)$ is a positive integer depending on n and p and not on G (universal estimate).

The proof of this theorem will be a rather lengthy induction on s. We start with a very simple remark.

Remark 2.2. Let $G \in \mathcal{M}_n(K[[X]])$ and assume that the eigenvalues $\alpha_1, \ldots, \alpha_n$ of $G(0)$ are prepared and satisfy $|\alpha_i| < 1$. The uniform part Y_G of the matrix solution at zero of the system $\delta - G$ satisfies the differential equation

$$\delta Y_G + Y_G G(0) - G Y_G = 0 . \tag{2.2}$$

If we write

$$G(X) = G(0) + G_1 X + G_2 X^2 + \cdots$$

with the $G_i \in \mathcal{M}_n(K)$ then to solve equation (2.2) we must inductively solve the equations

$$m Y_m + Y_m G(0) - G(0) Y_m = \sum_{j=1}^m G_j Y_{m-j} . \tag{2.3}$$

The eigenvalues of the linear map

$$\theta = \psi_{G(0)} : \mathcal{M}_n(K) \longrightarrow \mathcal{M}_n(K)$$
$$A \longmapsto G(0)A - AG(0)$$

are the differences of the eigenvalues of $G(0)$ (see Lemma III.8.4). Therefore $|\det(mI - \theta)| = 1$ for $1 \le m \le p - 1$, where I denotes the identity map of $\mathcal{M}_n(K)$. Assume $|G(0)| \le 1$ and $|G_j| \le 1$ for all $j \ge 1$. From $|G(0)| \le 1$ it follows that $|\operatorname{adj}(mI - \theta)| \le 1$ so that $|(mI - \theta)^{-1}| \le 1$ for $1 \le m \le p - 1$. Now from (2.3) we obtain

$$Y_m = (mI - \theta)^{-1} \sum_{j=1}^m G_j Y_{m-j}$$

and so we may conclude that

$$|Y_m| \leq \sup_{0 \leq j \leq m-1} |Y_j| \,.$$

Since $|Y_0| = |I_n| = 1$ by induction we obtain

$$|Y_m| \leq 1 \quad \text{for} \quad 1 \leq m \leq p-1 \,.$$

If $G \in \mathcal{R}$ then we clearly have the hypotheses required to obtain this last result.

The main point in the proof will be the following:

If $G \in \mathcal{R}$ then there exists a matrix $G^{(1)} \in \mathcal{R}$ such that

$$T(G, s) \leq T(G^{(1)}, s/p) B_n$$

where B_n is a constant independent of G and s and

$$T(G, s) = \sup_{0 \leq j \leq s} |Y_j| \,.$$

This suffices to obtain the desired estimate. Indeed, for $s \in \mathbf{N}$ let l be such that $1 \leq s/p^l < p$. Then

$$T(G, s) \leq T(G^{(1)}, s/p) B_n \leq T(G^{(2)}, s/p^2) B_n^2$$
$$\leq \cdots \leq T(G^{(l)}, s/p^l) B_n^l \leq B_n^l \,.$$

Now since $p^l \leq s < p^{l+1}$,

$$p^l = \sup_{k \leq s} \frac{1}{|k|} = T(s)$$

so that

$$B_n^l = (T(s))^{\log B_n / \log p}$$

and there remains to find a good estimate for B_n.

We divide the proof into five steps which we now sketch, but before proceeding we introduce some lemmas and terminology which will be of frequent use in the sequel.

An element $H \in G\ell(n, E)$ will be said to be *unimodular* if

$$|H|_{\text{gauss}} \leq 1 \quad \text{and} \quad |H^{-1}|_{\text{gauss}} \leq 1 . \tag{2.4}$$

We recall that for $G \in \mathcal{M}_n(E)$ and $H \in G\ell(n, E)$

$$G_{[H],\delta} = G_{[H]} = HGH^{-1} + (\delta H)H^{-1} .$$

We also recall the definition of $\log X$ and X^α for $\alpha \in K$ given in Section III.8. We remind the reader that $\log X$ and X^α are elements of a differential field extension $\mathcal{F}$ of $K((X))$. We let $\mathcal{F}_0 = \ker(\delta, \mathcal{F}) \supseteq K$ be the field of constants of $\mathcal{F}$ with respect to δ. Clearly $\mathcal{F}_0 \cap K((X)) = K$.

Lemma 2.3. *Let $P(X)$ be a non-zero element of $\mathcal{F}_0[X]$ and let $\alpha \in K$. Then $X^\alpha P(\log X)$ belongs to $K((X))$ if and only if $\alpha \in \mathbf{Z}$ and $\deg P = 0$.*

Proof.   Suppose $X^\alpha P(\log X) = w \in K((X))$. Then

$$P(\log X) = (X^\alpha)^{-1}w = X^{-\alpha}w .$$

Let $n = \deg P$ and put $D = d/dX$; since

$$\delta P(\log X) = (DP)(\log X) ,$$

we have

$$\delta^{n+1} P(\log X) = (D^{n+1}P)(\log X) = 0 ,$$

so that

$$\left(X^\alpha \circ \delta^{n+1} \circ X^{-\alpha}\right) w = (\delta - \alpha)^{n+1}w = 0 .$$

Therefore, if

$$w = \sum_j A_j X^j$$

with all $A_j \in K$ we obtain

$$\sum_j A_j (j - \alpha)^{n+1} X^j = 0 ;$$

hence, either $w = 0$ or $\alpha \in \mathbf{Z}$ and $X^{-\alpha}w = A_\alpha$. We can conclude that in either case $X^{-\alpha}w \in K$, and so we have $P(\log X) \in K$. If $n > 1$, then on applying δ we can reduce to the case in which P is of degree one, and so $\log X$ would be a constant, which is impossible.   **Q.E.D.**

Let $F, G \in \mathcal{M}_n(K[[X]])$, $H \in G\ell(n, K((X)))$ and assume that

$$G = F_{[H]} \; .$$

Assume moreover that the eigenvalues of $F(0)$ and $G(0)$ are prepared. Let

$$\overline{H} = Y_G^{-1} H Y_F \in G\ell(n, K((X))) \; .$$

We claim that $\overline{H}$ is actually an element of $G\ell\left(n, K[X, X^{-1}]\right)$, that is $\overline{H}$ and $\overline{H}^{-1}$ are both Laurent polynomials. Indeed, there exists a matrix $C \in G\ell(n, \mathcal{F}_0)$ such that

$$H Y_F X^{F(0)} = Y_G X^{G(0)} C \; ,$$

so that

$$\overline{H} X^{F(0)} = X^{G(0)} C \; .$$

If we put $A = F(0)$ and $B = G(0)$ this last identity is equivalent to

$$B = A_{[\overline{H}]} \; .$$

Therefore,

$$\delta \overline{H} = B \overline{H} - \overline{H} A \; ,$$

and putting $\overline{H} = \sum_s H_s X^s$ we find

$$s H_s = B H_s - H_s A$$

for all $s \in \mathbf{Z}$. Hence, if $H_s \neq 0$ then s is an eigenvalue of the map $\psi_{B,A}$ defined in Lemma III.8.4. Since by the same Lemma the eigenvalues of this map are precisely the differences of the eigenvalues of B and A we must have that $H_s = 0$ for almost all s, so that $\overline{H}$ is a Laurent polynomial. By symmetry $(A = B_{[\overline{H}^{-1}]})$ the same holds for $\overline{H}^{-1}$.

We can draw some consequences of the fact that the matrix $\overline{H}$ lies in $G\ell\left(n, K[X, X^{-1}]\right)$. First, since $\det \overline{H}$ must be a unit in $K[X, X^{-1}]$, we see that $\det \overline{H}$ is a monomial, that is $\det \overline{H} = cX^\mu$ for some non-zero $c \in K$ and some integer μ. Second, since

$$\delta(\det \overline{H}) = (\operatorname{Tr} G(0) - \operatorname{Tr} F(0)) \det \overline{H}$$

we have that $\operatorname{Tr} G(0) - \operatorname{Tr} F(0)$ is a rational integer.

Using Lemma 2.3 we can obtain more precise results.

Lemma 2.4. *Let* F, $G \in \mathcal{M}_n(K((X)))$ *and let* $H \in G\ell(n, K((X)))$. *Assume* $G = F_{[H]}$. *Let zero be a regular singularity of both systems* $\delta - F$ *and* $\delta - G$, *that is, let the systems* $\delta - F$ *and* $\delta - G$ *have matrix solutions of the form* $Y_F X^A$ *and* $Y_G X^B$ *with* Y_F, $Y_G \in G\ell(n, K((X)))$ *and* $A, B \in \mathcal{M}_n(K)$. *Let* $\alpha_1, \ldots, \alpha_n$ *and* $\beta_1, \ldots, \beta_n$ *be the eigenvalues of* A *and* B *respectively. Then there exist* C_1, $C_2 \in G\ell(n, K)$ *such that*

$$Y_G^{-1} H Y_F = C_1 M C_2$$

where

$$M = \left(C_{ij} X^{\beta_i - \alpha_j} \right)$$

with $C_{ij} \in K$. *Moreover, if* $C_{ij} \neq 0$ *then* $\beta_i - \alpha_j \in \mathbf{Z}$ *so that* $M \in G\ell\left(n, K[X, X^{-1}]\right)$. *In particular, for each* $i = 1, \ldots, n$ *there exists an index* j *such that* $\beta_i \equiv \alpha_j \mod \mathbf{Z}$.

Proof. As above we write $\overline{H} = Y_G^{-1} H Y_F$. Let

$$\Delta = \begin{pmatrix} \alpha_1 & & 0 \\ & \ddots & \\ 0 & & \alpha_n \end{pmatrix}, \qquad \Delta' = \begin{pmatrix} \beta_1 & & 0 \\ & \ddots & \\ 0 & & \beta_n \end{pmatrix}.$$

Moreover, let C_1, $C_2 \in G\ell(n, K)$ be such that

$$A = C_2^{-1}(\Delta + N) C_2$$
$$B = C_1(\Delta' + N') C_1^{-1}$$

where N and N' are nilpotent matrices in $\mathcal{M}_n(K)$ commuting respectively with Δ and Δ'. Since $B = A_{[\overline{H}]}$ there exists a matrix $\tilde{C} \in G\ell(n, \mathcal{F}_0)$ such that

$$\overline{H} X^A = X^B \tilde{C} .$$

From this it follows that

$$X^{\Delta'} X^{N'} \overline{C} X^{-N} X^{-\Delta} \in G\ell(n, K((X)))$$

where

$$\overline{C} = C_1^{-1} \tilde{C} C_2^{-1} \in G\ell(n, \mathcal{F}_0) .$$

Now

$$X^{N'} \overline{C} X^{-N} = (P_{ij}(\log X))$$

for suitable $P_{ij}(X) \in \mathcal{F}_0[X]$, so that

$$X^{\Delta'} X^{N'} \overline{C} X^{-N} X^{-\Delta} = \left(X^{\beta_i} X^{-\alpha_j} P_{ij}(\log X) \right) .$$

Therefore, from Lemma 2.3 it follows that each $P_{ij}(X)$ is a constant polynomial, so that

$$X^{N'} \overline{C} X^{-N} = \overline{C}$$

and

$$X^{\Delta'} \overline{C} X^{-\Delta} = \left(C_{ij} X^{\beta_i - \alpha_j} \right) = M \in G\ell(n, K((X))) .$$

Again applying Lemma 2.3 we find that if $C_{ij} \neq 0$ then $\beta_i - \alpha_j \in$ **Z**. Therefore $M \in G\ell\left(n, K[X, X^{-1}]\right)$ and $\overline{H} = C_1 M C_2$, as desired. **Q.E.D.**

We now sketch the five steps into which the proof of Theorem 2.1 is divided.

Step I.   Construction of Frobenius.

If $f = f(X) \in E$ we denote $f(X^p)$ by f^ϕ.

Let $G \in \mathcal{R}$. Then there exists $H \in G\ell(n, E_0)$, H unimodular, and a matrix F satisfying $\mathcal{R}1$, $\mathcal{R}2$, $\mathcal{R}3$, but not necessarily $\mathcal{R}4$ such that

$$G_{[H]} = pF^\phi ,$$

$$G(0) = pF(0) ,$$

$$Y_G = H^{-1} Y_F^\phi .$$

We observe that from the first equality it follows that $|F|_{\text{gauss}} \leq p$.

Even if we started with $G \in \mathcal{M}_n(K(X))$ (as is the case in many applications) the matrix H will not necessarily be in $G\ell(n, K(X))$, and so F too is not necessarily rational. This is the reason for introducing the ring E_0.

Step I is the basis of the theorem to be proved. Indeed, Steps II-V serve to overcome the failure of F to satisfy $\mathcal{R}4$.

Step II.   Choice of the cyclic vector.

Now F satisfies $\mathcal{R}1$, $\mathcal{R}2$ and $\mathcal{R}3$ and we also know that $|F|_{\text{gauss}} \leq p$. Then we can find $H_1 \in G\ell(n, E_0')$ such that

$$F_{[H_1]} = \begin{pmatrix} 0 & 1 & 0 & \dots & 0 \\ 0 & 0 & 1 & \dots & \vdots \\ \vdots & \vdots & & \ddots & 0 \\ 0 & 0 & & \dots & 1 \\ C_0 & C_1 & & \dots & C_{n-1} \end{pmatrix}$$

and

$$|H_1|_{\text{gauss}} \, |H_1^{-1}|_{\text{gauss}} \leq B_n \ .$$

Moreover, the matrix $F_{[H_1]}$ satisfies $\mathcal{R}2$ and therefore also $\mathcal{R}4$ by Theorem 1.2 (Dwork-Frobenius Theorem).

Since zero is a regular singularity of $\delta - F$ the same holds for $\delta - F_{[H_1]}$ which is a scalar equation and so, by Fuchs' condition (cf. Theorem III.8.9) $F_{[H_1]}$ is analytic at zero, though not necessarily in all of $D(0, 1^-)$.

The matrix $F_{[H_1]}$ no longer satisfies $\mathcal{R}1$.

Since F satisfies $\mathcal{R}1$, for every $b \in D(0, 1^-)\backslash\{0\}$ there exist a matrix solution $\mathcal{U}_{F,b}$ of $\delta - F$ analytic at b. Therefore there exists at b a meromorphic matrix solution of the system $\delta - F_{[H_1]}$.

Step III.   Removing Apparent Singularities.

There exists a unimodular matrix H_2 with $H_2 \in G\ell(n, K(X)) \cap G\ell(n, K[[X]])$ such that $F_{[H_2 H_1]} \in \mathcal{M}_n(E_0)$. Therefore, $F_{[H_2 H_1]}$ satisfies $\mathcal{R}1$, $\mathcal{R}2$ and $\mathcal{P}4$.

Step IV.   Shearing Transformations.

There exists a unimodular matrix $H_3 \in G\ell(n, K[X, X^{-1}])$ (so the only possible singularity of H_3 is at zero) such that $F_{[H_3 H_2 H_1]}$ is analytic at zero and the eigenvalues $\lambda_1, \ldots, \lambda_n$ of the matrix $F_{[H_3 H_2 H_1]}(0)$ are prepared.

Let

$$Q = H_3 H_2 H_1 \tag{2.5}$$

and

$$G^{(1)} = F_{[Q]} \ .$$

We claim that $G^{(1)} \in \mathcal{R}$.

Indeed, $F_{[H_2 H_1]}$ satisfies $\mathcal{R}1$, $\mathcal{R}2$ and $\mathcal{R}4$. Then $G^{(1)}$ satisfies $\mathcal{R}4$ because H_3 is unimodular. Since the only singularity of H_3 is at zero, $\mathcal{R}1$ also holds for $G^{(1)}$ (which is analytic at zero), and $\mathcal{R}2$ continues to hold as well since the transformation does not change the radius of convergence at t. It remains to prove that $G^{(1)}$ satisfies $\mathcal{R}3$.

By Lemma 2.4 there is a one to one correspondence between the eigenvalues of $F(0)$ and those of $F_{[Q]}(0)$ with the property that corresponding eigenvalues differ by integers. Hence all the eigenvalues of $F_{[Q]}(0)$ lie in $\mathbf{Z}$ and so are zero since they are prepared.

Step V.

In this last step we will prove that the matrix Q defined in (2.5) is an element of $G\ell(n, E_0)$ and that

$$QY_F = Y_{G^{(1)}}C .$$

Therefore we will have

$$C = Q(0)$$

and moreover

$$|C| = |Q(0)| \le |Q|_{\text{gauss}} = |H_3 H_2 H_1|_{\text{gauss}} \le |H_1|_{\text{gauss}} .$$

Assume that Steps I–V have been proved. Then from

$$Y_G = H^{-1} Y_F^\phi$$

it follows that

$$Y_G = H^{-1} \left(Q^{-1} Y_{G^{(1)}} \right)^\phi Q(0)$$
$$= W Y_{G^{(1)}}^\phi Q(0)$$

where

$$W = H^{-1} \left(Q^{-1} \right)^\phi \in G\ell(n, E_0)$$

(recall that $H, Q \in G\ell(n, E_0)$). From (2.5) we see that

$$W = H^{-1}(H_1^{-1})^\phi (H_2^{-1})^\phi (H_3^{-1})^\phi ,$$

each factor lies in $\mathcal{M}_n(E_0)$ and three of the factors are unimodular. Hence

$$|W|_{\text{gauss}} = \left| (H_1^{-1})^\phi \right|_{\text{gauss}} = \left| H_1^{-1} \right|_{\text{gauss}} .$$

Therefore if we write

$$W(X) = W_0 + W_1 X + W_2 X^2 + \cdots$$

with $W_j \in \mathcal{M}_n(K)$, then we have

$$|W_j| \le |H_1^{-1}|_{\text{gauss}} .$$

If we now write

$$Y_G = \sum_{s=0}^{\infty} Y_G^{(s)} X^s$$

and

$$Y_{G^{(1)}} = \sum_{s=0}^{\infty} Y_{G^{(1)}}^{(s)} X^s$$

then

$$Y_G^{(s)} = \sum_{pi+j=s} W_j Y_{G^{(1)}}^{(i)} Q(0)$$

and therefore

$$|Y_G^{(s)}| \leq \sup_{i \leq s/p} |Y_{G^{(1)}}^{(i)}| \, |H_1^{-1}|_{\mathrm{gauss}} \, |H_1|_{\mathrm{gauss}} \; .$$

It follows that

$$T(G,s) \leq T(G^{(1)}, s/p) |H_1^{-1}|_{\mathrm{gauss}} \, |H_1|_{\mathrm{gauss}} \; ,$$

so to find the constant B_n we have to estimate $|H_1^{-1}|_{\mathrm{gauss}} \, |H_1|_{\mathrm{gauss}}$.

3.   Proof of Step V.

Proposition 3.1. *Let* F, $G^{(1)} \in \mathcal{M}_n(E_0)$ *and* $Q \in G\ell(n, E_0')$ *with* $F(0)$, $G^{(1)}(0)$ *nilpotent and*

$$G^{(1)} = F_{[Q]} \; .$$

Then $Q \in \mathcal{M}_n(E_0)$, *and there exists* $C \in G\ell(n, K)$ *such that*

$$Q Y_F = Y_{G^{(1)}} C \; .$$

Proof.   Let $0 \neq b \in D(0, 1^-)$. Then the two systems $\delta - F$ and $\delta - G^{(1)}$ have normalized solution matrices $\mathcal{U}_{F,b}$ and $\mathcal{U}_{G^{(1)},b}$ analytic at b; since $Q \mathcal{U}_{F,b}$ is again a solution of $\delta - G^{(1)}$ there exists a matrix $C_b \in G\ell(n, K)$ such that

$$Q \mathcal{U}_{F,b} = \mathcal{U}_{G^{(1)},b} C_b \; .$$

This implies that Q and Q^{-1} are analytic at b.

On the other hand from Lemma 2.4 and since $G^{(1)}(0)$ and $F(0)$ are nilpotent, we obtain that

$$Y_{G^{(1)}}^{-1} Q Y_F = C \in G\ell(n, K) \ .$$

Q.E.D.

We note that the hypothesis that $G^{(1)}(0)$ and $F(0)$ be nilpotent is essential here as is shown by the example

$$\begin{pmatrix} X^{1/2} & 0 \\ 0 & X^{1/3} \end{pmatrix} C \begin{pmatrix} X^{1/2} & 0 \\ 0 & X^{2/3} \end{pmatrix}$$

which can be meromorphic at zero but not constant, with a suitable choice of C.

4.   Proof of Step IV. The Shearing Transformation.

Proposition 4.1. *Let $\mathcal{G} \in \mathcal{M}_n(E_0)$ and fix an eigenvalue λ of $\mathcal{G}(0)$. Then there exists a unimodular matrix $H \in G\ell\left(n, K[X, X^{-1}]\right)$ such that $\mathcal{G}_{[H]} \in \mathcal{M}_n(E_0)$ and the eigenvalues of $\mathcal{G}_{[H]}(0)$ are the same as those of $\mathcal{G}(0)$ (counting multiplicities), except that the eigenvalue λ is replaced by $\lambda + 1$. A similar statement holds for passing from λ to $\lambda - 1$.*

Proof.   The shearing transformation is constructed as in the proof of Lemma III.8.2. The only new point is to prove that we may choose our shearing transformation to be unimodular. But this transformation is obtained by taking a product of a matrix of the type

$$H = \begin{pmatrix} I_l X & 0 \\ 0 & I_{n-l} \end{pmatrix} ,$$

where l is the multiplicity of λ, with a matrix $C \in G\ell(n, K)$ such that $C\mathcal{G}(0)C^{-1}$ is upper triangular. The former matrices are obviously unimodular. The following Lemma shows that we may also choose C to be unimodular.   **Q.E.D.**

Lemma 4.2. *Let $A \in \mathcal{M}_n(K)$. Then there exists $C \in G\ell(n, \mathcal{O}_K)$ such that CAC^{-1} is upper (resp. lower) triangular.*

Proof. Let (recall that K is algebraically closed)

$$\begin{pmatrix} 1 \\ x_2 \\ \vdots \\ x_n \end{pmatrix} \in K^n$$

with $|x_i| \leq 1$ for $2 \leq i \leq n$, be an eigenvector of the matrix A corresponding to the eigenvalue λ (if necessary we perform a permutation of the basis which certainly is represented by a matrix in $G\ell(n, \mathcal{O}_K)$). If we now put

$$C_1 = \begin{pmatrix} 1 & 0 & \cdots & 0 \\ -x_2 & & & \\ \vdots & & I_{n-1} & \\ -x_n & & & \end{pmatrix}$$

then

$$C_1^{-1} = \begin{pmatrix} 1 & 0 & \cdots & 0 \\ x_2 & & & \\ \vdots & & I_{n-1} & \\ x_n & & & \end{pmatrix},$$

so that this $C_1 \in G\ell(n, \mathcal{O}_K)$. A simple calculation then shows that

$$C_1 A C_1^{-1} = \begin{pmatrix} \lambda & * & \cdots & * \\ 0 & * & \cdots & * \\ \vdots & \vdots & & \vdots \\ 0 & * & \cdots & * \end{pmatrix},$$

and so by induction we obtain the desired result. **Q.E.D.**

In our situation we apply Proposition 4.1 to the matrix $\mathcal{G} = F_{[H_2 H_1]}$, and so we see that we can find a unimodular $H_3 \in G\ell(n, K[X, X^{-1}])$ such that $F_{[H_3 H_2 H_1]}$ has prepared eigenvalues.

5. Proof of Step III. Removing Apparent Singularities.

Let $\mathcal{G} \in \mathcal{M}_n(E_0')$ (recall that E_0' is the quotient field of E_0). Let $a \in D(0, 1^-)$. If $\mathcal{G}$ is analytic at a we say that a is an *ordinary point* of the system $D - \mathcal{G}$ where $D = d/dX$. If a is not ordinary we say that a is a *singular point* of $D - \mathcal{G}$. The singular point a is said to be an *apparent singularity* of $D - \mathcal{G}$ if the system has a matrix solution at a belonging to $G\ell\,(n, K((X - a)))$. Finally, an apparent singularity, a, of $D - \mathcal{G}$ is said to be a *trivial singularity* if the system has a matrix solution analytic at a. As an example of a system with a trivial singularity, consider the system of order 2 defined by

$$\mathcal{G} = \begin{pmatrix} 0 & 1 \\ 0 & 1/X \end{pmatrix} \ .$$

This system has a singularity at zero and it has

$$\begin{pmatrix} 1 & X^2 \\ 0 & 2X \end{pmatrix}$$

as a matrix solution at zero and this is analytic (but not invertible) at zero.

Proposition 5.1. *Let $\mathcal{G} \in \mathcal{M}_n(E_0')$ and assume that $\mathcal{G}$ is analytic at zero.*

(i) If the system $\delta - \mathcal{G}$ has only apparent singularities in the punctured disk $D(0, 1^-)\backslash\{\,0\,\}$, then there exists a polynomial $P \in K[X]$ having $|P|_{\text{gauss}} = 1$ such that $\mathcal{G}_{[PI_n]}$ is analytic at zero and $\delta - \mathcal{G}_{[PI_n]}$ has only trivial singularities in $D(0, 1^-)\backslash\{\,0\,\}$.

(ii) If the system $\delta - \mathcal{G}$ has only trivial singularities in $D(0, 1^-)\backslash\{\,0\,\}$ then there exists a unimodular matrix $H \in G\ell\,(n, K(X))\cap G\ell\,(n, K[[X]])$ such that $\mathcal{G}_{[H]} \in \mathcal{M}_n(E_0)$.

Proof. (i) There exists $\xi \in E_0$ such that $\xi\mathcal{G} \in \mathcal{M}_n(E_0)$. By Proposition IV.4.2 the number of zeros of ξ in $D(0, 1^-)$ is finite. Therefore the number of poles of $\mathcal{G}$ in $D(0, 1^-)$ is finite.

Let $\mathcal{S}$ be the set of poles of $\mathcal{G}$ in $D(0, 1^-)$ and ν_b be the order of the pole of $\mathcal{U}_{\mathcal{G}, b}$ in b. Put

$$P(X) = \prod_{b \in \mathcal{S}}(X - b)^{\nu_b} \ .$$

Then $|P|_{\text{gauss}} = 1$ and the matrix $\mathcal{G}_{[PI_n]}$ is again analytic at zero. Moreover the system $\delta - \mathcal{G}_{[PI_n]}$ now has only trivial singularities in the punctured disk $D(0, 1^-)\backslash\{\,0\,\}$.

(ii) Let $\mathcal{F}$ be the set of all matrices $\mathcal{G}_{[H]}$ with H unimodular, $H \in Gl(n, K(X)) \cap Gl(n, K[[X]])$ and such that $\delta - \mathcal{G}_{[H]}$ has a solution matrix analytic at b for every $0 \neq b \in D(0, 1^-)$. Note that each $F \in \mathcal{F}$ is analytic at zero. For $F \in \mathcal{F}$ let

$$m(F) = \sum_{\substack{b \neq 0 \\ |b| < 1}} \mathrm{order}_b \, (\det \mathcal{U}_{F,b}) < \infty \, ,$$

where $\mathrm{order}_b \, (\det \mathcal{U}_{F,b})$ denotes the order of zero of $\det \mathcal{U}_{F,b}$ at b. Let $F \in \mathcal{F}$ be such that $m(F)$ is minimal. We claim that in fact $m(F) = 0$. Assume not. Then there exists a $b \in D(0, 1^-) \backslash \{0\}$ such that $\det \mathcal{U}_{F,b}(b) = 0$. Hence there exists a $(\lambda_1, \dots, \lambda_n) \in K^n$ such that

$$(\lambda_1, \dots, \lambda_n) \mathcal{U}_{F,b}(b) = 0$$

with $|\lambda_j| \leq 1$ for $j = 1, \dots, n$ and $|\lambda_i| = 1$ for some i. Let S be the matrix

$$\begin{pmatrix} I_{i-1} & & 0 & & & 0_{i-1,n-i} \\ \dfrac{\lambda_1}{X-b} & \cdots & \dfrac{\lambda_{i-1}}{X-b} & \dfrac{\lambda_i}{X-b} & \dfrac{\lambda_{i+1}}{X-b} & \cdots & \dfrac{\lambda_n}{X-b} \\ 0_{n-i,i-1} & & 0 & & & I_{n-i} \end{pmatrix} .$$

Then $\det S = \lambda_i/(X - b)$ so that $|\det S|_{\mathrm{gauss}} = |\lambda_i| = 1$, whence $|S^{-1}|_{\mathrm{gauss}} \leq 1$ and S is unimodular. Now $\mathcal{U}_{F_{[S]},b} = S\mathcal{U}_{F,b}$ is analytic at b. Indeed a possible problem can occur only in the i-th row, which is

$$\frac{1}{X-b}(\lambda_1, \dots, \lambda_n) \left(\mathcal{U}_{F,b}(b) + (X - b)(\text{something}) \right) \, .$$

Clearly in the other points $b' \in D(0, 1^-)$ analyticity is preserved so that $F_{[S]} \in \mathcal{F}$. Now

$$\det \mathcal{U}_{F_{[S]},b} = \frac{\lambda_i}{X-b} \det \mathcal{U}_{F,b}$$

and therefore $\mathrm{order}_b(\det \mathcal{U}_{F_{[S]},b}) = \mathrm{order}_b(\det \mathcal{U}_{F,b}) - 1$ so that $m(F_{[S]}) = m(F) - 1$ which contradicts the minimality of $m(F)$.   **Q.E.D.**

6.   The Operators ϕ and ψ.

Let $q = p^{l+1}$ where l is a non-negative integer, and let

$$\phi_q : \Omega \longrightarrow \Omega$$
$$x \longmapsto x^q . \tag{6.1}$$

Let D_1 and D_2 be subsets of Ω and assume that $\phi^{-1}D_2 \subseteq D_1$. Then for any map

$$f : D_1 \longrightarrow \Omega$$

we define

$$\psi_q f : D_2 \longrightarrow \Omega$$

by

$$(\psi_q f)(x) = \frac{1}{q} \sum_{z^q = x} f(z) . \tag{6.2}$$

If, for $D \subseteq \Omega$, we define

$$\mathcal{B}_\Omega(D) = \{ f \in \Omega^D \mid \sup_{x \in D} |f(x)| < \infty \}$$

then clearly

$$\psi_q : \mathcal{B}_\Omega(D_1) \longrightarrow \mathcal{B}_\Omega(D_2)$$

is Ω-linear and continuous with respect to the sup-norm since

$$\sup_{x \in D_2} |(\psi_q f)(x)| \leq q \sup_{z \in D_1} |f(z)| .$$

In the following t will denote a generic point in Ω with respect to K and $\pi \in K$ is such that $\pi^{p-1} = -p$.

Lemma 6.1. Let $1 > R \geq |\pi|^{p/q}$, where $q = p^{l+1}$, $l \geq 0$. The map

$$\phi_q : D(t, R^+) \longrightarrow D(t^q, (R^q)^+)$$
$$x \longmapsto x^q$$

is surjective and q–to–one. Hence, ϕ_q gives a q–to–one map of $D(t, 1^-)$ onto $D(t^q, 1^-)$.

Proof.   Let

$$(1 + X)^p = 1 + U , \qquad U = X^p + \binom{p}{1} X^{p-1} + \cdots + \binom{p}{p-1} X .$$

The Newton polygon of the polynomial U has a unique finite side; its slope is $-1/(p-1)$ since its endpoints are $(1, 1)$ and $(p, 0)$.

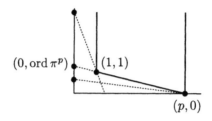

For $x \in \Omega$ we set $u = U(x)$. By the estimate for ord u which follows from the y–intercept method (cf. Section I.7) one easily verifies that

if $|x| > |\pi|$ then $|u| = |x|^p$,

if $|x| = |\pi|$ then $|u| \le |x|^p$,

if $|x| < |\pi|$ then $|u| = |p||x|$,

if $|u| \ge |\pi|^p$ then $|x| = |u|^{1/p}$,

if $|u| < |\pi|^p$ then $|x| = |u|/|p|$ or $|x| = |\pi|$ so that in either case $|x| \le |\pi|$.

Now let $r \ge |\pi|$ and $|a| = 1$. Then from the preceding statements it follows that the map

$$\phi_p : D(a, r^+) \longrightarrow D(a^p, (r^p)^+)$$
$$x \longmapsto x^p$$

is a p–to–one surjective map. Therefore if $R \ge |\pi|^{p/q}$ we see that ϕ_q is the composite map:

$$D(t, R) \xrightarrow{\phi_p} D(t^p, R^p) \xrightarrow{\phi_p} D(t^{p^2}, R^{p^2}) \longrightarrow \cdots$$
$$\cdots \longrightarrow D(t^{p^l}, R^{p^l}) \xrightarrow{\phi_p} D(t^q, R^q) .$$

Thus $\phi_q = (\phi_p)^{l+1}$ is clearly surjective and q–to–one.   **Q.E.D.**

Assume $f \in \mathcal{A}_t$, that is, f is an analytic function on the generic disk $D(t, 1^-)$. We claim that $\psi_q f \in \mathcal{A}_{t^q}$. Indeed, one checks easily that if

$$f = \sum_{s=0}^{N} A_s X^s$$

then

$$\psi_q f = \sum_{s=0}^{[N/q]} A_{qs} X^s \; .$$

Thus ψ_q sends polynomials into polynomials. Now if $f \in \mathcal{A}_t$ then f is a uniform limit of polynomials on $D(t, R^+)$ for every $R < 1$. Therefore, since by Lemma 6.1 we have $\phi_q^{-1} D(t^q, (R^q)^+) \subseteq D(t, R^+)$ for $1 > R > \varepsilon$ with a suitable $\varepsilon > 0$, it follows that $\psi_q f$ is itself a uniform limit of polynomials on each disk $D(t^q, R^{q+})$ and this suffices to prove our assertion.

Assume now that $f \in E_0'$. We wish to show that $\psi_q f$ is also an element of E_0'. In the particular case in which $f \in K(X)$ then

$$\psi_q f = \frac{1}{q} \mathrm{Tr}_{K(X^{1/q})/K(X)} f(X^{1/q})$$

and therefore $\psi_q f$ is a rational function. In general f is a uniform limit on $D(t, 1^-)$ of rational functions and so it follows that $\psi_q f$ is a uniform limit on $D(t^q, 1^-)$ of rational functions . Therefore, $\psi_q f \in E$. On the other hand, if $f \in E_0'$ then

$$f = \lim_i \frac{h_i}{P}$$

where $P \in K[X]$ and $h_i \in K(X) \cap E_0$. From this it easily follows that

$$(\psi_q f) \prod_{\zeta^q = 1} P(X^{1/q} \zeta) \in E_0 \; .$$

Here the product $\prod_{\zeta^q=1} P(X^{1/q} \zeta)$ is the norm from $K(X^{1/q})$ to $K(X)$ of $P(X^{1/q})$ so that it is a polynomial in $K[X]$ and $\psi_q f \in E_0'$ as desired.

Clearly if f has no pole in $D(0, 1^-)$ the same is true of $\psi_q f$ so that if $f \in E_0$ then $\psi_q f \in E_0$.

We also observe that if $f \in E_0$ then we may view f as a function on $\mathcal{D} = D(0, 1^-) \cup D(t, 1^-)$ and therefore $\psi_q f$ may be viewed as a function on $\mathcal{D}$.

Lemma 6.2. Let R be as in Lemma 6.1. If f is meromorphic in $D(t^q, (R^q)^+)$ (resp., in $D(t^q, 1^-)$) and $g \in E'_0$ is such that

$$f(x^q) = g(x)$$

for all $x \in D(t, R^+)$ (resp. in $D(t, 1^-)$) then there exists $h \in E'_0$ such that $h = f$ on $D(t^q, (R^q)^+)$ (resp. on $D(t, 1^-)$). Moreover if $g \in E_0$ then also $h \in E_0$.

Proof. Let

$$h(x) = (\psi_q g)(x) = \frac{1}{q} \sum_{z^q = x} g(z) .$$

Then, as we have just observed, $h \in E'_0$. We claim that $h = f$ on $D(t^q, (R^q)^+)$. Indeed, let $x \in D(t, R^+)$; then we have

$$h(x^q) = \frac{1}{q} \sum_{z^q = x^q} g(z) = \frac{1}{q} \sum_{\zeta^q = 1} g(x\zeta)$$

$$= \frac{1}{q} \sum_{\zeta^q = 1} f((x\zeta)^q) = f(x^q) ,$$

(the third equality holds because $x\zeta \in D(\zeta t, R^+)$ and $D(\zeta t, R^+) = D(t, R^+)$ since $|\zeta - 1| \le |\pi^{p/q}|$) so that $h = f$ on the image $\phi_q D(t, R) = D(t^q, R^q)$. The proof in the case of the open disk is now clear. **Q.E.D.**

7.   Proof of Step I. Construction of Frobenius.

We recall that for $f \in K[[X]]$ we define

$$f^\phi(X) = f(X^p) .$$

For later applications, we weaken condition $\mathcal{R}3$ in the following Proposition.

Proposition 7.1. Let G be a matrix satisfying conditions $\mathcal{R}1$, $\mathcal{R}2$ and $\mathcal{R}4$. Assume moreover that the eigenvalues of $G(0)$ are prepared and lie in $p\mathbf{Z}_p$. Then there exists a unimodular matrix $H \in G\ell(n, E_0)$ and a matrix $F \in \mathcal{M}_n(E_0)$ satisfying $\mathcal{R}1$ and $\mathcal{R}2$ such that

$$G_{[H]} = pF^\phi$$
$$G(0) = pF(0)$$
$$Y_G = H^{-1}Y_F^\phi .$$

In particular the eigenvalues of $F(0)$ are in $\mathbf{Z}_p$ and if $G \in \mathcal{R}$, then F also satisfies condition $\mathcal{R}3$.

Proof. For $s \geq 1$, let $\mathcal{G}_s$ be the matrix such that

$$\mathcal{G}_s \vec{u} = \frac{1}{s!} \frac{d^s}{dX^s} \vec{u}$$

for all solutions $\vec{u}$ of $\delta - G$. Consider now the power series in two variables

$$\mathcal{Y}(X, Z) = \sum_{s=0}^{\infty} \mathcal{G}_s(X)(Z - X)^s . \tag{7.1}$$

Since $|G|_{\text{gauss}} \leq 1$ we have $|s! \mathcal{G}_s|_{\text{gauss}} \leq 1$ and since G satisfies $\mathcal{R}2$ Remark IV.3.2 gives

$$|\mathcal{G}_s|_{\text{gauss}} \leq \{s, n - 1\}_p .$$

Let $\mathcal{D} = D(0, 1^-) \cup D(t, 1^-)$. Then for $x \in \mathcal{D}$ we have

$$|x^s \mathcal{G}_s(x)| \leq \{s, n - 1\}_p$$

so that for $0 \neq x \in \mathcal{D}$ the series

$$\mathcal{Y}(x, Z) = \sum_{s=0}^{\infty} x^s \mathcal{G}_s(x) \left(\frac{Z - x}{x} \right)^s$$

is convergent for

$$\left| \frac{z - x}{x} \right| < 1 .$$

We note that $\mathcal{Y}(x, X)$ is a normalized matrix solution to $\delta - G$ at $x \in \mathcal{D}$. Moreover, if $\mathcal{U}_b$ is a solution matrix of $\delta - G$ at some point b with radius of convergence $\rho(b)$, then for $x, z \in D(b, \rho(b)^-)$ such that $\mathcal{Y}(X, Z)$ converges at (x, z) we have

$$\mathcal{U}_b(z) = \mathcal{Y}(x, z) \mathcal{U}_b(x) .$$

Now let $0 \neq x \in \mathcal{D}$ and let $\zeta^p = 1$. From $|\zeta - 1| = |\pi|$ it follows that $\mathcal{Y}(x, x\zeta)$ converges and so for $x \in \mathcal{D}$ we can define

$$H(x) = \frac{1}{p} \sum_{\zeta^p = 1} \mathcal{Y}(x, x\zeta) .$$

From (7.1) it follows that

$$H(x) = \frac{1}{p} \sum_{s=0}^{\infty} x^s \mathcal{G}_s(x) \sum_{\zeta^p = 1} (\zeta - 1)^s$$

and so the s-th general term of this series is uniformly bounded on $\mathcal{D}$ by $p\{s, n-1\}_p |\pi|^s$. Since $X^s \mathcal{G}_s \in \mathcal{M}_n(E_0)$ it follows that $H(X)$ is also an element of $\mathcal{M}_n(E_0)$. We claim that

$$|H|_{\text{gauss}} \leq 1 .$$

Clearly $|s! X^s \mathcal{G}_s|_{\text{gauss}} \leq 1$ so that it suffices to prove that

$$z_s \stackrel{\text{def}}{=} \frac{1}{p} \sum_{\zeta^p = 1} \frac{(\zeta - 1)^s}{s!} \in \mathbf{Z}_p$$

for all $s \geq 0$. For $s = 0$ the result is clear. For $s \geq 1$ it suffices to observe that (cf. (II.4.6))

$$\text{ord}\left(\frac{1}{p}\frac{(\zeta - 1)^s}{s!}\right) = -1 + \frac{s}{p-1} - \text{ord } s! = -1 + \frac{S_s}{p-1} > -1$$

where S_s is the sum of the digits of s expressed in the base p. Indeed, surely $z_s \in \mathbf{Q}$ and ord $z_s > -1$ then implies that ord $z_s \geq 0$.

We now show that

$$H(0) = I_n .$$

The key point is that

$$\mathcal{G}_s(X) = \binom{G(0)}{s}\frac{1}{X^s} + \left(\frac{1}{X^{s-1}}\right)$$

so that

$$(X^s \mathcal{G}_s)(0) = \binom{G(0)}{s}$$

and

$$H(0) = \frac{1}{p} \sum_{\zeta^p = 1} \sum_{s=0}^{\infty} \binom{G(0)}{s}(\zeta - 1)^s = \frac{1}{p} \sum_{\zeta^p = 1} g_{G(0)}(\zeta - 1) = I_n$$

by Proposition IV.7.5. Therefore, since $\det H \in E_0$ and $|\det H|_{\text{gauss}} \leq 1$, we have

$$|\det H|_{\text{gauss}} = 1 ,$$

and $\det H$ has no zeros in $D(0, 1^-)$ (look at its Newton polygon), so that $\det H$ is a unit in E_0. Now $H^{-1} = (\det H)^{-1} \text{adj}\, H$ so that since $|\text{adj}\, H|_{\text{gauss}} \leq 1$ we have

$$\left| H^{-1} \right|_{\text{gauss}} \leq 1$$

which proves the unimodularity of H.

We now set

$$V = \psi_p \mathcal{U}_{G,t} .$$

By condition $\mathcal{R}2$ and the discussion in Section 6 it follows that $V \in \mathcal{M}_n(\mathcal{A}_{t^p})$. For $x \in D(t, 1^-)$ we then have

$$
\begin{aligned}
V^\phi(x) = V(x^p) &= \frac{1}{p} \sum_{z^p = x^p} \mathcal{U}_{G,t}(z) \\
&= \frac{1}{p} \sum_{\zeta^p = 1} \mathcal{U}_{G,t}(x\zeta) = \frac{1}{p} \sum_{\zeta^p = 1} \mathcal{Y}(x, x\zeta) \mathcal{U}_{G,t}(x) \\
&= H(x) \mathcal{U}_{G,t}(x)
\end{aligned}
$$

(note that $x\zeta \in D(\zeta t, 1^-) = D(t, 1^-)$). Now

$$\det V^\phi = (\det H)(\det \mathcal{U}_{G,t}) \neq 0 ,$$

so that $\det V \neq 0$, and therefore we can define

$$F = (\delta V)V^{-1} \in \mathcal{M}_n(\mathcal{A}'_{t^p}) .$$

Furthermore, since

$$\delta \circ \phi = p(\phi \circ \delta) , \tag{7.2}$$

we have

$$F^\phi = (\delta V)^\phi (V^{-1})^\phi = \frac{1}{p} (\delta V^\phi)(V^\phi)^{-1} .$$

Therefore

$$pF^\phi = \delta(H \mathcal{U}_{G,t})(H \mathcal{U}_{G,t})^{-1} = G_{[H]} .$$

This last equality holds in the disk $D(t, 1^-)$, but since $H \in G\ell\,(n, E_0)$ and $G \in \mathcal{M}_n(E_0)$ we have $G_{[H]} \in \mathcal{M}_n(E_0)$. We can now apply Lemma

6.2 to conclude that F can be extended to a matrix in $\mathcal{M}_n(E_0)$, which we still call F, such that

$$pF^\phi = G_{[H]}$$

on $\mathcal{D}$. In particular we obtain

$$pF(0) = G_{[H]}(0) = G(0) \ .$$

Clearly if the eigenvalues of $G(0)$ lie in $p\mathbf{Z}_p$ then the eigenvalues of F are in $\mathbf{Z}_p$, and if G satisfies condition $\mathcal{R}3$ the same holds for F.

There remains only to prove the last equation in the statement of the proposition. We first observe that from (7.2) and the uniqueness of Y_{pF^ϕ} it easily follows that

$$Y_{pF^\phi} = Y_F^\phi \ .$$

Let $\alpha_1, \dots, \alpha_n$ be the eigenvalues of $G(0)$. Now by Lemma 2.4 there exists C_1, $C_2 \in G\ell(n, K)$ such that

$$(Y_F^\phi)^{-1} H Y_G = C_1 M C_2 \tag{7.3}$$

where

$$M = \left(C_{ij} X^{\alpha_i - \alpha_j} \right)$$

with $C_{ij} \in K$. Since if $C_{ij} \neq 0$ we must have $\alpha_i - \alpha_j \in \mathbf{Z}$ and since $\alpha_1, \dots, \alpha_n$ are prepared, it follows that $M \in G\ell(n, K)$. On evaluating (7.3) at zero we obtain

$$C_1 M C_2 = I_n$$

as desired.   **Q.E.D.**

8.   Proof of Step II. Effective Form of the Cyclic Vector.

Proposition 8.1. *Let $G \in \mathcal{M}_n(E_0')$. Then there exists an $H \in G\ell(n, E_0')$ such that*

$$G_{[H]} = \begin{pmatrix} 0 & 1 & 0 & \dots & 0 \\ 0 & 0 & 1 & \dots & 0 \\ \multicolumn{5}{c}{\dots\dots\dots\dots\dots\dots\dots} \\ 0 & & & \dots & 1 \\ C_0 & C_1 & & \dots & C_{n-1} \end{pmatrix}$$

with $C_0, \ldots, C_{n-1} \in E'_0$, and such that

$$|H|_{\text{gauss}} |H^{-1}|_{\text{gauss}} \leq \frac{\sup(1, |G|_{\text{gauss}})^{2(n-1)}}{|(n-1)!|} .$$

If moreover $\mathcal{U}_{G,t}$, the normalized matrix solution of $\delta - G$ at the generic point t, converges on the disk $D(t, |\pi|^-)$ then H may be chosen such that

$$|H|_{\text{gauss}} |H^{-1}|_{\text{gauss}} \leq \frac{\sup(1, |G|_{\text{gauss}})^{n-1}}{|(n-1)! \prod_{j=1}^{n-1} \binom{n}{j}|} .$$

Proof. Let M be the differential module over E having basis $e_1, \ldots, e_n$ satisfying

$$\Delta \begin{pmatrix} e_1 \\ \vdots \\ e_n \end{pmatrix} = G \begin{pmatrix} e_1 \\ \vdots \\ e_n \end{pmatrix} .$$

Here we are assuming that for $u \in E$, $m \in M$ we have

$$\Delta(um) = u\Delta m + (\delta u)m$$

where $\delta = X d/dX$. We extend the norm on E to M by identifying M with E^n via the basis $e_1, \ldots e_n$; that is, if $m = a_1 e_1 + \cdots a_n e_n$ with $a_i \in E$ we set

$$|m|_{\text{gauss}} = \sup_{1 \leq i \leq n} |a_i|_{\text{gauss}} .$$

In the same way we can also extend the norm to $\bigwedge^s M$ by means of the basis

$$\{e_{j_1} \wedge e_{j_2} \wedge \cdots \wedge e_{j_s}\}_{1 \leq j_1 < j_2 < \cdots < j_s \leq n} .$$

In the sequel we will usually suppress the suffix "gauss" to simplify notation.

For $m \in M$ we set

$$\theta(m) = m \wedge \Delta m \wedge \cdots \wedge \Delta^{n-1} m .$$

By the Theorem of the Cyclic Vector (Theorem III.4.2) we know that there exists $m \in M$, which we can assume to satisfy $|m| \leq 1$, such that $\theta(m) \neq 0$. Moreover the map

$$m \longmapsto \theta(m)$$

is continuous, since the maps $m \longmapsto \Delta^i m$ are certainly continuous. Let

$$\tau = \sup_{|m| \leq 1} |\theta(m)| \leq \sup(1, |G|)^{n(n-1)/2} .$$

Since M is complete (recall that E is the completion of $K(X)$ with respect to the Gauss norm) we can choose the preceding m in such a way that

$$\tau = |\theta(m)| .$$

For any $\varepsilon > 0$, we may choose $m' \in K(X)e_1 + \cdots + K(X)e_n$ such that $|m - m'| < \varepsilon$ and indeed we may choose ε sufficiently small so that we have $|\theta(m') - \theta(m)| < \tau$ and hence $|\theta(m')| = \tau$. Therefore we may assume that $\tau = |\theta(m)|$ with $m \in \sum_{i=1}^{n} K(X)e_i$ such that $|m| \leq 1$.

We define the matrix H by

$$\begin{pmatrix} m \\ \Delta m \\ \vdots \\ \Delta^{n-1} m \end{pmatrix} = H \begin{pmatrix} e_1 \\ e_2 \\ \vdots \\ e_n \end{pmatrix} .$$

Since $G \in \mathcal{M}_n(E_0')$ the matrix $H \in G\ell(n, E_0')$. Clearly since $|m| \leq 1$ we have

$$|H| \leq \sup(1, |G|)^{n-1} .$$

We must now find an estimate for $|H^{-1}|$. Let

$$A_i(m) = (-1)^{n-i-1} m \wedge \Delta m \wedge \cdots \wedge \widehat{\Delta^i m} \wedge \cdots \wedge \Delta^{n-1} m \in \bigwedge^{n-1} M .$$

We claim that

$$\left| A_j(m) \wedge z + \binom{j+1}{j} A_{j+1}(m) \wedge \Delta z + \cdots \right.$$
$$\left. + \binom{n-1}{j} A_{n-1}(m) \wedge \Delta^{n-1-j} z \right| \leq \frac{\tau}{|(n-1)!|} \qquad (8.1)_j$$

for $j = 0, 1, \ldots, n-1$ and for every $z \in M$ with $|z| \leq 1$.

Indeed, let $\lambda \in K$, $z \in M$ be such that $|\lambda| \leq 1$ and $|z| \leq 1$. Then, since

$$|\theta(m + \lambda z)| = \left| \theta(m) + \lambda \sum_{j=0}^{n-1} A_j(m) \wedge \Delta^j z + (\lambda^2) \right| \leq \tau$$

for every $\lambda \in K$ with $|\lambda| \leq 1$, we obtain

$$\left| \sum_{j=0}^{n-1} A_j(m) \wedge \Delta^j z \right| \leq \tau$$

for every $z \in M$ with $|z| \leq 1$ (the residue field $\overline{K}$ being infinite we can actually find $\lambda_0, \dots, \lambda_{n-1} \in K$ with $|\lambda_i| \leq 1$ having Vandermonde determinant V with $|V| = 1$).

We now make the substitution $z \longmapsto X^s z$, for $s \in \mathbf{Z}$. From

$$\Delta^i(X^s z) = X^s \sum_{j=0}^{i} \binom{i}{j} s^j \Delta^{i-j} z$$

we obtain

$$\left| \sum_{i=0}^{n-1} A_i(m) \wedge \sum_{j=0}^{i} \binom{i}{j} s^j \Delta^{i-j} z \right| = \left| \sum_{j=0}^{n-1} s^j \sum_{i=j}^{n-1} A_i(m) \binom{i}{j} \wedge \Delta^{i-j} z \right| \leq \tau$$

for every $s \in \mathbf{Z}$. We now set

$$F_j(m, z) = \sum_{i=j}^{n-1} \binom{i}{j} A_i(m) \wedge \Delta^{i-j} z$$

and

$$Q(Y) = \sum_{j=0}^{n-1} F_j(m, z) Y^j .$$

If we take $h(Y) = Y(Y - 1) \cdots (Y - n + 1)$ then by the Lagrange interpolation formula we have

$$Q(Y) = \sum_{s=0}^{n-1} \frac{h(Y)Q(s)}{(Y - s)h'(s)} .$$

Now $|Q(s)| \leq \tau$ so that

$$|Q(Y)|_{\text{gauss}} \leq \tau \sup_{0 \leq s \leq n-1} \frac{1}{|h'(s)|} .$$

But $h'(s) = s(s-1) \cdots (s-s+1)(s-s-1) \cdots (s-n+1) = \pm s!(n-s-1)!$
whence

$$\frac{1}{|h'(s)|} = \left| \binom{n-1}{s} \frac{1}{(n-1)!} \right| \leq \frac{1}{|(n-1)!|} .$$

Therefore we have

$$|Q(Y)|_{\text{gauss}} \leq \frac{\tau}{|(n-1)!|}$$

so that

$$|F_j(m, z)| \leq \frac{\tau}{|(n-1)!|} ,$$

as desired.

We now prove that from (8.1) it follows that

$$|A_j(m)| \leq \frac{\tau}{|(n-1)!|} \sup(1, |G|)^{n-1-j} . \tag{8.2}$_j$$

We proceed by descending induction on j. Let $j = n - 1$. Then $(8.1)_{n-1}$ is

$$|A_{n-1}(m) \wedge z| \leq \frac{\tau}{|(n-1)!|}$$

for every $z \in M$ with $|z| \leq 1$. On varying z over the elements $e_1, \dots, e_n$ we conclude that

$$|A_{n-1}(m)| \leq \frac{\tau}{|(n-1)!|} .$$

Next let $j = n - 2$. Now $(8.1)_{n-2}$ takes the form

$$\left| A_{n-2}(m) \wedge z + \binom{n-1}{n-2} A_{n-1}(m) \wedge \Delta z \right| \leq \frac{\tau}{|(n-1)!|} .$$

But

$$|A_{n-1}(m) \wedge \Delta z| \leq |A_{n-1}(m)| \sup(1, |G|)$$

so that by $(8.2)_{n-1}$ we obtain

$$|A_{n-1}(m) \wedge \Delta z| \leq \frac{\tau}{|(n-1)!|} \sup(1, |G|)$$

and therefore

$$|A_{n-2}(m) \wedge z| \leq \frac{\tau}{|(n-1)!|} \sup(1, |G|) .$$

In the general case, assume $(8.2)_{j+l}$ true for $l = 1, 2, \ldots, n-1-j$. Then

$$\left| \sum_{l=1}^{n-1-j} \binom{j+l}{j} A_{j+l}(m) \wedge \Delta^l z \right|$$

$$\leq \frac{\tau}{|(n-1)!|} \sup_{1 \leq l \leq n-1-j} \sup(1, |G|)^{n-1-j-l} \sup(1, |G|)^l$$

$$= \frac{\tau}{|(n-1)!|} \sup(1, |G|)^{n-1-j}$$

whence from $(8.1)_j$ we obtain the desired estimate for $|A_j(m)|$.

If we write $z = y_0 m + y_1 \Delta m + \cdots + y_{n-1} \Delta^{n-1} m$ then

$$z \wedge A_i(m) = \pm y_i \theta(m) \ .$$

On the other hand, for $i = 1, \ldots, n$ we have

$$e_i = \sum_{j=0}^{n-1} (H^{-1})_{ij} \Delta^j(m)$$

so that

$$e_i \wedge A_j(m) = \pm (H^{-1})_{ij} \theta(m)$$

whence

$$\tau |(H^{-1})_{ij}| = |(H^{-1})_{ij} \theta(m)| = |e_i \wedge A_j(m)| \leq |A_j(m)|$$

$$\leq \frac{\tau}{|(n-1)!|} \sup(1, |G|)^{n-1-j}$$

and

$$|(H^{-1})_{ij}| \leq \frac{\sup(1, |G|)^{n-1-j}}{|(n-1)!|} \ ,$$

which gives the first estimate for $|H||H^{-1}|$.

We now establish the second estimate. We have the same estimate as given above for $|H|$ and we must establish that

$$|H^{-1}| \leq \frac{1}{\left| (n-1)! \prod_{j=1}^{n-1} \binom{n}{j} \right|} \ . \tag{8.3}$$

Note that this estimate is independent of the matrix G.

We have

$$\Delta^n m = \sum_{i=0}^{n-1} C_i \Delta^i m$$

where $C_0, C_1, \ldots, C_{n-1} \in E'_0$ satisfy $|C_i| \leq 1$ by Theorem 1.2. We now show by induction on s that for each $s \geq 0$ we may write

$$\Delta^s m = \sum_{i=0}^{n-1} C_{s,i} \Delta^i m$$

with $|C_{s,i}| \leq 1$ for $i = 0, \ldots, n-1$. Indeed,

$$\sum_{i=0}^{n-1} C_{s+1,i} \Delta^i m = \sum_{i=0}^{n-1} C_{s,i} \Delta^{i+1} m + \sum_{i=0}^{n-1} (\delta C_{s,i}) \Delta^i m$$

$$= \sum_{i=1}^{n-1} C_{s,i-1} \Delta^i m + C_{s,n-1} \sum_{i=0}^{n-1} C_i \Delta^i m$$

$$+ \sum_{i=0}^{n-1} (\delta C_{s,i}) \Delta^i m \,,$$

whence

$$C_{s+1,0} = \delta C_{s,0} + C_{s,n-1} C_0$$

and for $i \geq 1$

$$C_{s+1,i} = \delta C_{s,i} + C_{s,i-1} + C_{s,n-1} C_i \,,$$

which, by induction and the fact that $|\delta| \leq 1$, gives $|C_{s+1,i}| \leq 1$ for $i = 0, \ldots, n-1$.

Now let $y = (y_0, y_1, \ldots, y_{n-1}) \in E^n$; if

$$z = y_0 m + y_1 \Delta m + \cdots + y_{n-1} \Delta^{n-1} m$$

we can define $f_j(m, y) \in E$ by

$$f_j(m, y) \theta(m) = \sum_{s=0}^{n-1-j} \binom{j+s}{j} A_{j+s}(m) \wedge \Delta^s z \,.$$

We now estimate $|y|$. We shall prove that under the assumption $|z| \leq 1$ one has

$$|y| \leq \frac{1}{\left| (n-1)! \prod_{j=1}^{n-1} \binom{n}{j} \right|} \,. \tag{8.4}$$

Indeed by $(8.1)_j$ and since $|\theta(m)| = \tau$, we have

$$|f_j(m,y)| \leq \frac{1}{|(n-1)!|} \cdot \tag{8.5}$$

If we calculate

$$\Delta^s(y_l \Delta^l m) = \sum_{\alpha=0}^{s} \binom{s}{\alpha} (\delta^{s-\alpha} y_l) \Delta^{l+\alpha} m$$

$$= \sum_{\alpha=0}^{n-1-l} \binom{s}{\alpha} (\delta^{s-\alpha} y_l) \Delta^{l+\alpha} m$$

$$+ \sum_{\alpha=n-l}^{s} \binom{s}{\alpha} (\delta^{s-\alpha} y_l) \sum_{i=0}^{n-1} C_{l+\alpha,i} \Delta^i m$$

we find that the coefficient of $\Delta^{j+s} m$ is

$$\binom{s}{j+s-l} \delta^{l-j} y_l + \sum_{\alpha=n-l}^{s} \binom{s}{\alpha} \delta^{s-\alpha} y_l C_{l+\alpha,j+s} ;$$

therefore, since by definition

$$A_j(m) \wedge \Delta^j m = \theta(m)$$

for $j = 0, \ldots, n-1$, while

$$A_j(m) \wedge \Delta^i m = 0$$

for $i = 0, \ldots, n-1$, $i \neq j$, we obtain

$$A_{j+s}(m) \wedge \Delta^s(y_l \Delta^l m)$$

$$= \left[\binom{s}{j+s-l} \delta^{l-j} y_l + \sum_{\alpha=n-l}^{s} \binom{s}{\alpha} \delta^{s-\alpha} y_l C_{l+\alpha,j+s} \right] \theta(m)$$

and so

$$f_j(m,y) = \sum_{s=0}^{n-1-j} \binom{j+s}{j} \sum_{l=0}^{n-1} \binom{s}{j+s-l} \delta^{l-j} y_l$$

$$+ \sum_{s=0}^{n-1-j} \binom{j+s}{j} \sum_{l=0}^{n-1} \sum_{\alpha=n-l}^{s} \binom{s}{\alpha} C_{l+\alpha,j+s} \delta^{s-\alpha} y_l$$

$$= \sum_{s=0}^{n-1-j} \binom{j+s}{j} \sum_{l=j}^{j+s} \binom{s}{j+s-l} \delta^{l-j} y_l$$

$$+ \sum_{s=0}^{n-1-j} \binom{j+s}{j} \sum_{l=j+1}^{n-1} \sum_{\alpha=n-l}^{s} \binom{s}{\alpha} C_{l+\alpha,j+s} \delta^{s-\alpha} y_l ,$$

where the first change of indices is clear, while the second follows from $n - l \leq s \leq n - 1 - j$ which implies $l \geq j + 1$. Therefore we can write

$$f_j(m, y) = \left(\sum_{s=0}^{n-1-j} \binom{j+s}{j} \right) y_j + \sum_{h=1}^{n-1-j} \phi_{j,h}(\delta) y_{j+h} \qquad (8.6)_j$$

where $\phi_{j,h}(\delta)$ are elements of $\mathcal{O}_E[\delta]$ (we remind the reader that $\mathcal{O}_E$ is the ring $\{\xi \in E \,|\, |\xi| \leq 1\}$).

We now compute

$$q_j = \sum_{s=0}^{n-1-j} \binom{j+s}{j} .$$

We observe that q_j is the coefficient of X^j in

$$\sum_{h=j}^{n-1} (1+X)^h = (1+X)^j \sum_{h=0}^{n-1-j} (1+X)^h$$

$$= (1+X)^j \frac{1 - (1+X)^{n-j}}{1 - (1+X)}$$

$$= \frac{(1+X)^n - (1+X)^j}{X}$$

and therefore q_j is the coefficient of X^j in $(1+X)^n/X$, that is

$$q_j = \binom{n}{j+1} .$$

We now prove by descending induction on j that

$$|y_j| \leq \frac{1}{\left| (n-1)! \prod_{h=j+1}^{n-1} \binom{n}{h} \right|} .$$

For $j = n - 1$ equations $(8.6)_{n-1}$ and (8.5) give

$$|y_{n-1}| \leq \frac{1}{|(n-1)!|} .$$

For $j = n - 2$ we have

$$\left| \binom{n}{n-1} y_{n-2} + \phi_{n-2,1} y_{n-1} \right| \leq \frac{1}{|(n-1)!|} ,$$

and since

$$|\phi_{n-2,1} y_{n-1}| \leq \frac{1}{|(n-1)!|}$$

we obtain the desired inequality for $|y_{n-2}|$.

In the general case, for $1 \leq h \leq n-1-j$ we have

$$|\phi_{j,h} \, y_{j+h}| \leq |y_{j+h}|$$

$$\leq \frac{1}{\left| (n-1)! \prod_{k=j+h+1}^{n-1} \binom{n}{k} \right|}$$

$$\leq \frac{1}{\left| (n-1)! \prod_{k=j+2}^{n-1} \binom{n}{k} \right|}$$

and since the last quantity is $\geq |(n-1)!|^{-1}$ by (8.5) and (8.6) we obtain the desired estimate for $|y_j|$. Then (8.4) follows immediately and we obtain (8.3) by taking $z = e_i = \sum_{j=1}^{n} (H^{-1})_{ij} \Delta^{j-1} m$, for $i = 1, \ldots, n$. **Q.E.D.**

We can now finally conclude the proof of Christol–Dwork Theorem. In step I, given the matrix G we found a matrix F such that $pF^\phi = G_{[H]}$ which implies that $|F|_{\text{gauss}} \leq p$. We now apply Proposition 8.1 to this F. If we let H_1 be the matrix provided by Proposition 8.1 then we have the estimate

$$|H_1|_{\text{gauss}} \left| H_1^{-1} \right|_{\text{gauss}} \leq B_n$$

where

$$B_n = \frac{p^{n-1}}{|(n-1)!|} \inf \left(p^{n-1}, \frac{1}{\left| \prod_{j=1}^{n-1} \binom{n}{j} \right|} \right) . \tag{8.7}$$

Clearly, we can choose the sharper of the two estimates given by Proposition 8.1 because F satisfies condition $\mathcal{R}2$.

9. Effective Bounds. The Case of Unipotent Monodromy.

The goal of this section is to generalize the Christol-Dwork Theorem. We are mainly interested in systems $\delta - G$, with $\delta = X d/dX$, related to globally nilpotent equations. By Katz's theorem we know that all eigenvalues of $G(0)$ are rational numbers. One way to reduce this situation to the preceding one in which $G(0)$ is nilpotent is to take a common denominator N for the eigenvalues and then to make the substitution

$X \longrightarrow X^N$. The new system so obtained has integral exponents and by a shearing transformation we can actually assume that all exponents are zero. The weakness of this method lies in the fact that we have no *a priori* bound on the size of the denominators. Therefore we generalize the approach already used to cover the case of rational exponents.

Let G be a matrix satisfying conditions $\mathcal{R}1$, $\mathcal{R}2$, and $\mathcal{R}4$ of Section 2. We replace condition $\mathcal{R}3$ with the following

$\mathcal{R}3'$ The eigenvalues $\alpha_1, \ldots, \alpha_n$ of $G(0)$ are rational numbers which belong to $\mathbf{Z}_p \cap [0, 1)$.

Clearly the condition $\mathcal{R}3'$ implies that the eigenvalues $\alpha_1, \ldots, \alpha_n$ are prepared. We use $\mathcal{R}'$ to denote the set of matrices satisfying the conditions $\mathcal{R}1$, $\mathcal{R}2$, $\mathcal{R}3'$, and $\mathcal{R}4$.

Let

$$Y_G = \sum_{s=0}^{\infty} Y_s X^s \, .$$

Then we set

$$T(G, s) = \sup_{0 \le j \le s} |Y_j| \, .$$

For $\alpha \in \mathbf{Z}_p$ let $\mu \in \mathbf{Z}$, with $0 \le \mu \le p - 1$, be the representative of the residue class of $-\alpha$ modulo $p\mathbf{Z}_p$. Then we define the element $\alpha' \in \mathbf{Z}_p$ by

$$\alpha = -\mu + p\alpha' \, .$$

If $\alpha \in [0, 1)$ then one checks easily that $\alpha' \in [0, 1)$.

Theorem 9.1. *Let $G \in \mathcal{R}'$ and let $\alpha_1, \ldots, \alpha_n$ be the eigenvalues of $G(0)$. Then there exists a $G^{(1)} \in \mathcal{R}'$ such that the eigenvalues of $G^{(1)}(0)$ are $\alpha'_1, \ldots, \alpha'_n$ and such that*

$$T(G, s) \le B_n T(G, p - 1) \, T(G^{(1)}, (s + p - 1)/p) \qquad (9.1)$$

where B_n is the constant defined in (8.7).

Remark. When the matrix $G(0)$ is nilpotent, we have the inequality $T(G, p - 1) \le 1$. This no longer holds in the present case.

Proof. We will follow the steps of the proof of Theorem 2.1, but we must add a new initial step.

Step 0.   Preliminary Shearing.

There exists a unimodular matrix

$$H_0 \in G\ell\left(n, K[X, X^{-1}]\right) \cap \mathcal{M}_n(K[X])$$

such that the matrix

$$G' = G_{[H_0]}$$

satisfies $\mathcal{R}1$, $\mathcal{R}2$ and $\mathcal{R}4$ and is such that the eigenvalues of $G'(0)$ are $p\alpha'_1, \ldots, p\alpha'_n$. Moreover

$$\deg_X H_0 \le \mu_1 + \cdots + \mu_n \le n(p-1)$$

$$H_0^{-1} \in \mathcal{M}_n(K[X^{-1}]) \quad \text{and} \quad \deg_{1/X} H_0^{-1} \le \sup_i(\mu_i) \le p - 1 \ .$$

Furthermore there exists a matrix $J \in \mathcal{M}_n(K[X])$ such that

$$H_0 Y_G = Y_{G'} J \quad \text{and} \quad |J|_{\text{gauss}} \le T(G, p-1) \ .$$

We can be more precise with regard to the matrix J. With a slight change of notation, let $\alpha_1, \ldots, \alpha_l$ be the distinct eigenvalues of $G(0)$ with multiplicities $m_1, \ldots, m_l$. Then there exist matrices C_1, $C_2 \in G\ell(n, K)$ such that

$$J = C_2 \begin{pmatrix} C_{11}X^{\mu_1} & 0 & \cdots & 0 \\ 0 & C_{22}X^{\mu_2} & \cdots & 0 \\ \vdots & \vdots & \ddots & \vdots \\ 0 & 0 & \cdots & C_{ll}X^{\mu_l} \end{pmatrix} C_1^{-1}$$

where $C_{jj} \in \mathcal{M}_{m_j}(K)$.

Proof of Step 0.   To pass from the matrix $G(0)$ and its eigenvalues α_i to the matrix $G'(0)$ with eigenvalues $p\alpha'_i$ it suffices to increase the α_i by μ_i using a suitable sequence of shearing transformations. Thus from the results of Section 4 it follows that H_0 is a finite product of unimodular constant matrices and matrices of the form

$$\begin{pmatrix} XI_s & 0 \\ 0 & I_t \end{pmatrix} \ .$$

Therefore $\deg_X H_0 \le \mu_1 + \cdots + \mu_l \le n(p-1)$, and clearly we have $H_0^{-1} \in \mathcal{M}_n(K[X^{-1}])$.

The matrix G' certainly satisfies conditions $\mathcal{R}1$, $\mathcal{R}2$ and $\mathcal{R}4$, and by definition of α_i' the eigenvalues of $G'(0)$ are $p\alpha_1', \ldots, p\alpha_n'$. Observe also that the eigenvalues $p\alpha_i'$ are prepared.

Now by Lemma 2.4 there exist C_1, $C_2 \in Gl(n, K)$ such that

$$Y_{G'}^{-1} H_0 Y_G = C_1 M C_2$$

where

$$M = \left(C_{ij} X^{p\alpha_i' - \alpha_j} \right)$$

with

$$C_{ij} \in \mathcal{M}_{m_i, m_j}(K)$$

(i.e., C_{ij} is a block of size $m_i \times m_j$ with elements in K).

If $C_{ij} \neq 0$ then $p\alpha_i' - \alpha_j = p\alpha_i' - \alpha_i + \alpha_i - \alpha_j = \mu_i + \alpha_i - \alpha_j \in \mathbf{Z}$. Hence $\alpha_i - \alpha_j \in \mathbf{Z}$ and so $\alpha_i = \alpha_j$ (by condition $\mathcal{R}3'$). Thus $C_{ij} = 0$ for $i \neq j$ and we have

$$M = \begin{pmatrix} C_{11} X^{\mu_1} & 0 & \cdots & 0 \\ 0 & C_{22} X^{\mu_2} & \cdots & 0 \\ \vdots & \vdots & \ddots & \vdots \\ 0 & 0 & \cdots & C_{ll} X^{\mu_l} \end{pmatrix}.$$

If we put $J = C_1 M C_2$ then it is clear that $J \in \mathcal{M}_n(K[X])$ and that

$$\deg_X J \leq \sup_i(\mu_i) \leq p - 1$$

$$\deg_{X^{-1}} J^{-1} \leq \sup_i(\mu_i) \leq p - 1.$$

Now from $H_0 Y_G = Y_{G'} J$ we find

$$H_0^{-1} = Y_G J^{-1} Y_{G'}^{-1}$$

and since Y_G and $Y_{G'}^{-1}$ are analytic at zero, the order of pole at zero of H_0^{-1} is at most $p - 1$ and we conclude that $\deg_{X^{-1}} H_0^{-1} \leq p - 1$.

There remains to establish the bound on $|J|_{\text{gauss}}$. We observe that since the eigenvalues of $G'(0)$ are in $p\mathbf{Z}_p$, the eigenvalues of $(-G')^t$ are also in $p\mathbf{Z}_p$; moreover

$$\left| (-G')^t \right|_{\text{gauss}} = |G'|_{\text{gauss}} \leq 1.$$

Therefore
$$T((-G')^t, p - 1) = 1 \; ;$$
indeed, since by Lemma III.8.4 the map $\psi_{(-G')^t(0)}$ has eigenvalues in $p\mathbf{Z}_p$ and since G' satisfies condition $\mathcal{R}4$, we can apply Remark 2.2. Now we observe that

$$Y_{(-G')^t} X^{(-G')^t(0)} \quad \text{and} \quad \left(\left(Y_{G'} X^{G'(0)} \right)^{-1} \right)^t$$

are solution matrices at the origin of $\delta + (G')^t$. By the uniqueness of $Y_{(-G')^t}$ we have

$$Y_{G'}^{-1} = \left(Y_{(-G')^t} \right)^t \; .$$

The result now follows from consideration of the degree of J, the identity

$$J = Y_{G'}^{-1} H_0 Y_G \; ,$$

and the unimodularity of H_0.

Step I.   Construction of Frobenius.

There exist a unimodular $H \in G\ell(n, E_0)$ and $F \in \mathcal{M}_n(E_0)$ satisfying $\mathcal{R}1$, $\mathcal{R}2$ and $\mathcal{R}3'$ (the eigenvalues of $F(0)$ are $\alpha_1', \ldots, \alpha_n'$) such that

$$pF^\phi = G'_{[H]}$$
$$pF(0) = G'(0)$$
$$Y_{G'} = H^{-1} Y_F^\phi \; .$$

This is precisely the content of Proposition 7.1.

Step II.   Choice of the Cyclic Vector.

There exists $H_1 \in G\ell(n, E_0')$ such that

$$F_{[H_1]} = \begin{pmatrix} 0 & 1 & 0 & \ldots & 0 \\ 0 & 0 & 1 & \ldots & 0 \\ \multicolumn{5}{c}{\ldots\ldots\ldots\ldots\ldots} \\ 0 & \multicolumn{3}{c}{\ldots\ldots\ldots} & 1 \\ C_1 & \multicolumn{3}{c}{\ldots\ldots\ldots} & C_n \end{pmatrix}$$

and

$$\left| H_1 \right|_{\text{gauss}} \left| H_1^{-1} \right|_{\text{gauss}} \leq B_n \; .$$

Since F satisfies condition $\mathcal{R}2$ the same holds for $F_{[H_1]}$ and therefore, by the Dwork-Frobenius Theorem, $F_{[H_1]}$ also satisfies $\mathcal{R}4$. Moreover, $F_{[H_1]}$ is analytic at zero. See Proposition 8.1.

Step III.   Removing Apparent Singularities.

There exists a unimodular $H_2 \in G\ell\,(n, K(X)) \cap G\ell\,(n, K[[X]])$ such that

$$F_{[H_2 H_1]} \in \mathcal{M}_n(E_0) \,.$$

The matrix $F_{[H_2 H_1]}$ satisfies $\mathcal{R}1$, $\mathcal{R}2$ and $\mathcal{R}4$.

Step IV.   Shearing Transformation.

There exists a unimodular $H_3 \in G\ell\,(n, K[X, X^{-1}])$ such that the matrix $F_{[H_3 H_2 H_1]} \in \mathcal{M}_n(E_0)$ is such that the eigenvalues of $F_{[H_3 H_2 H_1]}(0)$ are not only prepared but also satisfy the further condition that if $\lambda \in \mathbf{R}$ is an eigenvalue, then $\lambda \in [0, 1)$. Let $\beta_1, \dots, \beta_n$ be these eigenvalues.

Step V.

Let

$$Q = H_3 H_2 H_1, \qquad (9.2)$$

and

$$G^{(1)} = F_{[Q]} \,.$$

Then $G^{(1)}$ satisfies $\mathcal{R}1$, $\mathcal{R}2$ and $\mathcal{R}4$. We claim that $G^{(1)}$ also satisfies $\mathcal{R}3'$, has eigenvalues $\alpha_1', \dots, \alpha_n'$ and is such that

$$Q Y_F = Y_{G^{(1)}} Q(0) \,.$$

Indeed, by Lemma 2.4, there exist $C_1,\ C_2 \in G\ell\,(n, K)$ such that

$$Y_{G^{(1)}}^{-1} Q Y_F = C_1 M C_2$$

where

$$M = \left(C_{ij} X^{\beta_i - \alpha_j'} \right)$$

with $C_{ij} \in K$. If $C_{ij} \neq 0$ then $\beta_i - \alpha_j' \in \mathbf{Z}$ so that, by Step IV, $\beta_i \in \mathbf{Q} \cap [0, 1)$ and therefore $\beta_i = \alpha_j'$. Thus $M \in G\ell\,(n, K)$ and we can conclude that

$$Q Y_F = Y_{G^{(1)}} C$$

where $C \in G\ell(n, K)$. From this equality it follows that both Q and Q^{-1} are analytic at zero. Therefore $C = Q(0)$ whence

$$QY_F = Y_{G^{(1)}}Q(0) \, ,$$

as claimed.

Finally, by Proposition 3.1, since $G^{(1)}$ and F satisfy $\mathcal{R}1$ it follows that the matrix Q lies in $G\ell(n, E_0)$. Clearly from (9.2) and the unimodularity of H_2 and H_3 it follows that

$$|Q|_{\text{gauss}} \leq |H_1|_{\text{gauss}} \, , \qquad \left|Q^{-1}\right|_{\text{gauss}} \leq \left|H_1^{-1}\right|_{\text{gauss}} \, .$$

Conclusion of the proof.

Putting the preceding steps together we obtain

$$\begin{aligned}
Y_G &= H_0^{-1} Y_{G'} J \\
&= H_0^{-1} H^{-1} Y_F^\phi J \\
&= H_0^{-1} H^{-1} \left(Q^{-1} Y_{G^{(1)}}\right)^\phi Q(0) J \\
&= H_0^{-1} W Y_{G^{(1)}}^\phi Q(0) J
\end{aligned}$$

where we set

$$W = H^{-1}(Q^{-1})^\phi = \sum_{i=0}^{\infty} W_i X^i \, .$$

Note that $W \in G\ell(n, E_0)$ since H and $Q \in G\ell(n, E_0)$. Moreover,

$$|W|_{\text{gauss}} = \left|H^{-1}(Q^{-1})^\phi\right|_{\text{gauss}} \leq \left|H_1^{-1}\right|_{\text{gauss}} \, ,$$

$$|Q(0)| \leq |Q|_{\text{gauss}} \leq |H_1|_{\text{gauss}} \, .$$

We now write

$$J = \sum_{i=0}^{p-1} J_i X^i \, , \qquad H_0^{-1} = \sum_{i=0}^{p-1} A_i X^{-i}$$

so that

$$|J_i| \leq T(G, p-1) \, , \qquad |A_i| \leq 1 \, ,$$

and

$$(Y_G)_s = \sum_{s = -i + j + ps' + l} A_i W_j (Y_{G^{(1)}})_{s'} Q(0) J_l \, .$$

From

$$s' \le \frac{s+i}{p} \le \frac{s+p-1}{p}$$

it follows that

$$T(G,s) \le T(G,p-1)\left|H_1^{-1}\right|_{\text{gauss}} |Q(0)| T\left(G^{(1)}, \frac{s+p-1}{p}\right)$$

$$\le B_n T(G,p-1) T\left(G^{(1)}, \frac{s+p-1}{p}\right),$$

as claimed. **Q.E.D.**

If we now iterate relation (9.1) we obtain

$$T(G,s) \le B_n^l T(G,p-1) T(G^{(1)},p-1) \cdots$$

$$T(G^{(l-1)},p-1) T\left(G^{(l)}, \frac{s}{p^l} + 1 - \frac{1}{p^l}\right)$$

so to find an estimate for $T(G,s)$ it suffices to take l sufficiently large so that

$$\frac{s}{p^l} + 1 - \frac{1}{p^l} < p$$

and to estimate $T(G,p-1)$ where G is any matrix satisfying conditions $\mathcal{R}1$, $\mathcal{R}2$, $\mathcal{R}3'$ and $\mathcal{R}4$. Note that the matrices $G, G^{(1)}, \ldots, G^{(l)}$ all satisfy these conditions: the eigenvalues of $G^{(i)}$ are $\alpha_1^{(i)}, \ldots, \alpha_n^{(i)}$ where $\alpha_j^{(i)} = (\alpha_j^{(i-1)})'$.

Estimate of $T(G,p-1)$.

Using the argument of Remark 2.2 and condition $\mathcal{R}4$ we obtain

$$T(G,p-1) \le \prod_{m=1}^{p-1} \prod_{i=1}^{n} \prod_{j=1}^{n} \frac{1}{|m - \alpha_i + \alpha_j|}.$$

In estimating this product we may clearly discard the terms in which $\mu_i = \mu_j$. Fix indices i and j with $\mu_i \ne \mu_j$. Recall that

$$\alpha_j - \alpha_i = \mu_i - \mu_j + p(\alpha_j' - \alpha_i').$$

Assume now $\mu_i < \mu_j$. Then

$$\prod_{m=1}^{p-1} |m - \alpha_i + \alpha_j| = |\mu_j - \mu_i - \alpha_i + \alpha_j| = |p||\alpha_j' - \alpha_i'|,$$

and

$$\prod_{m=1}^{p-1} |m + \alpha_i - \alpha_j| = \prod_{m=1}^{p-1} |m + \mu_j - \mu_i + p(\alpha_i' - \alpha_j')|$$
$$= |(p - \mu_j + \mu_i) + \mu_j - \mu_i + p(\alpha_i' - \alpha_j')|$$
$$= |p||1 + \alpha_i' - \alpha_j'| .$$

In order to avoid the distinction between the cases $\mu_i < \mu_j$ and $\mu_i > \mu_j$ it suffices to observe that for $\mu_i \neq \mu_j$ we now surely have

$$\prod_{m=1}^{p-1} |m - \alpha_i + \alpha_j||m - \alpha_j + \alpha_i| \geq |p|^2 |\alpha_j' - \alpha_i'||1 + \alpha_i' - \alpha_j'||1 + \alpha_j' - \alpha_i'| > 0 .$$

From this it follows that

$$T(G, p-1) \leq p^{n^2 - n} \prod_{\substack{\mu_i \neq \mu_j \\ i > j}} \frac{1}{|\alpha_j' - \alpha_i'||1 + \alpha_i' - \alpha_j'||1 + \alpha_j' - \alpha_i'|}$$
$$\leq p^{n^2 - n} \theta_p(\alpha_1, \dots, \alpha_n)$$

where

$$\theta_p(\alpha) = \theta_p(\alpha_1, \dots, \alpha_n) = \prod_{\substack{\alpha_i \neq \alpha_j \\ i > j}} \frac{1}{|\alpha_j' - \alpha_i'||1 + \alpha_i' - \alpha_j'||1 + \alpha_j' - \alpha_i'|} .$$

If we now choose l large enough so that

$$1 \leq \frac{s}{p^l} < p$$

then

$$T(G, s) \leq T(G, p-1)T(G^{(1)}, p-1) \cdots T(G^{(l-1)}, p-1)T(G^{(l)}, p)B_n^l ;$$

moreover

$$T(G^{(l)}, p) \leq T(G^{(l)}, p-1) \left(\prod_{i,j=1}^{n} |p - \alpha_i + \alpha_j| \right)^{-1} .$$

Now $|p - \alpha_i + \alpha_j| = 1$ if $\mu_i \neq \mu_j$, so that

$$\prod_{i,j=1}^{n} |p - \alpha_i + \alpha_j| = p^h \prod_{\mu_i = \mu_j} |1 + \alpha_j' - \alpha_i'||1 + \alpha_i' - \alpha_j'|$$

where $h = \#\{(i,j) \,|\, \mu_i = \mu_j\} \leq n^2$. Since for $p > n$ equation (8.7) becomes $B_n = p^{n-1}$, we have

$$T(G, s) \leq p^{l(n-1)} (p^{n(n-1)})^l B p^{n^2} \leq s^{n^2-1} B p^{n^2} \tag{9.3}$$

where

$$B = \theta_p(\alpha) \theta_p(\alpha') \cdots \theta_p(\alpha^{(l)}) .$$

Another method is the following. Let l be such that

$$p^{l-1} \leq s < p^l$$

so that

$$p^{l+1} \leq sp^2 \qquad \text{and} \qquad \frac{s}{p^l} + 1 - \frac{1}{p^l} < 1 + 1 - \frac{1}{p^l} < p .$$

Therefore, for $p > n$,

$$\begin{aligned}
T(G, s) &\leq B_n^l T(G, p-1) \cdots T(G^{(l)}, p-1) \\
&\leq B_n^l \left(p^{n(n-1)}\right)^{l+1} B \\
&\leq \left(p^{(n-1)+n(n-1)}\right)^{l+1} B \\
&\leq p^{n^2(l+1)} B \leq \left(sp^2\right)^{n^2} B .
\end{aligned}$$

Remark 9.2. Let N be a common denominator for $\alpha_1, \ldots, \alpha_n$. Then we have $\alpha_1, \ldots, \alpha_n \in \mathcal{N}$ where

$$\mathcal{N} = \{0, \frac{1}{N}, \ldots, \frac{N-1}{N}\} .$$

If $\alpha \in \mathcal{N}$ then it is easily seen that $\alpha' \in \mathcal{N}$, so that the map $\alpha \longmapsto \alpha'$ gives a permutation of the set $\mathcal{N}$. Moreover for $\alpha, \beta \in \mathcal{N}$ with $\alpha \neq \beta$, we have $|\alpha - \beta| = |1 + \alpha - \beta| = 1$ provided that $p > 2N$. Thus we can conclude that the constants $\theta_p(\alpha^{(j)})$ and so too the constant B defined above are equal to 1 for sufficiently large p.

CHAPTER VI

TRANSFER THEOREMS INTO DISKS WITH ONE SINGULARITY

We maintain the hypotheses of Chapter V on the fields K and Ω.

1. The Type of a Number.

In considering formal solutions of the hypergeometric equation in Section IV.8 we encountered the power series

$$f_\alpha(X) = \sum_{\substack{s=0 \\ s \neq \alpha}}^{\infty} \frac{X^s}{s - \alpha} \qquad (1.1)$$

with $\alpha \in K$. We define the *type* of α to be the radius of convergence of $f_\alpha(X)$. Let us observe in the first place that

$$\text{type}\, \alpha \leq 1 \qquad (1.2)$$

for every α. Indeed we have $|s-\alpha| \leq \sup(1, |\alpha|)$ so that the series can not converge for $|x| > 1$. If $\alpha \notin \mathbf{Z}_p$, that is if $d(\alpha, \mathbf{Z}) = \inf_{n \in \mathbf{Z}} |\alpha - n| > 0$, then clearly type $\alpha = 1$. On the contrary, for any fixed r, $0 \leq r < 1$, there are examples of $\alpha \in \mathbf{Z}_p$ with type $\alpha = r$. Indeed, let

$$\alpha = \sum_{j=0}^{\infty} p^{g(j)}$$

where $g(j)$ is a sequence of positive integers with $g(0) = 0$ and $g(j) \to \infty$ as $j \to \infty$. From $1/\text{type}\, \alpha = \limsup_{s \to \infty} |s - \alpha|^{-1/s}$, recalling that $\text{ord}\, (s - \alpha) = -\log |s - \alpha| / \log p$ and taking $s = 1 + p^{g(1)} + \cdots + p^{g(j-1)}$, we obtain

$$-\frac{\log \text{type}\, \alpha}{\log p} = \limsup_{s \to \infty} \frac{1}{s} \text{ord}\, (s - \alpha)$$

$$= \limsup_{j \to \infty} \frac{g(j)}{1 + p^{g(1)} + \cdots + p^{g(j-1)}} \ .$$

199

Note that the types of α and $-\alpha$ may be different. If type $\alpha < 1$ or type $(-\alpha) < 1$ we say that α is a p-adic Liouville number.

Proposition 1.1. *If α is algebraic over $\mathbf{Q}$ then type $\alpha = 1$.*

Proof.   The case in which $\alpha \in \mathbf{Q}$ is easily handled. Otherwise, let

$$g(X) = A(X - \alpha)(X - \alpha_2)\cdots(X - \alpha_n) \in \mathbf{Z}[X]$$

be the irreducible polynomial for α over $\mathbf{Q}$, with $\alpha_2, \dots, \alpha_n \in \mathbf{Q}^{\mathrm{alg}}$. Let

$$\varepsilon = \min_{2 \le j \le n} |\alpha - \alpha_j| .$$

Let $s \in \mathbf{N}$ be such that $|s - \alpha| < \varepsilon$. Then $|s - \alpha_j| = |\alpha - \alpha_j|$ so that

$$|g(s)| = |s - \alpha||g'(\alpha)| .$$

Thus for all $s \in \mathbf{N}$ we have

$$|s - \alpha| \ge \inf\left(\varepsilon, \frac{|g(s)|}{|g'(\alpha)|}\right) .$$

Since $g(s) \in \mathbf{Z}$, by the product formula for $\mathbf{Q}$ we have $|g(s)| \ge 1/|g(s)|_\infty$, where $|-|_\infty$ denotes the usual absolute value on $\mathbf{R}$. Thus

$$\frac{1}{|s - \alpha|} \le \sup\left(\frac{1}{\varepsilon}, |g'(\alpha)||g(s)|_{\mathbf{R}}\right) \le \frac{1}{\varepsilon} + |g'(\alpha)||g(s)|_{\mathbf{R}} \le h(s)$$

where h lies in $\mathbf{R}[s]$ and has positive coefficients. Thus

$$\sum_{s=0}^{\infty} \frac{X^s}{s - \alpha} \in \mathbf{Q}_p^{\mathrm{alg}}[[X]]$$

is dominated in the sense of Cauchy by

$$\sum_{s=0}^{\infty} h(s)X^s \in \mathbf{R}[[X]]$$

and this last is $h(X\,d/dX)\dfrac{1}{1 - X}$. Hence type $\alpha = 1$.   **Q.E.D.**

The following is a special case of a Kummer transformation formula for $_1F_1(a, b; X)$.

Lemma 1.2. *If $\alpha \in \Omega$ and $\alpha \notin \mathbf{N}$ then*

$$\sum_{s=0}^{\infty} \frac{X^s}{(1-\alpha)_s} = \alpha e^X \sum_{s=0}^{\infty} \frac{(-X)^s}{s!} \frac{1}{\alpha - s} \ .$$

Proof.   We recall that

$$(1-\alpha)_s = (1-\alpha)(1-\alpha+1)\cdots(1-\alpha+s-1)$$
$$= (1-\alpha)(2-\alpha)\cdots(s-\alpha) \ .$$

Let

$$L = \delta \circ (\delta - \alpha - X)$$

and let

$$y = \sum_{s=0}^{\infty} \frac{X^s}{(1-\alpha)_s} \ .$$

Then $L = \delta \circ (\delta - \alpha) - \delta \circ X$ and an easy calculation gives

$$Ly = \sum_{s=1}^{\infty} \left(\frac{s(s-\alpha)}{(1-\alpha)_s} - \frac{s}{(1-\alpha)_{s-1}} \right) X^s = 0 \ .$$

The indicial polynomial of L at zero is $s(s-\alpha)$ so that, since $\alpha \notin \mathbf{N}$, y is the unique analytic solution of the equation $Ly = 0$ at the origin satisfying the initial condition $y(0) = 1$. Let

$$z = \alpha e^X \sum_{s=0}^{\infty} \frac{(-X)^s}{s!} \frac{1}{\alpha - s} \ .$$

Then, since $z(0) = 1$, to establish the identity it suffices to show that $Lz = 0$.
      But

$$e^{-X} \circ \delta \circ e^X = \delta + X$$

and

$$e^{-X} \circ (\delta - \alpha - X) \circ e^X = \delta - \alpha \ ,$$

so that

$$e^{-X} \circ L \circ e^X = (\delta + X) \circ (\delta - \alpha)$$
$$= X \circ (D + 1) \circ (\delta - \alpha)$$

whence

$$\frac{1}{\alpha}e^{-X}Lz = e^{-X}Le^X \sum_{s=0}^{\infty} \frac{(-X)^s}{s!} \frac{1}{\alpha-s}$$

$$= X \circ (D+1) \circ (\delta - \alpha) \sum_{s=0}^{\infty} \frac{(-X)^s}{s!} \frac{1}{\alpha-s}$$

$$= -X \circ (D+1) \sum_{s=0}^{\infty} \frac{(-X)^s}{s!} = -X \circ (D+1)e^{-X} = 0 \ .$$

Q.E.D.

Let

$$\xi = \sum_{s=0}^{\infty} A_s X^s \ , \qquad \eta = \sum_{s=0}^{\infty} B_s X^s \ ;$$

the *Hadamard product* of ξ and η is the formal power series

$$\zeta = \sum_{s=0}^{\infty} A_s B_s X^s \ .$$

Let R_1 and R_2 be the radii of convergence of ξ and η respectively. We show that the radius of convergence of ζ is at least $R_1 R_2$. Indeed let $\varepsilon > 0$. Then there exists a positive constant M such that

$$|A_s| < \frac{M}{(R_1 - \varepsilon)^s} \ , \qquad |B_s| < \frac{M}{(R_2 - \varepsilon)^s}$$

for all s. Therefore for all s

$$|A_s B_s| \leq \frac{M'}{(R_1 R_2 - \varepsilon')^s}$$

where M' is a suitable positive constant and ε' may be chosen arbitrarily small. This proves that ζ converges for $|x| < R_1 R_2$.

Proposition 1.3. *If $\alpha \in K$, $\alpha \notin \mathbf{N}$, has type 1 then the series*

$$\sum_{s=0}^{\infty} \frac{X^s}{(1-\alpha)_s}$$

converges for $|x| < |\pi|$.

Proof. Recall that the series e^X has radius of convergence $|\pi|$. **Q.E.D.**

Proposition 1.4. *Let* $\alpha, \alpha' \in K$ *and assume that* $p\alpha' - \alpha \in \mathbf{Z}$ *and that* $|p\alpha'| < 1$. *Then*

$$\text{type } \alpha' = (\text{type } \alpha)^p .$$

Proof. Since $\text{type } \alpha = \text{type } (\alpha + \mu)$ for all $\mu \in \mathbf{Z}$ we may assume that $p\alpha' = \alpha$, $\alpha \notin \mathbf{Z}$ and $|\alpha| < 1$. Now

$$\sum_{s=0}^{\infty} \frac{X^s}{s - \alpha} = \sum_{s=0}^{\infty} \frac{(X^p)^s}{ps - p\alpha'} + \sum_{(s,p)=1} \frac{X^s}{s - \alpha}$$

$$= \frac{1}{p} \sum_{s=0}^{\infty} \frac{(X^p)^s}{s - \alpha'} + \sum_{(s,p)=1} \frac{X^s}{s - \alpha} .$$

In the last line the second series has radius of convergence 1 since $|s-\alpha| = 1$ if $(s, p) = 1$, while the first series has radius of convergence $(\text{type } \alpha')^{1/p}$. We now claim that the sum of the two series has radius of convergence $(\text{type } \alpha')^{1/p}$. This is certainly true if $(\text{type } \alpha')^{1/p} < 1$. If on the contrary $(\text{type } \alpha')^{1/p} = 1$ then the sum of the two series certainly converges for $|x| < 1$, but in fact neither series can converge for $|x| = 1$ so that the radius of convergence of the sum is again 1. To conclude it suffices to observe that the series on the left side has radius of convergence type α, by definition.   **Q.E.D.**

2.   Transfer into Disks with one Singularity: a First Estimate.

Let

$$L = D^n - C_1 D^{n-1} - \cdots - C_n \in E[D] .$$

By the Dwork-Frobenius Theorem we know that if all solutions of L at the generic point t converge on $D(t, |\pi|^-)$ then necessarily $|C_j|_{\text{gauss}} \leq 1$. The converse is obvious.

Let $D - G$, where

$$G = \begin{pmatrix} 0 & 1 & 0 & \cdots & 0 \\ 0 & 0 & 1 & \cdots & 0 \\ \cdots\cdots\cdots\cdots\cdots\cdots\cdots \\ 0 & & & \cdots & 1 \\ C_n & C_{n-1} & & \cdots & C_1 \end{pmatrix} ,$$

be the differential system associated to L. We recall (cf. Section III.5) that the radius of convergence of L or of $D-G$ is by definition the radius of convergence of the normalized matrix solution $\mathcal{U}_{G,t}$ of $D - G$ at the generic point t.

Lemma 2.1. *Let*

$$L = D^n - C_1 D^{n-1} - \cdots - C_n \in E[D]$$

and assume $\sup_j |C_j|_{\text{gauss}} > 1$. *Then the radius of convergence of* L *is* $|\pi|/\beta$ *where*

$$\beta = \sup_j |C_j|_{\text{gauss}}^{1/j} .$$

Proof.   Let $\gamma \in K$ be such that $|\gamma| = \beta$ and let

$$H = \begin{pmatrix} 1 & & & 0 \\ & 1/\gamma & & \\ & & \ddots & \\ 0 & & & 1/\gamma^{n-1} \end{pmatrix} .$$

We set

$$T = G_{[H]} = \gamma W ,$$

where

$$W = \begin{pmatrix} 0 & 1 & \cdots & 0 \\ \cdots\cdots\cdots\cdots\cdots\cdots \\ \cdots\cdots\cdots\cdots\cdots\cdots \\ 0 & 0 & \cdots & 1 \\ C_n/\gamma^n & & \cdots & C_1/\gamma \end{pmatrix} ,$$

and define the matrices T_s recursively by $T = T_1$ and

$$T_{s+1} = T_s T + \frac{dT_s}{dX} .$$

Then we find that

$$T_s \equiv \gamma^s W^s \mod \gamma^{s-1} ,$$

for every $s \geq 1$. Now, from

$$\left| \frac{T_s}{s!} \right|_{\text{gauss}} \leq \frac{|\gamma|^s}{|s!|} = \frac{\beta^s}{|s!|}$$

it follows that the matrix solution at t

$$U_{T,t} = \sum_{s=0}^{\infty} \frac{T_s(t)}{s!} (X - t)^s$$

converges for $|x - t| < |\pi|/\beta$, and the same holds for $\mathcal{U}_{G,t} = H^{-1}\mathcal{U}_{T,t}$. Therefore the radius of convergence R of L satisfies $R \geq |\pi|/\beta$.

To see the inequality in the other direction we note that by the definition of W, the matrix $W(t)$ has integral eigenvalues at least one of which is a unit. So let λ be an eigenvalue of $W(t)$ with $|\lambda| = 1$ and let $\vec{v}$ be a corresponding eigenvector such that $|\vec{v}| = 1$. We then have

$$T_s(t)\vec{v} \equiv \gamma^s W^s \vec{v} = \gamma^s \lambda^s \vec{v} \qquad \mathrm{mod}\ \gamma^{s-1}$$

so that

$$|T_s(t)\vec{v}| = |\gamma|^s$$

which implies that

$$\sum_{s=0}^{\infty} \frac{T_s(t)\vec{v}}{s!}(X - t)^s$$

can converge only for $|x - t| < |\pi|/\beta$. **Q.E.D.**

Corollary 2.2. *If $G \in \mathcal{M}_n(E)$ and the radius of convergence of $\mathcal{U}_{G,t}$ is R then there exists $H \in G\ell(n, E)$ such that*

$$\left|G_{[H]}\right|_{\mathrm{gauss}} \leq \sup\left(1, \frac{|\pi|}{R}\right) \ .$$

Proof.   Let $H \in G\ell(n, E)$ be such that

$$G_{[H]} = \begin{pmatrix} 0 & 1 & 0 & \ldots & 0 \\ 0 & 0 & 1 & \ldots & 0 \\ \multicolumn{5}{c}{\cdots\cdots\cdots\cdots\cdots} \\ 0 & \multicolumn{3}{c}{\cdots\cdots\cdots} & 1 \\ C_n & \multicolumn{3}{c}{\cdots\cdots\cdots} & C_1 \end{pmatrix} \ .$$

If $R \geq |\pi|$, we know that $\left|G_{[H]}\right|_{\mathrm{gauss}} \leq 1$.

If $R < |\pi|$ we know that $R = |\pi|/\sup_j |C_j|_{\mathrm{gauss}}^{1/j}$. We take $\gamma \in K$ such that $|\gamma| = \sup_j |C_j|_{\mathrm{gauss}}^{1/j}$; if

$$H_1 = \begin{pmatrix} 1 & & & 0 \\ & 1/\gamma & & \\ & & \ddots & \\ 0 & & & 1/\gamma^{n-1} \end{pmatrix}$$

then

$$G_{[H_1 H]} = \left(G_{[H]}\right)_{[H_1]} = \gamma W$$

where

$$W = \begin{pmatrix} 0 & 1 & \cdots & 0 \\ \cdots\cdots\cdots\cdots\cdots\cdots \\ \cdots\cdots\cdots\cdots\cdots\cdots \\ 0 & \cdots\cdots & 1 \\ C_n/\gamma^n & \cdots\cdots & C_1/\gamma \end{pmatrix} .$$

Then

$$\left|G_{[H_1 H]}\right|_{\text{gauss}} = |\gamma| \, |W|_{\text{gauss}} = |\gamma| = \frac{|\pi|}{R} .$$

Q.E.D.

If $Y \in \mathcal{M}_n(\Omega[[X - a]])$ with $a \in \Omega$ we define

$$R_a(Y) = \text{radius of convergence of } Y$$

and

$$r_a(Y) = \inf(1, R_a(Y)) .$$

We shall sometimes suppress the subscripts when the center of the disk is clear from the context.

We now consider another variant of condition $\mathcal{R}3$, namely

$\mathcal{R}3''$ The eigenvalues $\alpha_1, \ldots, \alpha_n$ of $G(0)$ are prepared and $\alpha_i - \alpha_j$ has type one for all i, j.

Proposition 2.3. *Let G satisfy $\mathcal{R}1$ and $\mathcal{R}3''$. Let Y_G be the uniform part of the solution $Y_G X^{G(0)}$ of $\delta - G$. Then we have*

$$r_0(Y_G) \geq \inf\left(|\pi|, R\right)^{n^2} \quad \left(= R^{n^2} \text{ if } R \leq |\pi| \right)$$

where $R = r_t(\mathcal{U}_{G,t})$ and $\mathcal{U}_{G,t}$ is the normalized matrix solution of $\delta - G$ at the generic point t.

Proof.

Step 1.

By the Theorem of the Cyclic Vector (Theorem III.4.2) there exists $H_1 \in G\ell(n, E_0')$ such that

$$G_{[H_1]} = \begin{pmatrix} 0 & 1 & \ldots & 0 \\ \cdots\cdots\cdots\cdots\cdots\cdots \\ \cdots\cdots\cdots\cdots\cdots\cdots \\ 0 & 0 & \ldots & 1 \\ C_n & C_{n-1} & \ldots & C_1 \end{pmatrix} .$$

If $R \geq |\pi|$ then $\left| G_{[H_1]} \right|_{\text{gauss}} \leq 1$ by Theorem V.1.2.

If $R < |\pi|$ we proceed as in the proof of Corollary 2.2 and take

$$H_2 = \begin{pmatrix} 1 & & & \\ & 1/\gamma & & \\ & & \ddots & \\ & & & 1/\gamma^{n-1} \end{pmatrix} , \qquad |\gamma| = \sup_j |C_j|_{\text{gauss}}^{1/j}$$

so that

$$\left| G_{[H_2 H_1]} \right|_{\text{gauss}} = |\gamma| = \frac{|\pi|}{R} .$$

We recall that because $\delta - G$ has a regular singularity at zero, Fuchs' condition (cf. Theorem III.8.9) implies that $G_{[H_1]}$ is analytic at zero and so the same holds for $G_{[H_2 H_1]}$.

Thus we may conclude that in both cases there exists a matrix $H \in G\ell(n, E_0')$ such that

$$\left| G_{[H]} \right|_{\text{gauss}} \leq \sup\left(1, \frac{|\pi|}{R} \right)$$

and $G_{[H]}$ is analytic at zero.

Step 2.   Removing Apparent Singularities.

By Proposition V.5.1 there exists a unimodular H_3 in $G\ell(n, K(X))$ with H_3, H_3^{-1} analytic at zero such that

$$G_{[H_3 H]} \in \mathcal{M}_n(E_0) .$$

Step 3.   Shearing.

By Proposition V.4.1 there exists a unimodular matrix H_4 belonging to $G\ell\left(n, K[X, X^{-1}]\right)$ such that

$$T = G_{[H_4 H_3 H]} \in \mathcal{M}_n(E_0) \,,$$

the eigenvalues of $T(0)$ are prepared and if an eigenvalue, β, of $T(0)$ lies in the same class mod $\mathbf{Z}$ as an eigenvalue, α, of $G(0)$ then $\alpha = \beta$. Furthermore

$$|T|_{\text{gauss}} \leq \sup\left(1, \frac{|\pi|}{R}\right) \,.$$

Let

$$Q = H_4 H_3 H$$

so that

$$T = G_{[Q]} \,.$$

Since $T \in \mathcal{M}_n(E_0)$ and $G \in \mathcal{M}_n(E_0)$ the proof of Proposition V.3.1 shows that Q and Q^{-1} have no poles in $D(0, 1^-)\backslash\{0\}$. Moreover from Lemma V.2.4 we obtain

$$QY_G = Y_T \tilde{M}$$

with $\tilde{M} \in G\ell\left(n, K[X, X^{-1}]\right)$. From this it easily follows that

$$r_0(Y_G) = r_0(Y_T) \,.$$

We now find an estimate for $r_0(Y_T)$.

We know that

$$|T|_{\text{gauss}} \leq \sup\left(1, \frac{|\pi|}{R}\right)$$

and, again by Lemma V.2.4, that the eigenvalues of $T(0)$ are $\alpha_1, \ldots, \alpha_n$. We set

$$T = T(0) + T_1 X + T_2 X^2 + \cdots \,,$$
$$Y_T = I_n + Y_1 X + Y_2 X^2 + \cdots \,.$$

Then

$$mY_m + Y_m T(0) - T(0)Y_m = T_1 Y_{m-1} + T_2 Y_{m-2} + \cdots + T_{m-1} Y_1 + T_m \,.$$

If θ_m is the map

$$A \longmapsto mA + AT(0) - T(0)A$$

then, since

$$|T(0)| \leq \sup\left(1, \frac{|\pi|}{R}\right) ,$$

we have

$$|\text{adj}\,\theta_m| \leq \sup\left(1, \frac{|\pi|}{R}\right)^{n^2-1}$$

so that, by Lemma III.8.4,

$$|\theta_m^{-1}| \leq \frac{\sup\left(1, \dfrac{|\pi|}{R}\right)^{n^2-1}}{|\det\theta_m|} \leq \frac{\sup\left(1, \dfrac{|\pi|}{R}\right)^{n^2-1}}{\displaystyle\prod_{i,j=1}^{n}|m-\alpha_i+\alpha_j|} .$$

From $Y_m = \theta_m^{-1}\left(T_1 Y_{m-1} + \cdots + T_{m-1}Y_1 + T_m\right)$ we obtain

$$|Y_m| \leq |\theta_m^{-1}|\,|T|_{\text{gauss}}\sup_{1\leq j\leq m-1}|Y_j|$$

$$\leq \frac{\sup\left(1, \dfrac{|\pi|}{R}\right)^{n^2}}{|\det\theta_m|}\sup_{1\leq j\leq m-1}|Y_j| .$$

Proceeding by induction on m we find

$$|Y_m| \leq \frac{\left(\sup\left(1, \dfrac{|\pi|}{R}\right)^{n^2}\right)^m}{\displaystyle\prod_{k=1}^{m}\prod_{i,j=1}^{n}|k-\alpha_i+\alpha_j|}$$

$$\leq \frac{\left(\sup\left(1, \dfrac{|\pi|}{R}\right)^{n^2}\right)^m}{\displaystyle\prod_{i,j=1}^{n}|(1+\alpha_i-\alpha_j)_m|} .$$

Therefore the series $\displaystyle\sum_{m=0}^{\infty}Y_m X^m$ has the same radius of convergence as
the series

$$\sum_{m=1}^{\infty}\frac{(zX)^m}{\displaystyle\prod_{i,j=1}^{n}|(1+\alpha_i-\alpha_j)_m|}$$

with $|z| = \sup(1, |\pi|/R)^{n^2}$. This last series is the Hadamard product of $n^2 + 1$ series n^2 of which converge for $|x| < |\pi|$ by Proposition 1.5, and one of which converges for $|zx| < 1$. This gives the desired result. **Q.E.D.**

Remark 2.4. If in Proposition 2.3 we make the further hypothesis that $|\alpha_i| \leq 1$ for $1 \leq i \leq n$, then we obtain the sharper estimate

$$r_0(Y) \geq \frac{|\pi|^{n^2}}{\sup\left(1, \dfrac{|\pi|}{R}\right)^n} \; .$$

To prove this estimate we insert the following extra steps in the proof of Proposition 2.3.

Step 4.

By Lemma V.4.2 there exists $C \in G\ell(n, K)$, with C unimodular, such that

$$T_{[C]}(0) = \begin{pmatrix} \alpha_1 & & * \\ & \ddots & \\ 0 & & \alpha_n \end{pmatrix}$$

is upper triangular.

Step 5.

Let $\gamma \in K$ be such that

$$|\gamma| = \sup\left(1, \frac{|\pi|}{R}\right)$$

and let

$$Z = T_{[\Gamma C]}$$

where

$$\Gamma = \begin{pmatrix} 1 & & & 0 \\ & \gamma & & \\ & & \ddots & \\ 0 & & & \gamma^{n-1} \end{pmatrix} \; .$$

Then

$$|Z|_{\text{gauss}} \leq |\gamma|^n$$

and

$$|Z(0)| = \left| \Gamma \begin{pmatrix} \alpha_1 & & * \\ & \ddots & \\ 0 & & \alpha_n \end{pmatrix} \Gamma^{-1} \right| \leq 1$$

since $|\gamma| \geq 1$ and $|\alpha_i| \leq 1$ for all i.

By Lemma V.2.4 there exists a matrix $\tilde{M} \in \mathcal{M}_n(K[X, X^{-1}])$ such that

$$\Gamma C Q Y_G = Y_Z \tilde{M} \; ;$$

therefore we have

$$r_0(Y_G) = r_0(Y_Z) \; .$$

Just as we have done for $r_0(Y_T)$ in the proof of Proposition 2.3 we find the estimate for $r_0(Y_Z)$. We follow the steps of the preceding estimate starting from the inequalities

$$|Z|_{\text{gauss}} \leq \sup \left(1, \frac{|\pi|}{R} \right)^n \; , \qquad |Z(0)| \leq 1 \; .$$

If θ_m is the map

$$A \longmapsto mA + AZ(0) - Z(0)A$$

as before we obtain

$$|\theta_m^{-1}| \leq \frac{1}{|\det \theta_m|} \leq \frac{1}{\displaystyle\prod_{i,j=1}^{n} |m - \alpha_i + \alpha_j|} \; ,$$

the point being that in the present case $|\text{adj } \theta_m| \leq 1$. Then if we set

$$Y_Z = I_n + Z_1 X + Z_2 X^2 + \cdots$$

we find

$$|Z_m| \leq \frac{\left(\sup \left(1, \dfrac{|\pi|}{R} \right)^n \right)^m}{\displaystyle\prod_{i,j=1}^{n} |(1 + \alpha_i - \alpha_j)m|} \; .$$

It is now easy to conclude as in the previous case.

3. The Theorem of Transfer of Radii of Convergence.

Theorem 3.1. *Let $G \in \mathcal{M}_n(E_0)$ satisfy $\mathcal{R}3''$. Assume that $\mathcal{U}_{G,t}$ converges in the disk $D(t, R^+)$ with*

$$1 > R \geq |\pi|^{p/q}, \quad q = p^{1+l}, \quad l \geq 0.$$

Then

$$r(Y_G) \geq \inf\left(R, |\pi|^{1/q}\right)^{n^2}$$

(so if $|\pi|^{1/q} \geq R$ then $r(Y_G) \geq R^{n^2}$).

Proof.

Step 0.

For $s \geq 1$, let $\mathcal{G}_s \in \mathcal{M}_n(E_0')$ be the matrix satisfying

$$\frac{1}{s!} D^s \vec{u} = \mathcal{G}_s \vec{u}$$

for every $\vec{u}$ such that $\delta\,\vec{u} = G\vec{u}$. Then

$$\mathcal{U}_{G,t} = \sum_{s=0}^{\infty} \mathcal{G}_s (X - t)^s .$$

We recall (cf. proof of Proposition IV.7.7) that

$$\mathcal{G}_s(X) = \binom{G(0)}{s} \frac{1}{X^s} + \left(\frac{1}{X^{s-1}}\right)$$

whence

$$X^s \mathcal{G}_s(X) = \binom{G(0)}{s} + (X)$$

so that

$$\left|\binom{G(0)}{s}\right| \leq |X^s \mathcal{G}_s(X)|_{\text{gauss}} = |\mathcal{G}_s(X)|_{\text{gauss}} .$$

Since by hypothesis we have

$$\lim_{s \to \infty} |\mathcal{G}_s(t)| \, R^s = \lim_{s \to \infty} |\mathcal{G}_s|_{\text{gauss}} R^s = 0$$

we obtain that the series $(1 + X)^{G(0)}$ converges in $D(0, R^+)$. Hence, by Proposition IV.7.5, if $\alpha_1, \ldots, \alpha_n$ are the eigenvalues of $G(0)$ then $\alpha_i \equiv \beta_i$

mod $\mathbf{Z}$ for suitable β_i such that $|\beta_i| < p^{-l}$. Clearly we may assume that the β_i are prepared; moreover we have $\mathrm{type}\,(\beta_i - \beta_j) = \mathrm{type}\,(\alpha_i - \alpha_j)$ for all i, j so that the β_i satisfy condition $\mathcal{R}3''$.

We now apply a shearing transformation. There exists a matrix H_0 in $G\ell\,(n, K[X, X^{-1}])$ such that

$$G' = G_{[H_0]} \in \mathcal{M}_n(E_0)$$

and such that the eigenvalues of $G'(0)$ are $\beta_1, \ldots, \beta_n$. Clearly the matrix solution $\mathcal{U}_{G',t}$ of $\delta - G'$ at the generic point t is convergent in the disk $D(t, R^+)$.

Step 1.   Frobenius.

For $f \in \Omega[[X]]$ we define

$$f^{\phi_q}(X) = f(X^q) \,.$$

We have a matrix $G' \in \mathcal{M}_n(E_0)$ and we know that $\mathcal{U}_{G',t}$ is convergent in $D(t, R^+)$ where $|\pi|^{1/q} > R \geq |\pi|^{p/q}$, with $q = p^{l+1}$ and $l \geq 0$; moreover we know that the eigenvalues $\beta_1, \ldots, \beta_n$ of $G'(0)$ satisfy $|\beta_i| < p^{-l}$. We wish to show that there exist:

— a matrix $H \in G\ell\,(n, E_0') \cap \mathcal{M}_n(E_0)$ with

$$H(0) = I_n \,,$$

— a matrix $F \in \mathcal{M}_n(E_0')$, analytic at zero and with $\delta - F$ having only apparent singularities in $D(0, 1^-)\backslash\{0\}$, which satisfies

$$qF^{\phi_q} = G'_{[H]} \,, \tag{3.1}$$

$$qF(0) = G'(0) \,, \tag{3.2}$$

$$r(\mathcal{U}_{F,t^q}) \geq R^q \,.$$

From (3.2) it follows that the eigenvalues of $F(0)$ are $\alpha_1' = \beta_1/q$, $\ldots$, $\alpha_n' = \beta_n/q$; these eigenvalues $\alpha_1', \ldots, \alpha_n'$ are prepared and are such that $\mathrm{type}\,(\alpha_i' - \alpha_j') = 1$ for all i, j by Proposition 1.4.

For $s \geq 1$ let $\mathcal{G}_s$ be the matrix such that

$$\mathcal{G}_s \vec{u} = \frac{1}{s!}\frac{d^s}{dX^s}\vec{u}$$

for every $\vec{u}$ satisfying $\delta\vec{u} = G'\vec{u}$. Let

$$\mathcal{Y}(X, Z) = \sum_{s=0}^{\infty} \mathcal{G}_s(X)(Z - X)^s \; .$$

This series converges if x and z are such that $x \neq 0$ and

$$\lim_{s \to \infty} \left| x^s \mathcal{G}_s(x) \left(\frac{z - x}{x} \right)^s \right| = 0 \; .$$

By hypothesis we know that

$$\lim_{s \to \infty} |\mathcal{G}_s|_{\text{gauss}} \, R^s = 0 \; ,$$

so that, taking into account the fact that in this case

$$|x^s \mathcal{G}_s(x)| \leq |\mathcal{G}_s|_{\text{gauss}} \; ,$$

we will surely have convergence if $0 \neq x \in \mathcal{D} = D(0, 1^-) \cup D(t, 1^-)$ and

$$\left| \frac{z - x}{x} \right| \leq R \; .$$

Let

$$H(X) = \frac{1}{q} \sum_{\zeta^q = 1} \mathcal{Y}(X, X\zeta)$$

$$= \sum_{s=0}^{\infty} X^s \mathcal{G}_s(X) \sum_{\zeta^q = 1} \frac{(\zeta - 1)^s}{q}$$

which converges uniformly on $\mathcal{D}$ since

$$\left| \frac{x\zeta - x}{x} \right| = |\zeta - 1| \leq |\pi|^{1/p'} = |\pi|^{p/q} \leq R \; .$$

Since $X^s \mathcal{G}_s(X) \in \mathcal{M}_n(E_0)$ it follows that $H \in \mathcal{M}_n(E_0)$. Since

$$\mathcal{G}_s(X) = \binom{G'(0)}{s} \frac{1}{X^s} + \left(\frac{1}{X^{s-1}} \right) \; ,$$

we have

$$X^s \mathcal{G}_s(X) = \binom{G'(0)}{s} + (X)$$

so that

$$H(0) = \frac{1}{q} \sum_{\zeta^q=1} \sum_{s=0}^{\infty} \binom{G'(0)}{s}(\zeta - 1)^s$$

$$= \frac{1}{q} \sum_{\zeta^q=1} g_{G'(0)}(\zeta - 1)$$

$$= I_n$$

by Proposition IV.7.5. Since we have established that $H \in \mathcal{M}_n(E_0)$ and $\det H$ is not identically zero we can conclude that $H \in G\ell(n, E_0') \cap \mathcal{M}_n(E_0)$. Now let

$$V = \psi_q \mathcal{U}_{G',t} \ .$$

As we have seen in Section V.6 for $x \in D(t^q, (R^q)^+)$ we have

$$V(x) = \frac{1}{q} \sum_{z^q=x} \mathcal{U}_{G',t}(z)$$

(note that by Lemma V.6.1 necessarily $z \in D(t, R^+)$). Moreover we know from Section V.6 that V is analytic in the disk $D(t^q, R^q)$.

We now follow the proof of Proposition V.7.1. We claim that for $x \in D(t, R)$ we have

$$V(x^q) = H(x)\mathcal{U}_{G',t}(x) \ .$$

Indeed, if $x \in D(t, R)$ we have $x^q \in D(t^q, R^q)$ and

$$V(x^q) = \frac{1}{q} \sum_{z^q=x^q} \mathcal{U}_{G',t}(z) = \frac{1}{q} \sum_{\zeta^q=1} \mathcal{U}_{G',t}(x\zeta)$$

$$= \frac{1}{q} \sum_{\zeta^q=1} \mathcal{Y}(x, x\zeta)\mathcal{U}_{G',t}(x) = H(x)\mathcal{U}_{G',t}(x) \ .$$

Therefore $\det V$ is not identically zero, so that we can define

$$F = (\delta V)V^{-1} \tag{3.3}$$

which is meromorphic in the disk $D(t^q, R^q)$. Moreover

$$F^{\phi_q} = (\delta V)^{\phi_q}\left(V^{\phi_q}\right)^{-1} = \frac{1}{q}\delta\left(V^{\phi_q}\right)\left(V^{\phi_q}\right)^{-1}$$

and therefore

$$qF^{\phi_q} = \delta\left(V^{\phi_q}\right)\left(V^{\phi_q}\right)^{-1} = \delta(HU_{G',t})(HU_{G',t})^{-1} = G'_{[H]} \ .$$

These equalities hold in the disk $D(t, R)$, but the right hand side is in $\mathcal{M}_n(E'_0)$ so that by Lemma V.6.2 we may conclude that F has an extension in $\mathcal{M}_n(E'_0)$ and that for this extension (which we still call F) we have

$$qF^{\phi_q} = G'_{[H]} \ .$$

Now from

$$G'_{[H]} = HG'H^{-1} + (\delta H)H^{-1}$$

and the fact that $H(0) = I_n$ (so that $H \in G\ell\,(n, K[[X]])$) we can easily conclude that $G'_{[H]}$ is analytic at zero and

$$G'_{[H]}(0) = G'(0) \ .$$

Therefore F is analytic at zero and, specializing at $X = 0$, we find that

$$qF(0) = G'_{[H]}(0) = G'(0) \ .$$

From (3.3) it follows that V is a matrix solution at t^q of $\delta - F$ such that $V(t^q) = H(t)$. Since $H \in \mathcal{M}_n(E_0)$ and $\det H \neq 0$ we have $\det H(t) \neq 0$. Therefore

$$\mathcal{U}_{F,t^q} = VH(t)^{-1}$$

so that

$$r(\mathcal{U}_{F,t^q}) = r_{t^q}(V) \geq R^q \ .$$

To conclude the proof of Step 1 we need only establish that $\delta - F$ has only apparent singularities in the disk $D(0, 1^-)\backslash\{0\}$. Let $0 \neq a \in D(0, 1^-)$. We wish to show that a^q is at worst a trivial singularity of $\delta - F$, that is, there exists a matrix solution of $\delta - F$ at $X = a^q$ which is analytic for x close to a^q. We know that

$$\mathcal{U}_{G',a}(Z) = \mathcal{Y}(a, Z)$$

is analytic for

$$\left|\frac{z - a}{a}\right| \leq R \ .$$

Then, for x sufficiently close to a^q, i.e. for

$$\left| \frac{x - a^q}{a^q} \right| \le R^q \; ,$$

we can define

$$W(x) = \frac{1}{q} \sum_{z^q = x} \mathcal{U}_{G',a}(z)$$

since in view of Lemma V.6.1

$$\left| \frac{z^q - a^q}{a^q} \right| \le R^q$$

implies that

$$\left| \frac{z - a}{a} \right| \le R$$

so that $\mathcal{U}_{G',a}(z)$ is defined. Thus we obtain, for x sufficiently close to a, i.e. for

$$\left| \frac{x - a}{a} \right| \le R$$

so that

$$\left| \frac{x^q - a^q}{a^q} \right| \le R^q \; ,$$

$$W(x^q) = \frac{1}{q} \sum_{z^q = x^q} \mathcal{U}_{G',a}(z) = \frac{1}{q} \sum_{\zeta^q = 1} \mathcal{U}_{G',a}(x\zeta)$$

$$= \frac{1}{q} \sum_{\zeta^q = 1} \mathcal{Y}(x, x\zeta) \mathcal{U}_{G',a}(x)$$

$$= H(x) \mathcal{U}_{G',a}(x) \; .$$

The key point is that R is so defined that if

$$\left| \frac{x - a}{a} \right| \le R$$

then

$$\left| \frac{x\zeta - a}{a} \right| \le R$$

for all ζ satisfying $\zeta^q = 1$.

Then since $\det H$ is not identically zero, the same is true for $\det W$, so that W is analytic at x for x close to a^q (but it may well not be invertible at a^q, i.e. $\det H(a)$ may be zero). Moreover we have

$$\delta(W^{\phi_q}) = \delta(H \, U_{G',a}) = G'_{[H]} H \, U_{G',a}$$
$$= G'_{[H]} W^{\phi_q} = q F^{\phi_q} W^{\phi_q} \ .$$

On the other hand

$$\delta(W^{\phi_q}) = q(\delta W)^{\phi_q}$$

and therefore

$$\delta W = F W \ .$$

We conclude that W is a matrix solution of $\delta - F$ at $x = a^q$.

Step 2.   Removing the singularities of F.

There exists $H_1 \in G\ell(n, K(X))$ with H_1, H_1^{-1} analytic at zero such that

$$F_{[H_1]} \in \mathcal{M}_n(E_0) \ .$$

Step 3.   Shearing Transformation.

There exists $H_2 \in G\ell\left(n, K[X, X^{-1}]\right)$ such that

$$S = F_{[H_2 H_1]} \in \mathcal{M}_n(E_0)$$

and the eigenvalues of $S(0)$ are prepared. From (3.1) and

$$\delta \circ \phi_q = q(\phi_q \circ \delta)$$

it follows that

$$q S^{\phi_q} = q \left(F_{[H_2 H_1]}\right)^{\phi_q} = \left(q F^{\phi_q}\right)_{[(H_2 H_1)^{\phi_q}]}$$
$$= \left(G'_{[H]}\right)_{[H_2^{\phi_q} H_1^{\phi_q}]} = G'_{[H_2^{\phi_q} H_1^{\phi_q} H]}$$
$$= G_{[H_2^{\phi_q} H_1^{\phi_q} H H_0]} \ .$$

Let

$$Q = H_2^{\phi_q} H_1^{\phi_q} H H_0$$

so that

$$q S^{\phi_q} = G_{[Q]} \ .$$

Since G, $S \in \mathcal{M}_n(E_0)$, as in the proof of Proposition V.3.1 we find that the matrices Q and Q^{-1} are analytic in $D(0, 1^-) \backslash \{0\}$. Moreover $Y_S X^{S(0)}$ is a matrix solution of $\delta - S$ so that $Y_S^{\phi_q} X^{qS(0)}$ is a matrix solution of $\delta - qS^{\phi_q}$; on the other hand, by Lemma V.2.4 we have

$$\left(Y_S^{\phi_q} \right)^{-1} Q Y_G = V$$

where $V \in G\ell \left(n, K[X, X^{-1}] \right)$. The matrix Q may have a pole at $X = 0$. Now from $Y_G = Q^{-1} Y_S^{\phi_q} V$ and the fact that Q^{-1} is analytic at every point of $D(0, 1^-) \backslash \{0\}$ if follows that

$$r(Y_G) \geq r \left(Y_S^{\phi_q} \right) .$$

Assume that $r(Y_S) = \rho$. Then $Y_S(X^q)$ converges for $|x^q| < \rho$, that is for $|x| < \rho^{1/q}$, and

$$r(Y_G) \geq r(Y_S)^{1/q} . \tag{3.4}$$

Thus we have reduced to estimating $r(Y_S)$. We know from Step 0 and Step 1 that

$$r(\mathcal{U}_{F, t^q}) \geq R^q .$$

Moreover, since $S = F_{[H_2 H_1]}$ and H_1, H_2 do not alter the radius of convergence at the generic point we have

$$r(\mathcal{U}_{S, t^q}) = r(\mathcal{U}_{F, t^q}) \geq R^q .$$

Now, let $\gamma_1, \ldots, \gamma_n$ be the eigenvalues of $S(0)$; then, again by Lemma V.2.4, for each i there exists j such that $\gamma_i \equiv \alpha'_j \mod \mathbf{Z}$. Therefore, since type $(\alpha'_i - \alpha'_j) = 1$ for all i, j, we obtain

$$\text{type} \left(\gamma_i - \gamma_j \right) = 1 ,$$

for all i, j. Therefore we can apply Proposition 2.3 to obtain

$$r(Y_S) \geq \inf \left(R^q, |\pi| \right)^{n^2}$$

so that by (3.4)

$$r(Y_G) \geq \inf \left(R, |\pi|^{1/q} \right)^{n^2} .$$

Q.E.D.

As an immediate consequence of Theorem 3.1 we can now state

Theorem 3.2. *Let $G \in M_n(E_0)$ and suppose that the eigenvalues of $G(0)$ satisfy condition $\mathcal{R}3''$. Let $\mathcal{U}_{G,t}$ be the normalized matrix solution of the system $\delta Y = GY$ at the generic point t and suppose that $\mathcal{U}_{G,t}$ converges in the disk $D(t, R^-)$ with $R \leq 1$. Then the uniform part Y_G of the solution $Y_G X^{G(0)}$ at zero converges in the disk $D\left(0, (R^{n^2})^-\right)$.*

Proof. If $R \leq |\pi|$ then the assertion is an immediate consequence of Proposition 2.3. If $1 > R > |\pi|$, choose q such that $|\pi|^{1/q} \geq R > |\pi|^{p/q}$. Then by Theorem 3.1

$$r_0(Y_G) \geq \inf\left(R, |\pi|^{1/q}\right)^{n^2} = R^{n^2}.$$

If $R = 1$ then clearly $r_0(Y_G) > 1 - \varepsilon$ for all $\varepsilon > 0$.   **Q.E.D.**

The following result, due to Christol, was the first in the direction of Theorem 3.1. The final form of Theorem 3.1 is due to André, Baldassarri and Chiarellotto.

Corollary 3.3. *Let G be as in Theorem 3.2, and assume that $r(\mathcal{U}_{G,t}) = 1$. Then $r(Y_G) = 1$.*

We now extend Theorem 3.2 to the case in which the system has trivial singularities in the punctured disk.

Corollary 3.4. *Let $Y \in G\ell(n, K((X))) \cap M_n(K[[X]])$, $A \in M_n(K)$. Let $Y X^A$ be a solution matrix at the origin of $\delta - G$ where $G \in M_n(E_0')$. We assume that*
 (i) The differences $\alpha_i - \alpha_j$ of the eigenvalues $\alpha_1, \ldots, \alpha_n$ of the matrix A all have type one.
 (ii) The operator $\delta - G$ has only trivial singularities in the punctured disk $D(0, 1^-)\backslash\{0\}$.
 (iii) The solution matrix $\mathcal{U}_{G,t}$ at the generic point t converges in $D(t, R^-)$ with $R \leq 1$.
Then

$$r_0(Y) \geq R^{n^2}.$$

Proof. We repeat steps used in the proof of Proposition 2.3.

Step 1.

By the theorem of the cyclic vector there exists $H_1 \in G\ell(n, E_0)$ such that

$$G_{[H_1]} = \begin{pmatrix} 0 & 1 & \ldots & 0 \\ \cdot\cdot\cdot\cdot\cdot\cdot\cdot\cdot\cdot\cdot\cdot\cdot\cdot\cdot\cdot\cdot\cdot\cdot\cdot\cdot \\ \cdot\cdot\cdot\cdot\cdot\cdot\cdot\cdot\cdot\cdot\cdot\cdot\cdot\cdot\cdot\cdot\cdot\cdot\cdot\cdot \\ 0 & 0 & \ldots & 1 \\ C_n & C_{n-1} & \ldots & C_1 \end{pmatrix}.$$

By Theorem III.8.9, the matrix $G_{[H_1]}$ is analytic at the origin.

Step 2.   Removal of apparent singularities.

By Proposition V.5.1 there exists $H_2 \in G\ell(n, K(X))$ with H_2, H_2^{-1} analytic at zero such that $G_{[H_2 H_1]} \in \mathcal{M}_n(E_0)$.

Step 3.   Shearing.

By Proposition V.4.1 there exists $H_3 \in G\ell(n, K[X, X^{-1}])$ such that $T = G_{[H_3 H_2 H_1]} \in \mathcal{M}_n(E_0)$ and the eigenvalues of $T(0)$ are prepared.

Let $H = H_3 H_2 H_1$. Then $Y_T X^{T(0)}$ is solution matrix of $\delta - T$ with $Y_T \in G\ell(n, K[[X]])$, $Y_T(0) = I_n$. A solution matrix at t of $\delta - T$ is given by $H \mathcal{U}_{G,t}$ and hence $\mathcal{U}_{T,t}$ converges in $D(t, R^-)$.

By Lemma V.2.4 the eigenvalues of $T(0)$ are congruent modulo $\mathbf{Z}$ to eigenvalues of A and hence the eigenvalues of $T(0)$ satisfy condition $\mathcal{R}3''$. Hence by Theorem 3.2, $r_0(Y_T) \geq R^{n^2}$. Again by Lemma V.2.4 we find that $Y_T^{-1} H Y \in G\ell(n, K(X))$ (in fact it lies in $G\ell(n, K[X, X^{-1}])$). It follows that Y is meromorphic in $D(0, (R^{n^2})^-)$.

Up to this point we have made no use of hypothesis (ii). By that condition, if $b \in D(0, 1^-)\backslash\{0\}$ then $Y(X)(X/b)^A$ is a solution matrix at b which must be analytic at b. It follows that Y is analytic at b. **Q.E.D.**

With the preceding proof we have actually established the following:

Corollary 3.5. *Under the same hypotheses as in Corollary 3.4, but with hypothesis (ii) replaced by the weaker assumption that $\delta - G$ has only apparent singularities in the punctured disk, then Y_G is meromorphic in the disk $D\left(0, (R^{n^2})^-\right)$.*

CHAPTER VII

DIFFERENTIAL EQUATIONS OF ARITHMETIC TYPE

We now begin our discussion of some global properties of differential systems, that is we will study systems

$$D\vec{y} = G\vec{y}$$

where $G \in \mathcal{M}_n(K(X))$, $D = d/dX$, and K is an algebraic number field (a finite extension of $\mathbf{Q}$).

1. The Height.

We fix some notation which will be maintained in the sequel. Let K be an algebraic number field and let $d = [K : \mathbf{Q}]$. We denote by $\mathcal{O}_K$ the ring of integers of K. By a *prime* of K we mean a valuation of K. We say that the prime is *finite* if the corresponding valuation is non-archimedean, and *infinite* otherwise. We denote by $\mathcal{P}$ (resp. $\mathcal{P}_f$, $\mathcal{P}_\infty$) the set of primes of K (resp. the set of finite and infinite primes). If $v \in \mathcal{P}_f$ then it is the extension of a p-adic valuation of $\mathbf{Q}$ for a well-determined rational prime p, and in this case we write $v \mid p$. We use K_v to denote the completion of K with respect to $v \in \mathcal{P}$. If v is finite and $v \mid p$ then K_v is a finite extension of $\mathbf{Q}_p$ and for each fixed rational prime p we have

$$d = \sum_{v \mid p} d_v$$

where $d_v = [K_v : \mathbf{Q}_v]$.

If $v \in \mathcal{P}_f$ and $v \mid p$ we assume v to be normalized so that

$$|\alpha|_v = \left| \, \mathrm{N}_{K_v/\mathbf{Q}_p} \alpha \, \right|_p^{1/d}$$

222

for all $\alpha \in K$, or equivalently so that

$$|p|_v = p^{-d_v/d} .$$

Similarly, if $v \in \mathcal{P}_\infty$ then we normalize by setting

$$|\alpha|_v = |N_{K_v/\mathbf{R}}\alpha|_{\mathbf{R}}^{1/d} ;$$

thus

$$|\alpha|_v = |\alpha|_{\mathbf{R}}^{1/d}$$

if v is *real*, that is if $K_v = \mathbf{R}$ and

$$|\alpha|_v = |\alpha|_{\mathbf{C}}^{2/d}$$

if v is *complex*, that is if $K_v = \mathbf{C}$. Here $|\alpha|_{\mathbf{R}}$ and $|\alpha|_{\mathbf{C}}$ denote respectively the classical real and complex magnitudes. With these normalizations we have the *Product Formula*: for $0 \neq \alpha \in K$

$$\prod_{v \in \mathcal{P}} |\alpha|_v = 1 .$$

Let $a \in \mathbf{R}$, $a > 0$. We define

$$\log^+ a = \log \sup(1, a) .$$

If $a, b \in \mathbf{R}$, $a, b > 0$ then

$$\sup(1, ab) \leq \sup(1, a) \sup(1, b)$$

so that

$$\log^+(ab) \leq \log^+ a + \log^+ b .$$

If $\alpha \in K$ the (*logarithmic absolute*) *height* of α is defined by

$$h(\alpha) = \sum_{v \in \mathcal{P}} \log^+ |\alpha|_v . \tag{1.1}$$

In view of our normalizations it is not difficult to see that $h(\alpha)$ depends only on α and not on the extension field containing α. Moreover from the product formula it follows that, for each $\alpha \neq 0$,

$$h(\alpha) = h(1/\alpha) .$$

Similarly we can also define

$$h_f(\alpha) = \sum_{v \in \mathcal{P}_f} \log^+ |\alpha|_v \ ,$$

$$h_\infty(\alpha) = \sum_{v \in \mathcal{P}_\infty} \log^+ |\alpha|_v \ .$$

(1.2)

For $\alpha, \beta \in K$ we clearly have

$$h(\alpha\beta) \leq h(\alpha) + h(\beta)$$

and similarly for h_f and h_∞.

If $(\alpha_1, \dots, \alpha_n) \in K^n$ we define

$$h(\alpha_1, \dots, \alpha_n) = \sum_{v \in \mathcal{P}} \sup_{1 \leq i \leq n} \log^+ |\alpha_i|_v$$

(1.3)

and similar definitions hold for $h_f(\alpha_1, \dots, \alpha_n)$ and $h_\infty(\alpha_1, \dots, \alpha_n)$.

If $P(X) \in K[X_1, \dots, X_n]$ then $h(P)$ is defined to be the height of the coefficient vector defining P. From the Gauss lemma it follows that for $P, Q \in K[X_1, \dots, X_n]$ we have

$$h_f(PQ) \leq h_f(P) + h_f(Q) \ .$$

Let σ be an isomorphism of K into $\mathbf{Q}^{\text{alg}}$. Then for $\alpha_1, \dots, \alpha_n \in K$ and for any $P \in K[X_1, \dots, X_s]$ we have

$$h(\alpha_1^\sigma, \dots, \alpha_n^\sigma) = h(\alpha_1, \dots, \alpha_n)$$

$$h(P^\sigma) = h(P)$$

where P^σ is obtained by applying σ to the coefficients. Similar equalities hold for h_f and h_∞.

If $\alpha_1, \dots, \alpha_n \in K$ we write

$$(\alpha_i) = \frac{\mathfrak{a}_i}{\mathfrak{b}_i} \ , \quad i = 1, \dots, n$$

where $\mathfrak{a}_i$ and $\mathfrak{b}_i$ are integral ideals of K and $(\mathfrak{a}_i, \mathfrak{b}_i) = 1$. Let

$$\mathfrak{b} = \text{l.c.m.}(\mathfrak{b}_1, \mathfrak{b}_2, \dots, \mathfrak{b}_n) \ ;$$

the positive integer $\text{den}(\alpha_1, \dots, \alpha_n)$ defined by

$$(\text{den}(\alpha_1, \dots, \alpha_n)) = N_{K/\mathbf{Q}}\mathfrak{b}$$

(1.4)

is called the *denominator* of $\alpha_1, \dots, \alpha_n$. Consequently if we now set $M = \text{den}(\alpha_1, \dots, \alpha_n)$, we have $\mathfrak{b} \mid M$ so that $(M) = \mathfrak{b}\mathfrak{b}'$ for a suitable integral ideal $\mathfrak{b}'$, and

$$(M\alpha_i) = (M)(\alpha_i) = \mathfrak{b}\mathfrak{b}'\mathfrak{a}_i/\mathfrak{b}_i = \mathfrak{b}'\mathfrak{a}_i(\mathfrak{b}/\mathfrak{b}_i)$$

and therefore, the last ideal being integral, $M\alpha_i \in \mathcal{O}_K$.

Lemma 1.1. *Let* $\alpha_1, \ldots, \alpha_n \in K$. *Then*

$$h_f(\alpha_1, \ldots, \alpha_n) = \frac{1}{d} \log \operatorname{den}(\alpha_1, \ldots, \alpha_n) .$$

Proof. Let v be a prime of K dividing the rational prime p. Since v is discrete there is an order function ord_v associated to v and because of our normalizations, for $\alpha \in K$ we have

$$|\alpha|_v = p^{-\frac{1}{d}f_v \operatorname{ord}_v \alpha}$$

where f_v denotes the relative degree of v.

Now let $(\alpha_i) = \mathfrak{a}_i/\mathfrak{b}_i$ with $\mathfrak{a}_i$, $\mathfrak{b}_i$ relatively prime integral ideals. Then

$$\mathfrak{b}_i = \prod_{\substack{v \\ \operatorname{ord}_v \alpha_i < 0}} v^{-\operatorname{ord}_v \alpha_i}$$

so that

$$\mathfrak{b} = \operatorname{l.c.m.}(\mathfrak{b}_1, \ldots, \mathfrak{b}_n) = \prod_{\substack{v \\ \inf_i(\operatorname{ord}_v \alpha_i) < 0}} v^{-\inf_i(\operatorname{ord}_v \alpha_i)} .$$

Now if $v \mid p$, then $N_{K/\mathbf{Q}} v = p^{f_v}$ and so we have

$$N_{K/\mathbf{Q}} \mathfrak{b} = \prod_{\substack{p, v \mid p \\ \inf_i(\operatorname{ord}_v \alpha_i) < 0}} p^{-f_v \inf_i(\operatorname{ord}_v \alpha_i)} .$$

On the other hand

$$\begin{aligned}
h_f(\alpha_1, \ldots, \alpha_n) &= \sum_{v \in \mathcal{P}_f} \sup_{1 \le i \le n} \log^+ |\alpha_i|_v \\
&= \sum_{v \in \mathcal{P}_f} \log^+ \sup_{1 \le i \le n} |\alpha_i|_v \\
&= \sum_{\substack{v \\ \inf_i(\operatorname{ord}_v \alpha_i) < 0}} \log \sup_{1 \le i \le n} |\alpha_i|_v \\
&= \frac{1}{d} \log \prod_{\substack{p, v \mid p \\ \inf_i(\operatorname{ord}_v \alpha_i) < 0}} p^{-f_v \inf_i(\operatorname{ord}_v \alpha_i)} .
\end{aligned}$$

Q.E.D.

2.   The Theorem of Bombieri–André.

Let K be a number field and let $G \in \mathcal{M}_n(K(X))$. For every $v \in \mathcal{P}_f$ let

$$R_v = R_{t_v}(\mathcal{U}_{G,t_v}) = v\text{-adic radius of convergence of } \mathcal{U}_{G,t_v} \text{ at } t_v$$

where t_v is the generic point relative to the prime v. By definition R_v is the radius of convergence of the system $D - G$ with respect to the v-topology. We define the *generic global inverse radius* of $D - G$ to be

$$\rho(G) = \sum_{v \in \mathcal{P}_f} \log^+ \frac{1}{R_v} \,. \tag{2.1}$$

It may well happen that $\rho(G)$ is infinite. For example, take $K = \mathbf{Q}$ and $G = I_n$; then

$$R_p = \left(\frac{1}{p}\right)^{1/(p-1)}$$

so that

$$\rho(G) = \sum_p \frac{1}{p-1} \log p = \infty \,.$$

We say that G satisfies the *Bombieri condition* if $\rho(G) < \infty$ (that is, the generic radii of convergence can not be too small for "general" v).

Let

$$\mathcal{S} = \{ p \mid R_v \leq |p|_v^{1/(1-p)} \text{ for some } v \text{ which divides } p \} \,.$$

Then

$$\rho(G) \geq \sum_{\substack{p \in \mathcal{S} \\ v \mid p}} \log^+ \frac{1}{R_v} \geq \sum_{p \in \mathcal{S}} \frac{\log p}{p-1} \frac{1}{d}$$

so that if $\rho(G) < \infty$ then

$$\sum_{p \in \mathcal{S}} \frac{\log p}{p-1} < \infty \,;$$

therefore

$$\sum_{p \in \mathcal{S}} \frac{1}{p} < \infty$$

and so $\mathcal{S}$ is a set of primes of density zero. Therefore, by Proposition III.5.1 and Remark III.6.2, all the singularities of the system are regular and have rational exponents.

We now define a second invariant. Consider the sequence of matrices defined inductively by $G_1 = G$ and

$$G_{s+1} = G_s G + \frac{d}{dX} G_s \ .$$

Clearly $G_s \in \mathcal{M}_n(K(X))$. For $v \in \mathcal{P}_f$ let

$$h(s, v) = \sup_{m \le s} \log^+ \left| \frac{G_m}{m!} \right|_{v, \text{gauss}} \tag{2.2}$$

Then we define the *size* of the matrix G to be

$$\sigma(G) = \limsup_{s \to \infty} \frac{1}{s} \sum_{v \in \mathcal{P}_f} h(s, v) \ . \tag{2.3}$$

We say that G satisfies the *Galočkin condition* if $\sigma(G) < \infty$.

We now make some remarks to clarify the meaning of this condition.

Let $P \in K[X]$ be a common denominator of the elements of G, that is assume that $PG \in \mathcal{M}_n(K[X])$. Then, as is easily seen,

$$P^s \frac{G_s}{s!} \in \mathcal{M}_n(K[X]) \ ,$$

and so we can define

$$q_s = \text{den} \left(P^0 \frac{G_0}{0!}, P^1 \frac{G_1}{1!}, \dots, P^s \frac{G_s}{s!} \right) \ ,$$

where the last symbol denotes the denominator of all the coefficients appearing in the polynomials involved.

From Lemma 1.1 it then follows that

$$\log q_s = d \sum_{v \in \mathcal{P}_f} \sup_{m \le s} \log^+ \left| P^m \frac{G_m}{m!} \right|_{v, \text{gauss}} \ .$$

If we now observe that

$$\sigma(G) < \infty$$

if and only if

$$\limsup_{s \to \infty} \frac{1}{s} \sum_{v \in \mathcal{P}_f} \sup_{m \le s} \log^+ \left| P^m \frac{G_m}{m!} \right|_{v,\text{gauss}} < \infty \, ,$$

we see that the Galočkin condition is equivalent to the following

$$\limsup_{s \to \infty} \frac{1}{s} \log q_s < \infty \, ,$$

that is there exists a constant C such that

$$\text{den} \left(P^0 \frac{G_0}{0!}, P^1 \frac{G_1}{1!}, \dots, P^s \frac{G_s}{s!} \right) \le C^s$$

(the least common multiple of the denominators has geometric growth).

As in the case of $\rho(G)$ we may have $\sigma(G) = \infty$. Indeed, for $G = I_n$ we have

$$\sigma(G) = \limsup_s \frac{1}{s} \log s! \sim \limsup_s \log s = \infty \, .$$

The following theorem provides a fairly precise relation between $\rho(G)$ and $\sigma(G)$.

Theorem 2.1 (Bombieri, André). *Let K be an algebraic number field and let $G \in \mathcal{M}_n(K(X))$. Then G satisfies the Bombieri condition if and only if it satisfies the Galočkin condition. More precisely,*

$$\rho(G) \le \sigma(G) \le \rho(G) + n - 1 \, .$$

Proof. We first prove that

$$\log^+ \frac{1}{R_v} = \limsup_{s \to \infty} \frac{1}{s} h(s, v) \, . \tag{2.4}$$

Indeed, we know that

$$R_v = \liminf_{s \to \infty} \left| \frac{G_s}{s!} \right|_{v,\text{gauss}}^{-1/s}$$

so that

$$\log^+ \frac{1}{R_v} = \limsup_{s \to \infty} \frac{1}{s} \log^+ |G_s/s!|_{v,\text{gauss}} \le \limsup_{s \to \infty} \frac{1}{s} h(s, v) \, .$$

On the other hand by Remark IV.3.2 we have

$$\left| \frac{G_m}{m!} \right|_{v,\text{gauss}} \leq \frac{1}{R_v^m} \{m, n-1\}_v \sup_{j \leq n-1} |G_j|_{v,\text{gauss}}$$

where

$$\{m, n-1\}_v = \sup_{1 \leq \lambda_1 < \cdots < \lambda_{n-1} \leq m} \frac{1}{|\lambda_1 \cdots \lambda_{n-1}|_v} = \{m, n-1\}_p^{d_v/d} .$$

We recall that

$$\{m, n-1\}_p \quad \begin{cases} \leq m^{n-1} & \text{if} \quad p \leq m \\ = 1 & \text{if} \quad p > m . \end{cases}$$

Then

$$\log^+ \left| \frac{G_m}{m!} \right|_{v,\text{gauss}} \leq m \log^+ \frac{1}{R_v} + \left\{ \begin{array}{ll} \frac{d_v}{d}(n-1)\log m & \text{if} \quad p \leq m \\ 0 & \text{if} \quad p > m \end{array} \right\} + C_v$$

where

$$C_v = \log^+ \sup_{j \leq n-1} |G_j|_{v,\text{gauss}} \tag{2.5}$$

is a non-negative constant independent of m. Therefore

$$h(s, v) \leq s \log^+ \frac{1}{R_v} + \left\{ \begin{array}{ll} \frac{d_v}{d}(n-1)\log s & \text{if} \quad p \leq s \\ 0 & \text{if} \quad p > s \end{array} \right\} + C_v$$

and

$$\frac{1}{s} h(s, v) \leq \log^+ \frac{1}{R_v} + \left\{ \begin{array}{ll} \frac{d_v}{d}(n-1)\frac{\log s}{s} & \text{if} \quad p \leq s \\ 0 & \text{if} \quad p > s \end{array} \right\} + \frac{C_v}{s} \tag{2.6}$$

so that

$$\limsup_{s \to \infty} \frac{1}{s} h(s, v) \leq \log^+ \frac{1}{R_v} .$$

From (2.4) it now follows that

$$\rho(G) = \sum_v \limsup_{s \to \infty} \frac{1}{s} h(s, v) . \tag{2.7}$$

We now observe that $|G|_{v,\text{gauss}} \leq 1$ for almost all $v \in \mathcal{P}_f$ so that the same is true for the G_j for $j \leq n$. Therefore the constant C_v defined in (2.5) is zero for almost all v. Hence from (2.6), on summing over all $v \in \mathcal{P}_f$, we obtain

$$\sum_{v \in \mathcal{P}_f} \frac{1}{s} h(s, v) \leq \sum_{v \in \mathcal{P}_f} \log^+ \frac{1}{R_v} + \sum_{p \leq s} (n-1) \frac{\log s}{s} + \frac{C}{s}$$

for a suitable constant C. We recall that according to the Prime Number Theorem (cf. Hardy and Wright [1], p. 345) we have

$$\pi(s) \sim \frac{s}{\log s}$$

where $\pi(s)$ denotes the number of rational primes less than or equal to s. Therefore, passing to the limit on both sides of the last inequality we obtain

$$\sigma(G) \leq \rho(G) + n - 1 .$$

To establish the second inequality we first prove that

$$\log^+ \frac{1}{R_v} = \liminf_{s \to \infty} \frac{1}{s} h(s, v) .$$

From this and (2.4) it will follow that

$$\lim_{s \to \infty} \frac{1}{s} h(s, v)$$

exists.

For any vector $\vec{y}$ such that $D\vec{y} = G\vec{y}$ we have

$$D^s \vec{y} = G_s \vec{y}$$

and we have

$$\begin{aligned}
\frac{G_{s+m}}{(s+m)!} \vec{y} &= \frac{D^m}{(s+m)!} D^s \vec{y} = \frac{D^m}{(s+m)!} (G_s \vec{y}) \\
&= \frac{1}{(s+m)!} \sum_{i+j=m} \binom{m}{i} D^i G_s \, D^j \vec{y} \\
&= \frac{1}{(s+m)!} \sum_{i+j=m} \binom{m}{i} D^i G_s \, G_j \vec{y} .
\end{aligned}$$

Therefore

$$\frac{G_{s+m}}{(s+m)!} = \sum_{i+j=m} \frac{s!m!}{(s+m)!} \frac{D^i G_s}{i!s!} \frac{G_j}{j!} \, .$$

Since by equation (IV.1.3) we have

$$\left| \frac{D^i}{i!} \left(\frac{G_s}{s!} \right) \right|_{v,\text{gauss}} \leq \left| \frac{G_s}{s!} \right|_{v,\text{gauss}} \, ,$$

it follows that

$$\left| \frac{G_{s+m}}{(s+m)!} \right|_{v,\text{gauss}} \leq \left| \frac{G_s}{s!} \right|_{v,\text{gauss}} \left| \frac{s!m!}{(s+m)!} \right|_v \sup_{j \leq m} \left| \frac{G_j}{j!} \right|_{v,\text{gauss}}$$

and so

$$\log^+ \left| \frac{G_{s+m}}{(s+m)!} \right|_{v,\text{gauss}} \leq \log^+ \left| \frac{G_s}{s!} \right|_{v,\text{gauss}} + h(m,v) - \log \left| \binom{s+m}{s} \right|_v \, .$$

Therefore, proceeding by induction on k, we obtain

$$\log^+ \left| \frac{G_{s+km}}{(s+km)!} \right|_{v,\text{gauss}}$$

$$\leq \log^+ \left| \frac{G_s}{s!} \right|_{v,\text{gauss}} + kh(m,v)$$

$$- \log \left| \binom{s+m}{s} \binom{s+2m}{s+m} \cdots \binom{s+km}{s+(k-1)m} \right|_v$$

$$\leq \log^+ \left| \frac{G_s}{s!} \right|_{v,\text{gauss}} + kh(m,v) - \log \left| \frac{(s+km)!}{s!(m!)^k} \right|_v \, .$$

Assume now that $v \mid p$ and recall that if h is a positive integer with $p^{l-1} \leq h < p^l$ and $h = h_0 + h_1 p + \cdots + h_{l-1} p^{l-1}$ with $0 \leq h_i < p$ for $0 \leq i \leq l-1$, then (see equation (II.4.6))

$$-\frac{\log |h!|_p}{\log p} = \frac{h - S_h}{p - 1} \, ,$$

where

$$S_h = h_0 + \cdots + h_{l-1} \, .$$

Note that

$$\frac{S_h}{p-1} \leq l \leq 1 + \frac{\log h}{\log p} \, .$$

Returning to the case at hand we find that

$$-\log\left|\frac{(s+km)!}{s!(m!)^k}\right|_v = \frac{d_v}{d}\frac{S_s + kS_m - S_{s+km}}{p-1}\log p$$

and therefore

$$-\log\left|\frac{(s+km)!}{s!(m!)^k}\right|_v \le \log s + \log p + k(\log p + \log m)$$

$$\le (k+1)(\log p + \log m)$$

where the last inequality holds provided $s \le m$. Since, for $s \le m$, we have

$$\log^+\left|\frac{G_s}{s!}\right|_{v,\text{gauss}} \le h(m,v)\ ,$$

it follows that for $s \le m$

$$\log^+\left|\frac{G_{s+mk}}{(s+mk)!}\right|_{v,\text{gauss}} \le (k+1)(h(m,v) + \log p + \log m)\ . \qquad (2.8)$$

For $N \ge 1$, let

$$N = s + m\left[\frac{N}{m}\right]\ , \qquad \text{with} \quad s < m\ .$$

Then (2.8) becomes

$$\log^+\left|\frac{G_N}{N!}\right|_{v,\text{gauss}} \le \left(1 + \left[\frac{N}{m}\right]\right)(h(m,v) + \log p + \log m)\ ,$$

so that

$$h(N,v) \le \frac{N+m}{m}(h(m,v) + \log p + \log m)$$

and

$$\frac{1}{N}h(N,v) \le \left(\frac{1}{m} + \frac{1}{N}\right)(h(m,v) + \log p + \log m)\ .$$

Therefore

$$\log^+\frac{1}{R_v} = \limsup_{N\to\infty}\frac{1}{N}h(N,v) \le \frac{1}{m}(h(m,v) + \log p + \log m)$$

and taking a sequence of m such that

$$\frac{1}{m}h(m,v) \longrightarrow \liminf_{s\to\infty} \frac{1}{s}h(s,v)$$

we obtain the desired equality.

To conclude the proof it suffices to show that

$$\sum_{v\in\mathcal{P}_f} \liminf_{s\to\infty} \frac{1}{s}h(s,v) \le \liminf_{s\to\infty} \sum_{v\in\mathcal{P}_f} \frac{1}{s}h(s,v) . \tag{2.9}$$

Indeed we then have

$$\rho(G) = \sum_{v\in\mathcal{P}_f} \log^+ \frac{1}{R_v} = \sum_{v\in\mathcal{P}_f} \liminf_{s\to\infty} \frac{1}{s}h(s,v)$$

$$\le \liminf_{s\to\infty} \sum_{v\in\mathcal{P}_f} \frac{1}{s}h(s,v)$$

$$\le \limsup_{s\to\infty} \sum_{v\in\mathcal{P}_f} \frac{1}{s}h(s,v)$$

$$= \sigma(G) .$$

Inequality (2.9) is a consequence of Fatou's Lemma. We give a direct proof. Let $\varepsilon > 0$, and $N \in \mathbf{N}$. There exists an integer s_0 such that for all $p \le N$ and all v such that $v \mid p$ we have

$$\liminf_{s\to\infty} \frac{1}{s}h(s,v) \le \frac{1}{m}h(m,v) + \frac{\varepsilon}{dN}$$

for all $m > s_0$. Therefore

$$\sum_{p\le N}\sum_{v\mid p} \liminf_{s\to\infty} \frac{1}{s}h(s,v) \le \sum_{p\le N}\sum_{v\mid p} \left(\frac{1}{m}h(m,v) + \frac{\varepsilon}{dN}\right)$$

$$\le \varepsilon + \frac{1}{m}\sum_{p\le N}\sum_{v\mid p} h(m,v)$$

$$\le \varepsilon + \frac{1}{m}\sum_{v\in\mathcal{P}_f} h(m,v)$$

for every $m > s_0$. If we choose a sequence of integers m such that

$$\frac{1}{m}\sum_{v\in\mathcal{P}_f} h(m,v) \longrightarrow \liminf_{s\to\infty} \frac{1}{s}\sum_{v\in\mathcal{P}_f} h(s,v)$$

we obtain

$$\sum_{p \leq N} \sum_{v \mid p} \liminf_{s \to \infty} \frac{1}{s} h(s, v) \leq \varepsilon + \liminf_{s \to \infty} \frac{1}{s} \sum_{v \in \mathcal{P}_f} h(s, v) \ .$$

Now we let $N \to \infty$ and find

$$\sum_{v \in \mathcal{P}_f} \liminf_{s \to \infty} \frac{1}{s} h(s, v) \leq \varepsilon + \liminf_{s \to \infty} \frac{1}{s} \sum_{v \in \mathcal{P}_f} h(s, v) \ ;$$

finally letting $\varepsilon \to 0$ gives the desired result.   **Q.E.D.**

3.   Transfer Theorems for Differential Equations of Arithmetic Type.

Let K be a number field and let $G \in \mathcal{M}_n(K(X))$. We assume that $\rho(G) < \infty$. In this case, as we have already observed, the singularities of the system $D - G$ are regular with rational exponents. We know from Section III.8 that at each point $\zeta \in \mathbf{Q}^{\mathrm{alg}} \cup \{\infty\}$ the system $D - G$ has a matrix solution of the form $Y_\zeta (X - \zeta)^{A_\zeta}$ where $Y_\zeta \in Gl\left(n, \mathbf{Q}^{\mathrm{alg}}((X - \zeta))\right)$ and $A_\zeta \in \mathcal{M}_n(\mathbf{Q}^{\mathrm{alg}})$ (if $\zeta = \infty$ replace $X - \zeta$ by $1/X$). Replacing the matrix A_ζ by $A_\zeta - NI_n$ for a suitable positive integer N if necessary, we can assume that

$$Y_\zeta \in Gl\left(n, \mathbf{Q}^{\mathrm{alg}}((X - \zeta))\right) \cap \mathcal{M}_n\left(\mathbf{Q}^{\mathrm{alg}}[[X - \zeta]]\right) \ .$$

We will assume this to be the case in the following discussion. Moreover, we will assume for simplicity (if necessary replacing K by a finite extension) that all the singularities of $D - G$ lie in K and that Y_ζ and A_ζ have coefficients in K. For $v \in \mathcal{P}$ we denote the v–adic radius of convergence of Y_ζ at ζ by $R_v(Y_\zeta)$, and we define the *global inverse radius* of Y_ζ to be

$$\rho(Y_\zeta) = \sum_{v \in \mathcal{P}} \log^+ \frac{1}{R_v(Y_\zeta)} \ .$$

In the present context we say that $D - G$ satisfies the *local Bombieri condition* at ζ if $\rho(Y_\zeta) < \infty$. (More generally without the hypothesis that $\rho(G) < \infty$ we may say that the local Bombieri condition at ζ is satisfied if

(i) ζ is a regular singularity,

(ii) $\rho(Y_\zeta) < \infty$,

(iii) the eigenvalues of A_ζ lie in $\mathbf{Q}$.)

Let the matrices G_s be defined, as in Section 2, by

$$D^s \vec{y} = G_s \vec{y}$$

for each solution $\vec{y}$ of $D - G$. Let $\zeta \in K$ be any finite point. We now estimate the order of pole at ζ of any element f of G_s. We shall denote this order by $\nu_\zeta = \nu_\zeta(f)$. Our estimate for ν_ζ depends on the nature of the singularity of $D - G$ at the point ζ (see Section V.5):

1. If ζ is an ordinary point of $D - G$, then clearly $\nu_\zeta = 0$.

2. If ζ is a trivial singularity then

$$\nu_\zeta = O(1) \,,$$

that is there is a constant C independent of s such that

$$\nu_\zeta(f) \leq C$$

for all elements f of any G_s. Indeed, if ζ is a trivial singularity then $A_\zeta = 0$ and

$$G_s Y_\zeta = D^s(Y_\zeta)$$

so that

$$G_s = (D^s Y_\zeta) Y_\zeta^{-1} = (D^s Y_\zeta)(\mathrm{adj}\, Y_\zeta)(\det Y_\zeta)^{-1} \,.$$

Therefore $\nu_\zeta \leq$ the order of zero at ζ of $\det Y_\zeta$.

3. If ζ is a non-trivial singularity, then

$$\nu_\zeta \leq s + O(1) \,.$$

Let $Y = Y_\zeta$ and $A = A_\zeta$. Then

$$G_s = \left(D^s \left(Y(X - \zeta)^A \right) \right) (X - \zeta)^{-A} Y^{-1}$$

$$= \left(\sum_{i+j=s} \binom{s}{i} (D^i Y)(j!) \binom{A}{j} (X - \zeta)^{A-j} \right) (X - \zeta)^{-A} Y^{-1} \,,$$

so that $\nu_\zeta \left((X - \zeta)^s f \right) \leq$ the order of zero at ζ of $\det Y$.

We now show that if $\zeta = \infty$ then any element f of G_s has a zero of order $\geq s + O(1)$ at ζ, regardless of whether or not ∞ is a singular point. Let $T = 1/X$, $Y = Y_\infty(T)$, and $A = A_\infty$. We have

$$X^s \frac{d^s}{dX^s} = \delta(\delta - 1)\cdots(\delta - s + 1), \qquad \delta = X\frac{d}{dX},$$

so that at ∞

$$D^s = (-T)^s \delta'(\delta' + 1)\cdots(\delta' + s - 1)$$

where

$$\delta' = T\frac{d}{dT} = -\delta.$$

Then if we set

$$P_u(\delta') = \delta'(\delta' + 1)\cdots(\delta' + u - 1)$$

for $u = 1, 2, \ldots$, we obtain

$$G_s(T^{-1}) = D^s(YT^A)T^{-A}Y^{-1}$$

$$= \left(\sum_{i+j=s} \binom{s}{i}(D^iY)(D^jT^A)\right)T^{-A}Y^{-1}$$

$$= (-T)^s\left(\sum_{i+j=s} \binom{s}{i}P_i(\delta')YP_j(\delta')T^A\right)T^{-A}Y^{-1}$$

$$= (-T)^s\left(\sum_{i+j=s} \binom{s}{i}P_i(\delta')YA(A + I_n)\cdots(A + (j-1)I_n)\right)Y^{-1}$$

which gives the desired conclusion.

From the preceding discussion we see that for any element f of G_s we have that the total number of poles with multiplicities is bounded from above by $\nu s + O(1)$, where ν is the number of finite non-trivial singularities. In other words, the number of finite zeros of f counted with multiplicity is less than or equal to $(\nu - 1)s + C$ where C is a constant independent of s.

In the following lemma Ω is an algebraically closed complete field with a non–archimedean valuation, and $|f|_0(r)$ is as in (IV.1.1). This lemma is a p-adic analogue of the argument principle in complex variables.

Lemma 3.1. *Let $r \in \mathbf{R}$, $r > 0$. For $f \in \Omega(X)$ let $Z(r)$ (resp. $P(r)$) be the number of zeros (resp. poles) of f (with multiplicities) in $D(0,r)$. Then*

$$\frac{d}{d \log r} \log |f|_0(r) = Z(r) - P(r) .$$

Proof.   Take $f = X - \zeta$. Then

$$|X - \zeta|_0(r) = \begin{cases} |\zeta| & \text{if} \quad |\zeta| \geq r , \\ r & \text{if} \quad |\zeta| < r . \end{cases}$$

Thus

$$\log |f|_0(r) = \begin{cases} \log |\zeta| & \text{if} \quad r \leq |\zeta| \\ \log r & \text{if} \quad r \geq |\zeta| . \end{cases}$$

In the general case it suffices to write

$$f = \frac{\prod_i (X - \xi_i)}{\prod_j (X - \eta_j)} .$$

Q.E.D.

Remark 3.2.   Lemma 3.1 also holds for $f \in E_0'$. We will not use this fact.

   Let $0 < R_1 < R_2$, and let f be an element of G_s. Suppose $\Omega \supset K$. Then

$$\frac{d}{d \log r} \log |f|_0(r) \leq Z(r) \leq (\nu - 1)s + O(1)$$

from which it follows that

$$\frac{|f|_0(R_2)}{|f|_0(R_1)} \leq \left(\frac{R_2}{R_1}\right)^{(\nu-1)s} C(R_1, R_2) \tag{3.1}$$

where $C(R_1, R_2)$ is a constant independent of s.

   Let f be an element of $X^s G_s$. Then the polar divisor $(f)_\infty$ of f satisfies

$$(f)_\infty \leq \sum_{\substack{\zeta \neq 0 \\ \zeta \text{ non-trivial}}} s\zeta + \text{a divisor independent of } s .$$

Therefore if $0 < R_1 < R_2$, we have

$$\log \frac{|f|_0(R_2)}{|f|_0(R_1)} \geq -\int_{R_1}^{R_2} P(r) d \log r$$

$$\geq -s \sum_{\substack{\zeta \text{ non-trivial} \\ 0 < |\zeta| < R_2}} \int_{\sup(R_1, |\zeta|)}^{R_2} d \log r + \text{ constant}$$

$$\geq -s \sum_{\substack{\zeta \text{ non-trivial} \\ 0 < |\zeta| < R_2}} \log \frac{R_2}{\sup(R_1, |\zeta|)} + \text{ constant}.$$

Thus we obtain

$$|f|_0(R_1) \leq |f|_0(R_2) C \left(\prod_{\substack{\zeta \text{ non-trivial} \\ 0 < |\zeta| < R_2}} \frac{R_2}{|\zeta|} \right)^s \tag{3.2}$$

where C is a constant independent of s but depending on R_1, R_2.

If the origin is a trivial singularity of $D - G$ then (3.2) also holds for each coefficient, f, of G_s.

The following theorem shows that the global Bombieri condition implies the local Bombieri condition. The converse will be shown in Section 5.

Theorem 3.3. Let $G \in \mathcal{M}_n(K(X))$ with $\rho(G) < \infty$ and suppose that zero is a regular singularity of the system $D - G$. If $Y X^A$ with $Y \in \mathcal{M}_n(K[[X]]) \cap G\ell(n, K((X)))$ and $A \in \mathcal{M}_n(K)$, is a matrix solution of $D - G$ at zero, then we have

$$\rho(Y) \leq n^2 \rho(G) + (n^2 + 1) \sum_{\substack{\zeta \neq 0, \infty \\ \zeta \text{ non-trivial}}} h(\zeta).$$

Proof. Let v be a fixed finite prime of K and let α with $|\alpha|_v < 1$ be such that $D - G$ has only trivial singularities in the punctured disk $D(0, |\alpha|_v^-) \backslash \{0\}$ (the choice of α depends on v). Then the system

$$X \frac{d}{dX} - \alpha X G(\alpha X)$$

has only trivial singularities in the punctured disk $D(0, 1^-) \backslash \{0\}$ and $Y(\alpha X) X^A$ is a matrix solution at zero. Since Theorem III.6.1 implies that the exponents are rational, we can apply Corollary VI.3.4 to conclude that

$$R_v(Y(\alpha X)) \geq R_v \big(\mathcal{U}_{X\alpha G(\alpha X),t} \big)^{n^2} .$$

Since

$$R_v(Y(\alpha X)) = \frac{1}{|\alpha|_v} R_v(Y)$$

we obtain

$$R_v(Y) \geq |\alpha|_v R_v \big(\mathcal{U}_{X\alpha G(\alpha X),t} \big)^{n^2} . \tag{3.3}$$

We now calculate $\mathcal{U}_{X\alpha G(\alpha X),t}$. Let t_α be a generic point such that $|t_\alpha|_v = |\alpha|_v$. Then

$$\sum_{s=0}^{\infty} \frac{G_s(t_\alpha)}{s!} (X - t_\alpha)^s$$

is the matrix solution at t_α of the system

$$X \frac{d}{dX} - XG(X)$$

and therefore the matrix solution at t_α/α of the system $X\alpha \dfrac{d}{d(X\alpha)} - X\alpha G(X\alpha)$ is the matrix

$$\sum_{s=0}^{\infty} \frac{G_s(t_\alpha)}{s!} (\alpha X - t_\alpha)^s = \sum_{s=0}^{\infty} \alpha^s \frac{G_s(t_\alpha)}{s!} (X - t)^s ,$$

where $t = t_\alpha/\alpha$. Therefore

$$\mathcal{U}_{X\alpha G(\alpha X),t} = \sum_{s=0}^{\infty} \alpha^s \frac{G_s(t_\alpha)}{s!} (X - t)^s .$$

Now by (3.2) it follows that

$$\left| \frac{\alpha^s G_s(t_\alpha)}{s!} \right|_v \leq \left| \frac{G_s X^s}{s!} \right|_{v,0} (|\alpha|_v) \leq C \left| \frac{G_s X^s}{s!} \right|_{v,0} (1) \left(\prod_{\substack{0 < |\zeta|_v < 1 \\ \zeta \text{ non-trivial}}} \frac{1}{|\zeta|_v} \right)^s$$

$$\leq C \left| \frac{G_s}{s!} \right|_{v,0} (1) \left(\prod_{\substack{0 < |\zeta|_v < 1 \\ \zeta \text{ non-trivial}}} \frac{1}{|\zeta|_v} \right)^s$$

$$\leq C \frac{1}{R_v^s} \{s, n-1\}_v \left(\prod_{\substack{0 < |\zeta|_v < 1 \\ \zeta \text{ non-trivial}}} \frac{1}{|\zeta|_v} \right)^s ,$$

where C is a constant independent of s. Therefore

$$R_v\left(\mathcal{U}_{X\alpha G(\alpha X),t}\right) \geq R_v \prod_{\substack{0<|\zeta|_v<1 \\ \zeta \text{ non-trivial}}} |\zeta|_v \ ,$$

where $R_v = R_v(\mathcal{U}_{G,t})$. Then from (3.3) it follows that

$$R_v(Y) \geq |\alpha|_v \left(R_v \prod_{\substack{0<|\zeta|_v<1 \\ \zeta \text{ non-trivial}}} |\zeta|_v \right)^{n^2} \ ,$$

and that

$$\log^+ \frac{1}{R_v(Y)} \leq \log \frac{1}{|\alpha|_v} + n^2 \log \frac{1}{R_v} + n^2 \sum_{\substack{0<|\zeta|_v<1 \\ \zeta \text{ non-trivial}}} \log^+ \frac{1}{|\zeta|_v} \ .$$

If we take α such that

$$|\alpha|_v = \inf_{\substack{\zeta \neq 0,\infty \\ \zeta \text{ non-trivial}}} (1,|\zeta|_v) \geq \prod_{\substack{0<|\zeta|_v<1 \\ \zeta \text{ non-trivial}}} |\zeta|_v$$

then we have

$$\log^+ \frac{1}{|\alpha|_v} \leq \sum_{\substack{\zeta \neq 0,\infty \\ \zeta \text{ non-trivial}}} \log^+ \frac{1}{|\zeta|_v}$$

so that

$$\log^+ \frac{1}{R_v(Y)} \leq n^2 \log \frac{1}{R_v} + (n^2+1) \sum_{\substack{\zeta \neq 0,\infty \\ \zeta \text{ non-trivial}}} \log^+ \frac{1}{|\zeta|_v} \ . \qquad (3.4)$$

If v is an infinite prime of K then from the classical theory of regular singularities if follows that

$$R_v(Y) \geq \inf_{\substack{\zeta \neq 0,\infty \\ \zeta \text{ non-trivial}}} |\zeta|_v \geq \prod_{0<|\zeta|_v<1} |\zeta|_v \ . \qquad (3.5)$$

(The point is, cf. Appendix II, that $R_v(Y) > 0$ and that Y can be continued analytically to a meromorphic function in any simply connected neighborhood of zero containing only apparent singularities of

$D - G$. Hence the series representation of Y converges up to the nearest non-trivial singularity.)

Now, summing (3.4) and (3.5) over all primes of K, we obtain the desired inequality using the fact that $h(z) = h(1/z)$.   **Q.E.D.**

The preceding Theorem is of course valid if zero is an ordinary point or even an apparent singularity, but in that case we will obtain a stronger conclusion. With that goal in mind, we make the following remarks. Let $G \in \mathcal{M}_n(E)$ and $H \in G\ell(n, E)$. We recall that E is the completion of $\Omega(X)$ in the Gauss norm, where Ω is a p-adic field containing K. So, if $G_{[H]} = G_{[H],D}$ (see (III.8.4)), then

$$HU_{G,t} = U_{G_{[H]},t} H(t) .$$

Hence

$$r(U_{G,t}) = r(U_{G_{[H]},t})$$

(p-adic radii of convergence). Now suppose $H \in G\ell(n, K(X))$ and $G \in \mathcal{M}_n(K(X))$. Then

$$\rho(G) = \rho(G_{[H]}) . \tag{3.6}$$

For $Y \in \mathcal{M}_n(K((X)))$ and any prime $v \in \mathcal{P}$ we define $R_v(Y) = R_v(X^m Y)$, where m is any positive integer such that $X^m Y$ is an element of $\mathcal{M}_n(K[[X]])$. We set $r_v(Y) = \inf(1, R_v(Y))$. The *global inverse radius* of Y is defined by

$$\rho(Y) = \sum_{v \in \mathcal{P}} \log^+ \frac{1}{R_v(Y)} .$$

Similar definitions may be given for $\rho_f(Y)$ and $\rho_\infty(Y)$.

We now prove the stronger conclusion alluded to above.

Theorem 3.4. *Let $G \in \mathcal{M}_n(K(X))$ and assume that zero is an apparent singularity of the system $D - G$. Let $Y \in G\ell(n, K((X)))$ be a matrix solution of $D - G$ at zero. Then we have*

$$\rho(Y) \leq \rho(G) + \sum_{\substack{\zeta \neq 0, \infty \\ \zeta \text{ non-trivial}}} h(\zeta) .$$

Proof. We may assume $\rho(G) < \infty$. Let $m > 0$ be such that $X^m Y \in \mathcal{M}_n(K[[X]])$ and let $\tilde{G} = G_{[X^m I_n]}$. Then by (3.6)

$$\rho(\tilde{G}) = \rho(G) .$$

The non-trivial singularities of $\tilde{G}$ are the non-zero non-trivial singularities of $D - G$. Choose r such that $0 < r < 1$, and let t_r be a generic point such that $|t_r|_v = r$. Then, by (3.2) (applied to $\tilde{G}_s$ since the origin is a trivial singularity of $D - \tilde{G}$),

$$|\tilde{G}_s|_{v,0}(r) \leq C_r |\tilde{G}_s|_{v,0}(1) \left(\prod_{\substack{0 < |\zeta|_v < 1 \\ \zeta \text{ non-trivial}}} \frac{1}{|\zeta|_v} \right)^s$$

where C_r is a constant independent of s. It follows that, if $R_v = R_v(U_{G,t})$,

$$R_v(U_{\tilde{G},t_r}) \geq R_v \prod_{\substack{0 < |\zeta|_v < 1 \\ \zeta \text{ non-trivial}}} |\zeta|_v$$

(note that this lower bound is independent of r). If we take $r < R'$ with

$$R' = R_v \prod_{\substack{0 < |\zeta|_v < 1 \\ \zeta \text{ non-trivial}}} |\zeta|_v ,$$

then $U_{\tilde{G},t_r}$ converges in the disk $D(t_r, R'^-) = D(0, R'^-)$, so that, in particular, it converges at zero, and therefore

$$\log^+ \frac{1}{R_v(Y)} \leq \log^+ \frac{1}{R_v} + \sum_{\substack{\zeta \neq 0,\infty \\ \zeta \text{ non-trivial}}} \log^+ \left| \frac{1}{\zeta} \right|_v .$$

If $v \in \mathcal{P}_\infty$ then (3.5) holds again so that

$$\log^+ \frac{1}{R_v(Y)} \leq \sum_{\substack{\zeta \neq 0,\infty \\ \zeta \text{ non-trivial}}} \log^+ \frac{1}{|\zeta|_v} .$$

Summing over all v we obtain the desired inequality, as in Theorem 3.3. **Q.E.D.**

4. Size of Local Solution Bounded by its Global Inverse Radius.

Let $G \in \mathcal{M}_n(K(X))$ and assume that zero is a regular singularity with rational exponents of the system $D - G$. Let YX^A be a matrix solution of the system at zero such that $Y \in G\ell(n, K((X))) \cap \mathcal{M}_n(K[[X]])$ and $A \in \mathcal{M}_n(K)$. If we now write

$$Y = \sum_{m=0}^{\infty} Y_m X^m \, ,$$

then we define the *size* of Y to be

$$\sigma(Y) = \limsup_{s \to \infty} \frac{1}{s} \sum_{v \in \mathcal{P}} h(s, v)$$

where

$$h(s, v) = \sup_{m \le s} \log^+ |Y_m|_v \, .$$

We recall that

$$\rho(Y) = \sum_{v \in \mathcal{P}} \log^+ \frac{1}{R_v(Y)} \, .$$

We shall bound $\sigma(Y)$ in terms of $\rho(Y)$ subject to the hypothesis that the eigenvalues of A lie in $\mathbf{Q}$.

Lemma 4.1. *With the above notation let v be a prime (possibly infinite) of K. Then*

$$\log^+ \frac{1}{R_v(Y)} = \limsup_{s \to \infty} \frac{1}{s} h(s, v) \, .$$

Proof. We have

$$R_v(Y) = \liminf_{m \to \infty} |Y_m|_v^{-1/m}$$

so that

$$\log \frac{1}{R_v(Y)} = \limsup_{m \to \infty} \frac{1}{m} \log |Y_m|_v$$

and also

$$\log^+ \frac{1}{R_v(Y)} = \limsup_{m \to \infty} \frac{1}{m} \log^+ |Y_m|_v \le \limsup_{s \to \infty} \frac{1}{s} h(s, v) \, .$$

Fix $\varepsilon > 0$. Then there exists an s_0 such that

$$\frac{1}{m} \log^+ |Y_m|_v \leq \varepsilon + \log^+ \frac{1}{R_v(Y)}$$

for all $m > s_0$. Now

$$h(s, v) = \sup_{m \leq s} \log^+ |Y_m|_v = \sup \left(\sup_{m \leq s_0} \log^+ |Y_m|_v , \sup_{s_0 \leq m \leq s} \log^+ |Y_m|_v \right)$$

$$\leq \sup \left(\sup_{m \leq s_0} \log^+ |Y_m|_v , \left(\varepsilon + \log^+ \frac{1}{R_v(Y)} \right) s \right) .$$

Hence

$$\frac{1}{s} h(s, v) \leq \sup \left(\frac{1}{s} \sup_{m \leq s_0} \log^+ |Y_m|_v , \varepsilon + \log^+ \frac{1}{R_v(Y)} \right) ,$$

and therefore

$$\limsup_{s \to \infty} \frac{1}{s} h(s, v) \leq \varepsilon + \log^+ \frac{1}{R_v(Y)} .$$

Now let $\varepsilon \to 0$ to obtain the desired result.   **Q.E.D.**

Theorem 4.2. *In the preceding notation, assume that the eigenvalues of A are in $N^{-1}\mathbf{Z}$ where N is a positive integer. Then*

$$\sigma(Y) \leq \rho(Y) + N(n - 1) .$$

If, in particular, zero is an apparent singularity of $D - G$ then

$$\sigma(Y) \leq \rho(Y) + (n - 1) .$$

Proof.

Step 1.   By Theorem III.8.6 there exists a matrix $H_0 \in G\ell(n, K(X))$ such that the matrix $XG_{[H_0],D}$ (see equation (III.8.4)) is analytic at zero. By Lemma V.2.4 the eigenvalues of the specialization at zero of this matrix lie in $\mathbf{Q}$.

We apply a shearing transformation to this matrix so as to assure that the eigenvalues of the transformed matrix are prepared. More precisely, there exists a matrix $H_1 \in G\ell(n, K[X, 1/X])$ such that if we put

$$\hat{G} = XG_{[H_1 H_0],D}$$

then $\hat{G}$ is still analytic at zero, and the eigenvalues of $\hat{G}(0)$ are prepared. Let $\hat{Y}X^{\hat{A}}$ be the normalized matrix solution of $\delta - \hat{G}$ at zero, where $\hat{A} = \hat{G}(0)$ and $\hat{Y} \in \mathcal{M}_n(K[[X]])$. Thus, by Lemma V.2.4, there exists $M_1 \in G\ell\left(n, K[X, X^{-1}]\right)$ such that

$$\hat{Y} = H_1 H_0 Y M_1 . \tag{4.1}$$

The same lemma also shows that the eigenvalues of $\hat{A}$ lie in $(1/N)\mathbf{Z}$.

Recall $r_v(Y) = \inf(1, R_v(Y))$. Since the matrix H_0 has only a finite number of poles, $r_v(Y)$ is equal to $r_v(\hat{Y})$ for all but a finite number of finite primes v of K. Indeed, it suffices to discard the primes v for which there exist non-zero poles of either H_0 or of H_0^{-1} lying in the open unit v-adic disk with center at the origin.

Step 2.   We make the substitution $X \to X^N$ in the equation $\delta - \hat{G}$. Then

$$\hat{Y}(X^N)X^{N\hat{G}(0)}$$

is a matrix solution of the system $\delta - \overline{G}$ where

$$\overline{G}(X) = N\hat{G}(X^N) .$$

Now the matrix $\overline{G}(0)$ has integral eigenvalues.

Step 3.   There exists a matrix $H_2 \in G\ell(n, K[X, 1/X])$ such that the matrix

$$\tilde{G} = (\overline{G})_{[H_2], \delta}$$

has the property that $\tilde{G}(0)$ is nilpotent. Now let

$$\tilde{Y}X^{\tilde{G}(0)}$$

be the normalized matrix solution of the system $\delta - \tilde{G}$ at zero. Then, by Lemma V.2.4, there exists a matrix $M_2 \in G\ell\left(n, K[X, X^{-1}]\right)$ such that

$$H_2(X)\hat{Y}(X^N) = \tilde{Y}(X)M_2(X) . \tag{4.2}$$

Step 4.   We now fix a finite prime v of K (excluding the finite set for which $r_v(\hat{Y})$ fails to coincide with $r_v(Y)$) and choose α_v such that

$$|\alpha_v|_v = \inf(1, R_v(Y)^{1/N}) . \tag{4.3}$$

Then $\hat{Y}((\alpha_v X)^N)$ converges in the disk $D_v(0, 1^-)$ and therefore substituting $\alpha_v X$ for X in (4.2), we find that the same holds for $\tilde{Y}(\alpha_v X)$.

We define
$$\tilde{G}_v(X) = \tilde{G}(\alpha_v X) \, .$$

We now wish to establish that for almost all v the matrix $\tilde{G}_v$ satisfies conditions $\mathcal{R}1$, $\mathcal{R}2$, $\mathcal{R}3$ and $\mathcal{R}4$ listed in Section V.2. Indeed, since $\tilde{G} \in \mathcal{M}_n(E_{0,v})$ for almost all v, the same holds for $\tilde{G}_v$; since $|G|_{v,\text{gauss}} \leq 1$ for almost all v we also have $|\tilde{G}_v|_{v,\text{gauss}} \leq 1$ for almost all v; the matrix solution $\mathcal{U}_{\tilde{G}_v, t_v}$ of $\delta - \tilde{G}_v$ at the generic point t_v converges in the disk $D(t_v, 1^-)$ by Corollary IV.5.3 and by transfer from a disk with one regular singularity to a contiguous disk with no singularity (Proposition IV.5.2); finally $\tilde{G}_v(0)$ is nilpotent by construction.

We observe that $\tilde{Y}(\alpha_v X)$ is the uniform part of the matrix solution of $\delta - \tilde{G}_v$ at zero. If we write

$$\tilde{Y}(X) = \sum_{m=0}^{\infty} \tilde{Y}_m X^m$$

then by the Christol-Dwork Theorem (Theorem V.2.1) we have, for $v|p$, $p > n$

$$|\alpha_v^m \tilde{Y}_m|_v \leq \begin{cases} m^{(n-1)d_v/d} & \text{if } m \geq p \\ 1 & \text{if } m < p \end{cases} \tag{4.4}$$

where d_v and d have the same meanings as in Section 1.

Step 5.   If we set

$$H(X) = H_2(X) H_1(X^N) H_0(X^N)$$

and

$$M(X) = M_2(X) M_1(X^N)^{-1} \in G\ell(n, K[X, 1/X])$$

then, for almost all primes v, the matrices H and M are v–unimodular and H, H^{-1} are analytic in the punctured disk $D_v(0, 1^-) \backslash \{0\}$; from (4.1) and (4.2) it follows that

$$Y(X^N) = H^{-1}(X) \tilde{Y}(X) M(X) \, . \tag{4.5}$$

We choose an integer L greater than every rational prime p which is divisible by a prime of K discarded above. We fix $\mathcal{S}$ to be the set of all primes of K which are either infinite or are a finite prime divisor of a prime $\leq L$. If we write

$$Y = \sum_{m \geq 0} Y_m X^m$$

$$H^{-1} = \sum_{h \geq -h_0} (H^{-1})_h X^h \qquad (4.6)$$

$$M = \sum_{l \geq -l_0} M_l X^l$$

then we have the relations

$$Y_m = \sum_{\substack{h+k+l=mN \\ k \geq 0, h \geq -h_0, l \geq -l_0}} (H^{-1})_h \tilde{Y}_k M_l .$$

Thus, for $v \notin \mathcal{S}$,

$$|Y_m|_v \leq \sup_{k \leq mN + l_0 + h_0} |\tilde{Y}_k|_v$$

so that, when $v|p$, from (4.4) it follows that

$$|Y_m|_v \leq$$
$$\left(\frac{1}{|\alpha_v|_v}\right)^{mM + l_0 + h_0} \begin{cases} 1 & \text{if } mN + l_0 + h_0 < p \\ (mN + l_0 + h_0)^{(n-1)d_v/d} & \text{if } mN + l_0 + h_0 \geq p. \end{cases}$$

Therefore

$$\log^+ |Y_m|_v \leq (mN + l_0 + h_0) \log \frac{1}{|\alpha_v|_v}$$
$$+ \begin{cases} 0 & \text{if } mN + l_0 + h_0 < p \\ (n-1)\dfrac{d_v}{d} \log(mN + l_0 + h_0) & \text{if } mN + h_0 + l_0 \geq p \end{cases}$$

so that by (4.3)

$$h(s, v) = \sup_{m \leq s} \log^+ |Y_m|_v$$
$$\leq \left(s + \frac{l_0 + h_0}{N}\right) \log^+ \frac{1}{R_v(Y)}$$
$$+ (n-1)\frac{d_v}{d} \begin{cases} 0 & \text{if } sN + l_0 + h_0 < p \\ \log(sN + l_0 + h_0) & \text{if } sN + h_0 + l_0 \geq p \end{cases} .$$

Now, for L as above, we have

$$\frac{1}{s}\sum_{v\notin S}h(s,v) \le \sum_{v\notin S}\log^+\frac{1}{R_v(Y)} + \frac{l_0+h_0}{Ns}\sum_{v\notin S}\log^+\frac{1}{R_v(Y)}$$

$$+\frac{1}{s}(n-1)\log(sN+h_0+l_0)\sum_{L<p\le sN+h_0+l_0}\sum_{\substack{v\notin S\\ v|p}}\frac{d_v}{d}$$

$$=\sum_{v\notin S}\log^+\frac{1}{R_v(Y)} + \frac{l_0+h_0}{Ns}\sum_{v\notin S}\log^+\frac{1}{R_v(Y)}$$

$$+\frac{1}{s}(n-1)\log(sN+h_0+l_0)\left(\pi(sN+h_0+l_0)-\pi(L)\right)$$

where $\pi(t)$ is the number of rational primes less than or equal to t.
Now we have

$$\sigma(Y) = \limsup_{s\to\infty}\frac{1}{s}\sum_{v\in\mathcal{P}}h(s,v)$$

$$\le \sum_{v\in S}\limsup_{s\to\infty}\frac{1}{s}h(s,v) + \limsup_{s\to\infty}\frac{1}{s}\sum_{v\notin S}h(s,v)$$

$$\le \sum_{v\in S}\log^+\frac{1}{R_v(Y)} + \sum_{v\notin S}\log^+\frac{1}{R_v(Y)}$$

$$+(n-1)\limsup_{s\to\infty}\frac{1}{s}\log(sN+h_0+l_0)\left(\pi(sN+h_0+l_0)-\pi(L)\right) \ .$$

Since by the Prime Number Theorem (Hardy and Wright [1], p.345)

$$\pi(t) \sim \frac{t}{\log t}$$

we obtain the desired inequality.

In the particular case that zero is an apparent singularity we can clearly take $N = 1$ in the preceding proof. **Q.E.D.**

There is also a second method for estimating $\sigma(Y)$ in terms of $\rho(Y)$ which we now derive.

If N is a positive integer we define

$$z(N) = \frac{N}{\phi(N)}\sum_{(j,N)=1}\frac{1}{j}$$

where $\phi(N)$ is Euler's totient function.

Theorem 4.3. *Maintaining the hypotheses and notations of Theorem 4.2, we have*

$$\sigma(Y) \leq \rho(Y) + n^2 + n - 1 + (n^2 - n)z(N) .$$

Proof. We repeat the first step of Theorem 4.2 and again find matrices $H_0 \in G\ell(n, K(X))$ and $H_1 \in G\ell(n, K[X, 1/X])$ such that the matrix $X\hat{G} = XG_{[H_1 H_0], D}$ is analytic at zero and the eigenvalues of $\hat{G}(0)$ lie in $N^{-1}\mathbf{Z} \cap [0, 1)$ and are congruent modulo $\mathbf{Z}$ to the eigenvalues of A. As before we then find the matrix $M \in G\ell(n, K[X, 1/X])$ such that

$$\hat{Y}M = HY \qquad (4.7)$$

where $H = H_1 H_0$ and $\hat{Y}$ is the uniform part of the matrix solution $\hat{Y}X^{\hat{G}(0)}$ at zero. Then again $r_v(Y) = r_v(\hat{Y})$ for almost all $v \in \mathcal{P}_f$.

For each finite prime v of K let α_v be such that

$$|\alpha_v|_v = \inf(1, R_v(Y)) \qquad (4.8)$$

and set $\hat{G}_v = \hat{G}(\alpha_v X)$. Then $\hat{Y}(\alpha_v X)$ is the uniform part of the solution at zero of $\delta - \hat{G}_v$ and $\hat{Y}(\alpha_v X)$ converges in the disk $D(0, 1^-)$. Condition $\mathcal{R}3'$ (see Section V.9) is satisfied for all finite $v \notin \mathcal{S}_1$ where $\mathcal{S}_1$ is a finite subset of $\mathcal{P}_f$. By (V.9.3) and Remark V.9.2 we can find a finite subset $\mathcal{S}_2$ of $\mathcal{P}_f$ containing $\mathcal{S}_1$ such that if v is a finite prime not in $\mathcal{S}_2$ and $v|p$, then

$$|\alpha_v^s \hat{Y}_s|_v \leq \begin{cases} (s^{n^2-1}p^{n^2})^{d_v/d} & \text{if } p \leq s \\ T_v(\hat{G}_v, s) & \text{if } p > s . \end{cases} \qquad (4.9)$$

where the $\hat{Y}_s$ are defined by $\hat{Y} = \sum_{s=0}^{\infty} \hat{Y}_s X^s$.

As in Theorem 4.2 we find a finite set $\mathcal{S}_3$ in $\mathcal{P}_f$ containing $\mathcal{S}_2$ such that if v is a finite prime not in $\mathcal{S}_3$ then H and H^{-1} are analytic in the punctured disk $D_v(0, 1^-)\backslash\{0\}$ and H, M are v–unimodular. Finally we again choose an integer L greater than every rational prime p which is divisible by an element of $\mathcal{S}_3$. We fix $\mathcal{S}$ to be the set of primes of K which are either infinite or are a finite prime divisor of a prime $\leq L$. Hence using the notations of (4.6), by (4.7) we obtain

$$Y_s = \sum_{h+k+l=s} (H^{-1})_h \hat{Y}_k M_l$$

so that for $v \notin S$ and $v|p$, by (4.9) we have

$$|Y_s|_v \leq \sup_{k \leq s+h_0+l_0} |\hat{Y}_k|_v \leq$$

$$\left(\frac{1}{|\alpha_v|_v}\right)^{s+h_0+l_0} \begin{cases} \left((s+h_0+l_0)^{n^2-1} p^{n^2}\right)^{d_v/d} & \text{if } p \leq s+h_0+l_0 \\ T_v(\hat{G}_v, s+h_0+l_0) & \text{if } p > s+h_0+l_0 . \end{cases}$$

Hence by (4.8)

$$h(s,v) = \sup_{m \leq s} \log^+ |Y_m|_v$$

$$\leq (s+h_0+l_0) \log^+ \frac{1}{|R_v(Y)|_v}$$

$$+ \begin{cases} \dfrac{d_v}{d}\left[(n^2-1)\log(s+h_0+l_0)+n^2\log p\right] & \text{if } p \leq s+h_0+l_0 \\ \log T_v(\hat{G}_v, s+h_0+l_0) & \text{if } p > s+h_0+l_0 . \end{cases}$$

Thus, by our choice of L in the definition of S, we find that

$$\sigma(Y) \leq \sum_{v \in S} \log^+ \frac{1}{R_v(Y)} + \sum_{v \notin S} \log^+ \frac{1}{R_v(Y)}$$

$$+ (n^2-1) \limsup_{s \to \infty} \frac{1}{s} \log(s+h_0+l_0) \sum_{L<p\leq s+h_0+l_0} \sum_{\substack{v|p \\ v \notin S}} \frac{d_v}{d}$$

$$+ n^2 \limsup_{s \to \infty} \frac{1}{s} \sum_{L<p\leq s+h_0+l_0} \log p \sum_{\substack{v|p \\ v \notin S}} \frac{d_v}{d}$$

$$+ \limsup_{s \to \infty} \frac{1}{s} \sum_{\substack{v \notin S,\, v|p \\ p>s+h_0+l_0}} \log T_v(\hat{G}_v, s+h_0+l_0)$$

$$\leq \rho(Y) + (n^2-1) \limsup_{s \to \infty} \frac{1}{s} \log(s+h_0+l_0)\left(\pi(s+h_0+l_0)-\pi(L)\right)$$

$$+ n^2 \limsup_{s \to \infty} \frac{1}{s}\left(\theta(s+h_0+l_0)-\theta(L)\right)$$

$$+ \limsup_{s \to \infty} \frac{1}{s} \sum_{\substack{v \notin S,\, v|p \\ p>s+h_0+l_0}} \log T_v(\hat{G}_v, s+h_0+l_0) ,$$

where

$$\theta(t) = \sum_{\substack{p \leq t \\ p \text{ prime}}} \log p .$$

Now by the Prime Number Theorem and the fact that

$$\theta(t) \sim \pi(t) \log t$$

(cf. Hardy and Wright [1], p. 345) we obtain

$$\sigma(Y) \le \rho(Y) + (n^2 - 1) + n^2 + \limsup_{s \to \infty} \frac{1}{s} \sum_{\substack{v \notin S, \, v|p \\ p > s + h_0 + l_0}} \log T_v(\hat{G}_v, s + h_0 + l_0) \, .$$

There remains only to estimate the last term in the above inequality. Let $\alpha_1, \dots, \alpha_n \in \{0, 1/N, \dots, (N-1)/N\}$ be the eigenvalues of $\hat{G}(0)$. As we have seen at the end of Section V.9 and observing that $\hat{G}_v(0) = \hat{G}(0)$ we then have

$$T_v(\hat{G}_v, s) \le \prod_{i=1}^{n} \prod_{j=1}^{n} \prod_{m=1}^{s} \frac{1}{|\alpha_i - \alpha_j + m|_v}$$

$$\le \left[\prod_{\alpha_i \ne \alpha_j} \prod_{m=1}^{s} \frac{1}{|\alpha_i - \alpha_j + m|_p} \right]^{d_v/d} \, .$$

Thus

$$\frac{1}{s} \sum_{\substack{v \notin S, \, v|p \\ p > s + h_0 + l_0}} \log T_v(\hat{G}_v, s + h_0 + l_0)$$

$$= \frac{1}{s} \sum_{p > s + h_0 + l_0} \sum_{v \notin S, \, v|p} \log T_v(\hat{G}_v, s + h_0 + l_0)$$

$$\le \frac{1}{s} \sum_{p > s + h_0 + l_0} \sum_{\beta \in \mathcal{E}} \sum_{m=1}^{s + h_0 + l_0} \log |\beta + m|_p^{-1} \sum_{\substack{v \notin S \\ v|p}} \frac{d_v}{d}$$

$$= \frac{1}{s} \sum_{p > s + h_0 + l_0} \sum_{\beta \in \mathcal{E}} \sum_{m=1}^{s + h_0 + l_0} \log |\beta + m|_p^{-1} \, ,$$

where $\mathcal{E} = \{\alpha_i - \alpha_j \mid \alpha_i \ne \alpha_j\}$; clearly $\#\mathcal{E} \le n^2 - n$.

To conclude the proof we shall need the following two lemmas.

Lemma 4.4. *Let M be a positive integer and j an integer such that $(j, M) = 1$. Then*

$$\limsup_{s \to \infty} \frac{1}{s} \sum_{\substack{C \in \mathbb{Z}/M\mathbb{Z} \\ (C,M)=1}} \sum_{\substack{p \in C \\ p > s}} \sum_{m=1}^{s} \log \left| \frac{j}{M} + m \right|_p^{-1} \leq z(M) - 1$$

Proof. For each residue class C modulo M such that $(C, M) = 1$ let l_C be the integer such that $l_C p \equiv j \mod M$ for all primes p in the class C and such that $0 < l_C < M$. Then for sufficiently large s (for example $s \geq \sup(j, M)$) and $p > s$ with $p \in C$, we have

$$\prod_{m=1}^{s} \left| \frac{j}{M} + m \right|_p$$

$$= \begin{cases} 1 & \text{if } s < (l_C p - j)/M \\ \left| \frac{j}{M} + \frac{l_C p - j}{M} \right|_p = |p|_p \left| \frac{l_C}{M} \right|_p = |p|_p & \text{if } s \geq (l_C p - j)/M \ . \end{cases}$$

Therefore

$$\sum_{\substack{p \in C \\ p > s}} \sum_{m=1}^{s} \log \left| \frac{j}{M} + m \right|_p^{-1} = \sum_{s < p \leq (sM+j)/l_C} \log p$$

$$= \theta \left(\frac{sM + j}{l_C}, C \right) - \theta(s, C)$$

where

$$\theta(t, C) = \sum_{\substack{p \in C \\ p \leq t}} \log p \ .$$

Since (see the Remark following this proof)

$$\theta(t, C) \sim \frac{t}{\phi(M)} \tag{4.10}$$

we obtain

$$\limsup_{s \to \infty} \frac{1}{s} \sum_{\substack{p \in C \\ p > s}} \sum_{m=1}^{s} \log \left| \frac{j}{M} + m \right|_p^{-1} = \frac{1}{\phi(M)} \left(\frac{M}{l_C} - 1 \right) \ .$$

Finally, since $(j, M) = 1$ implies that $(l_C, M) = 1$, we have

$$\sum_{\substack{C \in \mathbb{Z}/M\mathbb{Z} \\ (C,M)=1}} \frac{1}{\phi(M)} \left(\frac{M}{l_C} - 1 \right) \leq z(M) - \sum_C \frac{1}{\phi(M)} = z(M) - 1 \ .$$

Q.E.D.

Remark. Formula (4.10) is a variant of the Prime Number Theorem for arithmetic progressions. Lacking a reference for (4.10) we refer to another variant of that theorem. We introduce the function

$$\psi(t, C) = \sum_{\substack{n \leq t \\ n \in C}} \Lambda(n)$$

where

$$\Lambda(n) = \begin{cases} \log p & \text{if } n = p^m , \\ 0 & \text{otherwise} . \end{cases}$$

Then, as in Hardy and Wright [1], p. 341, one proves that

$$\psi(t, C) = \theta(t, C) + O\{t^{1/2}(\log t)^2\} .$$

Now, since (cf. Davenport [1], p. 133)

$$\psi(t, C) \sim \frac{t}{\phi(M)} ,$$

we obtain (4.10).

Lemma 4.5. If M, $N \in \mathbf{N}$ and $M | N$ then $z(M) \leq z(N)$.

Proof. The result is clear if M and N have exactly the same prime factors with (possibly) different exponents, since in that case

$$\frac{M}{\phi(M)} = \frac{M}{M \prod_{p|M}(1 - 1/p)} = \frac{1}{\prod_{p|M}(1 - 1/p)} = \frac{N}{\phi(N)}$$

and the right sum contains all the terms of the left sum and other terms as well. Hence it suffices to treat the case $N = Mp$ where p is a prime with $(M, p) = 1$. If $p > M$ the result is again clear. Thus we may suppose that $p < M$ and that l is chosen so that $p^l < M < p^{l+1}$. Let

$$\sum = \sum_{j \in S} \frac{1}{j} \quad \text{where} \quad S = \{j \mid 1 \leq j < M, \quad (j, pM) = 1\} .$$

Then

$$z(M) = \frac{M}{\phi(M)} \left[\sum_{\substack{(j,pM)=1 \\ 1 \leq j < M}} \frac{1}{j} + \frac{1}{p} \sum_{\substack{(j,pM)=1 \\ 1 \leq j < M/p}} \frac{1}{j} + \cdots + \frac{1}{p^l} \sum_{\substack{(j,pM)=1 \\ 1 \leq j < M/p^l}} \frac{1}{j} \right]$$

$$\leq \frac{M}{\phi(M)} \left(1 + \frac{1}{p} + \cdots + \frac{1}{p^l}\right) \sum$$

$$< \frac{p}{p-1} \frac{M}{\phi(M)} \sum ,$$

while

$$z(pM) = \frac{p}{p-1} \frac{M}{\phi(M)} \left[\sum + \sum_{\substack{(j,pM)=1 \\ M \leq j < pM}} \frac{1}{j} \right] ,$$

and the result is now clear.   **Q.E.D.**

We can now conclude the proof of Theorem 4.3. Indeed, let

$$\{N_1, \ldots, N_r\} = \{\text{den } \beta \mid \beta \in \mathcal{E}\}$$

and let $\mathcal{E}_i = \{\beta \in \mathcal{E} \mid \text{den } \beta = N_i\}$ for $i = 1, 2, \ldots, r$. The N_i are positive integers which divide N. Furthermore, if $q_i = \#\mathcal{E}_i$ then

$$\sum_{i=1}^{r} q_i \leq n^2 - n .$$

From Lemmas 4.4 and 4.5 it easily follows that

$$\limsup_{s \to \infty} \frac{1}{s} \sum_{\substack{v \notin S, v \mid p \\ p > s + h_0 + l_0}} \log T_v(\hat{G}_v, s + h_0 + l_0) \leq \sum_{i=1}^{r} q_i \left(z(N_i) - 1 \right)$$

$$\leq (n^2 - n) \left(z(N) - 1 \right) .$$

Q.E.D.

5.   Generic Global Inverse Radius Bounded by the Global Inverse Radius of a Local Solution Matrix.

Let Ω be a p-adic field and let $Y \in \mathcal{M}_n\left(\Omega((X))\right)$. We define the *radius of convergence* $R(Y)$ of Y to be the radius of convergence of $X^m Y$ if m is a positive integer such that $X^m Y \in \mathcal{M}_n(\Omega[[X]])$. We also recall that $r(Y) = \inf(1, R(Y))$.

Let $C \in \mathcal{M}_n(\Omega)$. We denote the radius of convergence of the system $\delta - C$ (or equivalently of the system $D - C/X$) by $R(C)$ (although this usage conflicts with the above definition of $R(Y)$, this will surely cause no problems). We recall that by definition (cf. Section III.5) $R(C)$ is the radius of convergence of $(X/t)^C$ at the generic point t. Clearly $R(C)$ equals the radius of convergence at $x = 1$ of X^C. Moreover we set $r(C) = \inf(1, R(C))$.

Let $G \in \mathcal{M}_n(\Omega(X))$. We denote by $\nu'(G)$ the number of finite non-apparent singularities of the system $D - G$, by $\nu(G)$ the number of finite non-trivial singularities of $D - G$, and by $\kappa(G)$ the total number of poles of G, that is

$$\kappa(G) = - \sum_{\zeta, \mathrm{ord}_\zeta(G) < 0} \mathrm{ord}_\zeta G .$$

Lemma 5.1. *Let $G \in \mathcal{M}_n(\Omega(X))$ and let $Y X^C$ be a matrix solution of the system $D - G$ at zero with $Y \in G\ell(n, \Omega((X)))$ and $C \in \mathcal{M}_n(\Omega)$. Let R denote the radius of convergence of the system $D - G$. Then*

$$R \geq r(C) r(Y)^{\kappa(G)+1} .$$

Moreover, if $D - G$ has only regular singularities we have

$$R \geq r(C) r(Y)^{\nu'(G)} .$$

Proof. Let a be a generic point with $|a| < r(Y) = \inf(1, R(Y))$. Then $\mathcal{U} = Y(X)(X/a)^C$ is a matrix solution (not necessarily normalized) of $D - G$ at a. Clearly we have

$$R_a(\mathcal{U}) \geq |a| r(C) .$$

The normalized matrix solution of $D - G$ at a is

$$\mathcal{U}_{G,a}(X) = \sum_{s=0}^{\infty} \frac{G_s(a)}{s!} (X - a)^s$$

and $\mathcal{U}_{G,a} = \mathcal{U}\mathcal{U}(a)^{-1}$ so that

$$R(\mathcal{U}_{G,a}) \geq |a| r(C) .$$

The G_s are defined by the recursion:

$$G_{s+1} = DG_s + G_s G , \qquad G_0 = I_n$$

Therefore, by Remark IV.3.2, we have

$$\left| \frac{G_s(a)}{s!} \right| \leq C_1 \frac{1}{(|a| r(C))^s} \{s, n - 1\} \tag{5.1}$$

where C_1 is a positive constant independent of s.

Let ζ be a pole of G and let $\mathrm{ord}_\zeta G = -\mu_\zeta$ with $\mu_\zeta \geq 1$. We prove by induction on s that

$$\mathrm{ord}_\zeta G_s \geq -\mu_\zeta s .$$

Indeed, if ζ is finite we have

$$
\begin{aligned}
\mathrm{ord}_\zeta G_{s+1} &\geq \inf\left(\mathrm{ord}_\zeta DG_s, \mathrm{ord}_\zeta G_s + \mathrm{ord}_\zeta G\right) \\
&\geq \inf(-\mu_\zeta s - 1, -\mu_\zeta s - \mu_\zeta) \\
&= -\mu_\zeta(s+1) .
\end{aligned}
$$

A similar calculation holds for $\zeta = \infty$. Thus we can conclude that if f is an element of G_s then the number of finite zeros (counting multiplicities) of f is at most $\kappa(G)s + O(1)$.

If $D - G$ has only regular singularities and its only apparent singularities are trivial, then as we have seen in Section 3 we can replace $\kappa(G)$ by $\nu(G) - 1$.

An application of equation (3.1) yields,

$$\left|\frac{G_s}{s!}\right|_0 (1) \leq C_2 \left|\frac{G_s}{s!}\right|_0 (|a|) \left(\frac{1}{|a|}\right)^{s\kappa(G)} , \qquad (5.2)$$

where C_2 is a positive constant. Since

$$\left|\frac{G_s}{s!}\right|_0 (1) = \left|\frac{G_s(t)}{s!}\right|$$

and $\{s, n-1\} \leq s^{n-1}$ from (5.1) and (5.2) it follows that

$$
\begin{aligned}
R = \liminf_{s \to \infty} \left|\frac{G_s(t)}{s!}\right|^{-1/s} \\
\geq \liminf_{s \to \infty} \{s, n-1\}^{-1/s} r(C) |a|^{\kappa(G)+1} \\
= r(C) |a|^{\kappa(G)+1} .
\end{aligned}
$$

Taking $|a| \to r(Y)$ we obtain the desired inequality.

If $D - G$ has only trivial apparent singularities we can replace $\kappa(G)$ by $\nu(G) - 1$ in the above calculation so that we obtain

$$R \geq r(C) r(Y)^{\nu(G)} .$$

There remains only the case in which $D - G$ has only regular singularities but not all the apparent singularities are trivial.

As we have seen in the proof of Proposition V.5.1 we can find a polynomial P such that $D - \tilde{G}$, where $\tilde{G} = G_{[PI_n]}$, has the property that its only apparent singularities are trivial singularities. Then $\tilde{Y}X^C$, where $\tilde{Y} = PY$, is a matrix solution of $D - \tilde{G}$. Let $\tilde{R}$ denote the radius of convergence of the system $D - \tilde{G}$. Then we have $\tilde{R} = R$ and $\nu(\tilde{G}) = \nu'(G)$ so that

$$R = \tilde{R} \geq r(C)r(\tilde{Y})^{\nu(\tilde{G})}$$
$$= r(C)r(PY)^{\nu(\tilde{G})}$$
$$\geq r(C)r(Y)^{\nu'(G)} ,$$

as desired.   **Q.E.D.**

Let K be an algebraic number field. For $C \in \mathcal{M}_n(K)$ and $v \in \mathcal{P}_f$ we define $R_v(C)$ to be the radius of convergence of the system $\delta - C$ with respect to the topology induced by v. We then have (cf. (2.1))

$$\rho(C/X) = \sum_{v \in \mathcal{P}_f} \log^+ \frac{1}{R_v(C)} . \tag{5.3}$$

We observe that $\rho(C/X) < \infty$ if and only if the eigenvalues of C lie in $\mathbf{Q}$. Indeed, if $\rho(C/X) < \infty$ certainly the exponents of $\delta - C$ are rational, i.e. the eigenvalues of C are in $\mathbf{Q}$. The converse is trivial: in fact if $\alpha \in \mathbf{Z}_p$ then $\delta - \alpha$ has $R_v = 1$ and if $\alpha \in \mathbf{Q}$ then $R_v = 1$ for almost all v.

We now show that the local Bombieri condition at one point implies the global Bombieri condition.

Theorem 5.2. *Let $G \in \mathcal{M}_n(K(X))$ and let YX^C be a matrix solution at zero for $D - G$ with $Y \in G\ell(n, K((X)))$ and $C \in \mathcal{M}_n(K)$. Assume moreover that $\rho_f(Y)$ and $\rho(C/X)$ are finite. Then $\rho(G) < \infty$ so that all singularities are regular with rational exponents. Furthermore*

$$\rho(G) \leq \nu'(G)\rho_f(Y) + \rho(C/X) .$$

Proof.   By Lemma 5.1 for each finite place v of K (and denoting by R_v the v–adic radius of convergence of $D - G$) we have

$$\log^+ \frac{1}{R_v} \leq (\kappa(G) + 1) \log^+ \frac{1}{R_v(Y)} + \log^+ \frac{1}{R_v(C)} .$$

Summing over all finite primes of K we obtain

$$\rho(G) \le (\kappa(G) + 1)\rho_f(Y) + \rho(C/X)$$

so that $\rho(G) < \infty$. Therefore, by Katz's Theorem (Theorem III.6.1), Lemma 5.1 allows us to make the above calculation replacing $\kappa(G)$ by $\nu'(G) - 1$.   **Q.E.D.**

Example 5.3.   Consider the system associated to the hypergeometric equation, that is take

$$G = \begin{pmatrix} 0 & 1 \\ \dfrac{ab}{X(1-X)} & -\dfrac{c - (a+b+1)X}{X(1-X)} \end{pmatrix} ,$$

with a, b, $c \in \mathbf{Q}$. A matrix solution at zero of $D - G$ is given by YX^C where

$$Y = \begin{pmatrix} u & v \\ u' & (1-c)X^{-1}v + v' \end{pmatrix} ,$$

$$C = \begin{pmatrix} 0 & 0 \\ 0 & 1-c \end{pmatrix} ,$$

and

$$u = F(a, b, c, X) = \sum_{s=0}^{\infty} \frac{(a)_s (b)_s}{(c)_s s!} X^s ,$$

$$v = F(a+1-c, b+1-c, 2-c, X) .$$

Let $a = 1/p$, $b = 2/p$, $c = 3/p$ where $p > 3$ is a rational prime. Then $F(a, b, c, X)$ and $F(a+1-c, b+1-c, 2-c, X)$ converge for $|x|_{p'} < 1$ if $p' \ne p$ and for $|x|_p < |p\pi_p|_p$; hence

$$\rho_f(Y) = \left(1 + \frac{1}{p-1}\right) \log p$$

and

$$\rho(C/X) = \left(1 + \frac{1}{p-1}\right) \log p .$$

Therefore

$$\rho(G) \le 3\frac{p}{p-1} \log p .$$

Lemma 5.4. *Let $G \in \mathcal{M}_n(K(X))$ with $\rho(G) < \infty$ and let $\xi \in K$. If we define*

$$\tilde{G}(X) = G(X + \xi)$$

then

$$\rho(\tilde{G}) \leq \rho(G) + (2\nu' - 1)h_f(\xi) .$$

Proof. Let $P \in K[X]$ be such that the apparent singularities of $G_{[PI_n]}$ are trivial (see Proposition V.5.1). Then $\rho(G_{[PI_n]}) = \rho(G)$. Furthermore

$$G_{[PI_n]} = G + \frac{P'}{P}I_n$$

and if we set

$$\tilde{P}(X) = P(X + \xi)$$

then

$$\tilde{G}_{[\tilde{P}I_n]} = \tilde{G} + \frac{\tilde{P}'}{\tilde{P}}I_n$$

so that

$$\tilde{G}_{[\tilde{P}I_n]}(X) = G_{[PI_n]}(X + \xi) = \widetilde{G_{[PI_n]}}(X) .$$

Thus, if we assume the result to be true for system whose only apparent singularities are trivial then we obtain the general case as well. Indeed,

$$\begin{aligned}
\rho(\tilde{G}) &= \rho(\widetilde{G_{[PI_n]}}) \\
&\leq \rho(G_{[PI_n]}) + (2\nu(G_{[PI_n]}) - 1)h_f(\xi) \\
&= \rho(G) + (2\nu'(G) - 1)h_f(\xi) .
\end{aligned}$$

Assume that $\delta - G$ has only trivial apparent singularities. If $|\xi|_v \leq 1$ then

$$\left|\tilde{G}_s\right|_{\text{gauss}} = |G_s|_{\text{gauss}} \tag{5.4}$$

whence $\tilde{R}_v = R_v$ (where $\tilde{R}_v$ and R_v are the v–adic radii of convergence of $D - \tilde{G}$ and $D - G$ respectively) and so we suppose that $|\xi|_v > 1$. We assert that

$$|\tilde{G}_s(t)|_v = |G_s(t + \xi)|_v \leq |G_s(t)|_v |\xi|_v^{(2\nu-1)s} \tag{5.5}$$

where $\nu = \nu(G)$.

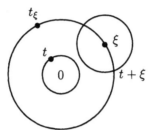

To see this let t_ξ be a generic point such that $|t_\xi| = |\xi|$ (we momentarily suppress the suffix v). Then

$$|G_s(t + \xi)| = |G_s|_\xi(1)$$

since $t + \xi$ is a generic point on the circle $|z - \xi| = 1$.

Then from (3.2) we obtain

$$|G_s|_\xi(1) \leq C_1 |G_s|_\xi(|\xi|) \prod_{\substack{\zeta \text{ non-trivial} \\ |\zeta - \xi| < |\xi|}} \left(\frac{|\xi|}{\sup(1, |\zeta - \xi|)} \right)^s ,$$

where C_1 is a constant independent of s. On the other hand

$$|G_s|_\xi(|\xi|) = |G_s(t_\xi)| = |G_s|_0(|\xi|) \leq C_2 |G_s|_0(1)|\xi|^{(\nu - 1)s} .$$

Therefore

$$|G_s(t + \xi)| \leq C_1 C_2 |G_s(t)| \left(|\xi|^{\nu - 1} \prod_{\substack{\zeta \text{ non-trivial} \\ |\zeta - \xi| < |\xi|}} \frac{|\xi|}{\sup(1, |\zeta - \xi|)} \right)^s$$

$$\leq C_1 C_2 |G_s(t)||\xi|^{(2\nu - 1)s} ,$$

as desired.

From (5.4) and (5.5) it now follows that

$$\tilde{R}_v \geq \frac{R_v}{\sup(1, |\xi|_v)^{2\nu - 1}}$$

so that

$$\log^+ \frac{1}{\tilde{R}_v} \leq \log^+ \frac{1}{R_v} + (2\nu - 1) \log^+ |\xi|_v .$$

To conclude it suffices to sum over all $v \in \mathcal{P}_f$.   **Q.E.D.**

Remark 5.5. Assume ξ to be at worst an apparent singularity of the system $D - G$ and let $Y_\xi \in \mathcal{M}_n\left(K((X - \xi))\right)$ be a solution matrix at ξ. If we put $\tilde{G}(X) = G(X + \xi)$ and $\tilde{Y}(X) = Y_\xi(X + \xi)$ then $\tilde{Y}$ is a matrix solution at zero of the system $D - \tilde{G}$ and $\rho(Y_\xi) = \rho(\tilde{Y})$. Therefore from the preceding result and Theorem 3.4 we have

$$\rho_\xi(Y_\xi) = \rho(\tilde{Y}) \leq \rho(\tilde{G}) + \sum_{\substack{\zeta \neq \xi, \infty \\ \zeta \, \text{non-trivial}}} h(\zeta - \xi)$$

$$\leq \rho(G) + (2\nu' - 1)h_f(\xi) + \sum_{\substack{\zeta \neq \xi, \infty \\ \zeta \, \text{non-trivial}}} h(\zeta - \xi) \, .$$

We now observe that for x, $y \in K$, we have

$$h(x + y) \leq \log 2 + h(x) + h(y) \, .$$

Indeed, if $v \in \mathcal{P}_f$, then

$$\sup(1, |x + y|_v) \leq \sup(1, |x|_v) \sup(1, |y|_v) \, ,$$

while, if $v \in \mathcal{P}_\infty$, then

$$\sup(1, |x + y|_v) \leq 2^{d_v/d} \sup(1, |x|_v) \sup(1, |y|_v) \, ;$$

finally recall that $\sum_{v \in \mathcal{P}_\infty} d_v = d$. Therefore we obtain

$$\rho_\xi(Y_\xi) - \rho(G) \leq (2\nu' - 1)h_f(\xi) + \nu h(\xi) + \nu \log 2 + \sum_{\substack{\zeta \neq \xi, \infty \\ \zeta \, \text{non-trivial}}} h(\zeta)$$

$$\leq (2\nu' + \nu - 1)h(\xi) + \nu \log 2 + \sum_{\substack{\zeta \neq \infty \\ \zeta \, \text{non-trivial}}} h(\zeta) \, .$$

Bombieri [1] found the following inequality:

$$\rho_\xi(Y_\xi) - \rho(G) \leq (\nu - 1)h(\xi) + \sum_{\substack{\zeta \, \text{non-trivial} \\ \zeta \neq \infty}} (h(\zeta) + h(\xi - \zeta))$$

which gives

$$\rho_\xi(Y_\xi) - \rho(G) \leq (2\nu - 1)h(\xi) + \nu \log 2 + 2 \sum_{\substack{\zeta \, \text{non-trivial} \\ \zeta \neq \infty}} h(\zeta) \, .$$

The following diagram summarizes the various implications regarding finiteness of $\rho(G)$, $\sigma(G)$, $\rho(Y)$ and $\sigma(Y)$.

$$\boxed{\sigma(G) < \infty} \; \overset{2.1}{\rightleftharpoons} \; \boxed{\rho(G) < \infty} \; \overset{5.2}{\underset{3.3}{\rightleftharpoons}} \; \boxed{\rho(Y) < \infty, \; \rho(C/X) < \infty}$$

$$\Downarrow{4.2}$$

$$\boxed{\sigma(Y) < \infty}$$

Here we assume that YX^C is a solution matrix of $D - G$ at $X = 0$.

It is interesting to observe that there seems at present to be no direct demonstration that $\sigma(G) < \infty$ implies that $\sigma(Y) < \infty$.

In the next chapter we show that the condition $\sigma(Y) < \infty$ together with $\rho(C/X) < \infty$ implies that $\sigma(G) < \infty$.

CHAPTER VIII

G–SERIES. THE THEOREM OF CHUDNOVSKY

1. Definition of G-series. Statement of Chudnovsky's Theorem.

Let K be a number field with $[K : \mathbf{Q}] = d$. Let

$$y = \sum_{s=0}^{\infty} A_s X^s \in K[[X]] .$$

We define

$$\sigma(y) = \limsup_{s \to \infty} \frac{1}{s} \sum_{v \in \mathcal{P}} h(s, v)$$

where

$$h(s, v) = \sup_{m \leq s} \log^+ |A_m|_v ,$$

and

$$\rho(y) = \sum_{v \in \mathcal{P}} \log^+ \frac{1}{R_v(y)}$$

where, for $v \in \mathcal{P}$, $R_v(y)$ denotes the radius of convergence of y as an element of $K_v[[X]]$. Similar definitions hold for $\sigma_\infty(y)$, $\sigma_f(y)$, $\rho_\infty(y)$ and $\rho_f(y)$. According to Lemma VII.4.1 we have

$$\rho(y) = \sum_{v \in \mathcal{P}} \limsup_{s \to \infty} \frac{1}{s} h(s, v)$$

and similar identities hold for $\rho_\infty(y)$ and $\rho_f(y)$. Moreover,

$$\rho(y) = \rho_\infty(y) + \rho_f(y)$$
$$\sigma(y) \leq \sigma_\infty(y) + \sigma_f(y)$$
$$\sigma_f(y) \leq \sigma(y) \tag{1.1}$$
$$\sigma_\infty(y) \leq \sigma(y)$$
$$\sigma_\infty(y) \leq \rho_\infty(y).$$

263

We observe that in view of our normalizations $\sigma(y)$ and $\rho(y)$ depend only on y and not on the field K such that $y \in K[[X]]$.

We say that $y \in K[[X]]$ is a *G-series* if
1) $\sigma(y) < \infty$,
2) $Ly = 0$ for some $L \in K(X)[d/dX]$, that is y is a solution of a linear differential equation with rational functions over K as coefficients.

Proposition 1.1. *Let $y \in K[[X]]$. Then $\sigma(y) < \infty$ if and only if the following two conditions hold:*

a) y has a non-zero radius of convergence for each imbedding of K into **C***;*

b) for each $s \geq 0$ there exists an $N_s \in$ **N** *such that $N_s A_j \in \mathcal{O}_K$ for all $j \leq s$ and there exists a positive constant C such that $N_s \leq C^s$ for all $s \geq 0$.*

Proof. Clearly y satisfies condition a) if and only if $\rho_\infty(y) < \infty$. We also assert that y satisfies condition b) if and only if $\sigma_f(y) < \infty$. To establish this, let $q_s = \text{den}(A_0, \ldots, A_s)$ for all $s \geq 0$ (cf. (VII.1.4)). Then from Lemma VII.1.1 it follows that

$$\sum_{v \in P_f} h(s, v) = \sum_{v \in P_f} \sup_{j \leq s} \log^+ |A_j|_v$$
$$= h_f(A_0, \ldots, A_s)$$
$$= \frac{1}{d} \log q_s .$$

Therefore $\sigma_f(y) < \infty$ if and only if there exists a positive constant C such that $q_s \leq C^s$ for every $s \geq 0$. Thus, if $\sigma_f(y) < \infty$ then y satisfies condition b) on taking $N_s = q_s$. Conversely, let $N_s \in$ **N** for $s = 0, 1, \ldots$, be a sequence satisfying the hypothesis of condition b). Then from the definition of q_s it follows easily that $q_s | N_s^d$ for all $s \geq 0$, so that $q_s \leq C^{sd}$ for all $s \geq 0$, and therefore $\sigma_f(y) < \infty$.

We now establish that $\sigma(y) < \infty$ if and only if $\rho_\infty(y) < \infty$ and $\sigma_f(y) < \infty$. Indeed, if $\sigma(y) < \infty$ then from (1.1) it follows that $\sigma_f(y) < \infty$; moreover, since for any fixed $v_0 \in \mathcal{P}$ we have

$$\limsup_{s \to \infty} \frac{1}{s} h(s, v_0) \leq \limsup_{s \to \infty} \frac{1}{s} \sum_{v \in \mathcal{P}} h(s, v) = \sigma(y) < \infty ,$$

we obtain that $\rho_\infty(y) < \infty$ (the infinite primes are finite in number). Conversely, if $\rho_\infty(y) < \infty$ and $\sigma_f(y) < \infty$, from (1.1) it follows immediately that $\sigma(y) < \infty$. **Q.E.D.**

Remark 1.2. This proposition establishes the equivalence of the definition of G-series given above with Siegel's original one, according to which a G-series is a Taylor series satisfying conditions a), b) and 2).

Lemma 1.3.

(i) If τ is an automorphism of $K/\mathbf{Q}$ then $\sigma(\tau y) = \sigma(y)$ where τy is defined by the natural action on the coefficients of y.

(ii) If $y_1, \ldots, y_m \in K[[X]]$ then

$$\sigma(y_1 + \cdots + y_m) \leq \sigma(y_1) + \cdots + \sigma(y_m).$$

(iii) If $y \in K[[X]]$ then for every non-zero $\alpha \in K$,

$$\sigma(\alpha y) = \sigma(y) \, .$$

Proof. Statement (i) is obvious since τ permutes the valuations of K.

Statement (ii) is also easy to verify, since if we set

$$y_i = \sum_{s=0}^{\infty} A_{i,s} X^s$$

then for $v \in \mathcal{P}_f$ we have

$$
\begin{aligned}
\sup_{j \leq s} \log^+ |A_{1,j} + \cdots + A_{m,j}|_v &\leq \sup_{j \leq s} \log^+ \sup_{1 \leq i \leq m} |A_{i,j}|_v \\
&= \sup_{1 \leq i \leq m} \sup_{j \leq s} \log^+ |A_{i,j}|_v \\
&\leq \sum_{i=1}^{m} \sup_{j \leq s} \log^+ |A_{i,j}|_v
\end{aligned}
$$

while for $v \in \mathcal{P}_\infty$ we certainly have

$$
\begin{aligned}
\sup_{j \leq s} \log^+ |A_{1,j} + \cdots + A_{m,j}|_v &\leq \sup_{j \leq s} \log^+ \left(m \sup_{1 \leq i \leq m} |A_{i,j}|_v \right) \\
&\leq \log m + \sup_{1 \leq i \leq m} \sup_{j \leq s} \log^+ |A_{i,j}|_v \\
&\leq \log m + \sum_{i=1}^{m} \sup_{j \leq s} \log^+ |A_{i,j}|_v \, ,
\end{aligned}
$$

so that on summing over $v \in \mathcal{P}$ and dividing by s we obtain

$$\frac{1}{s} \sum_v \sup_{j \le s} \log^+ |A_{1,j} + \cdots + A_{m,j}|_v \le \frac{1}{s}\text{const.} + \frac{1}{s} \sum_{i=1}^{m} \sum_v h_i(s, v)$$

where

$$h_i(s, v) = \sup_{j \le s} \log^+ |A_{i,j}|_v \ .$$

Taking the lim sup as $s \to \infty$ we obtain (ii).

As to (iii), for $v \in \mathcal{P}$ we have

$$\sup_{j \le s} \log^+ |\alpha A_j|_v \le \sup_{j \le s} \left(\log^+ |\alpha|_v + \log^+ |A_j|_v\right)$$

$$\le \log^+ |\alpha|_v + \sup_{j \le s} \log^+ |A_j|_v$$

so that

$$\frac{1}{s} \sum_v \sup_{j \le s} \log^+ |\alpha A_j|_v \le \frac{1}{s}h(\alpha) + \frac{1}{s} \sum_v h(s, v)$$

whence

$$\sigma(\alpha y) \le \sigma(y) \ .$$

Using α^{-1} in place of α and αy for y then shows that (iii) holds. **Q.E.D.**

We give an elementary criterion for $y \in K[[X]]$ to be a G-series.

Proposition 1.4.   Let $\{\omega_1, \ldots, \omega_d\}$ be a basis of $K/\mathbf{Q}$. Assume that $y_i \in \mathbf{Q}[[X]]$ $(1 \le i \le d)$ and let $y = y_1\omega_1 + \cdots + y_d\omega_d$. Then y is a G-series if and only if each y_i is a G-series.

Proof.   That condition 2) is satisfied for y if and only if it is satisfied for each y_i is an immediate consequence in one direction of ii) and in the other direction of iv) of Proposition 2.1 which follows. As to condition 1), we prove that $\sigma(y) < \infty$ if and only if $\sigma(y_i) < \infty$ for all i. In one direction this is an immediate consequence of ii) and iii) of the preceding Lemma 1.3. Conversely, let $\omega_1', \ldots, \omega_d'$ be the dual basis to the basis $\omega_1, \ldots, \omega_d$ of K over $\mathbf{Q}$, that is let $\omega_1', \ldots, \omega_d' \in K$ satisfy

$$\text{Tr}_{K/\mathbf{Q}}(\omega_i'\omega_j) = \delta_{ij} \ .$$

Then, with the obvious meaning of the symbols, we can write $y_i = \text{Tr}_{K/\mathbf{Q}}(\omega_i' y)$ and so by Lemma 1.3 we can estimate $\sigma(y_i)$ in terms of $\sigma(y)$. **Q.E.D.**

For $\vec{y} = \sum_{m=0}^{\infty} \vec{y}_m X^m \in K[[X]]^n$ we define

$$\tilde{h}(s, v) = \sup_{m \le s} \log^+ |\vec{y}_m|_v$$

and

$$\sigma(\vec{y}) = \limsup_{s \to \infty} \frac{1}{s} \sum_{v \in \mathcal{P}} \tilde{h}(s, v) .$$

The aim of this Chapter will be to prove the following theorem due to Chudnovsky:

Theorem 1.5. *Let K be a number field and let $G \in \mathcal{M}_n(K(X))$. Suppose that*

$$\vec{y} = \begin{pmatrix} y_0 \\ \vdots \\ y_{n-1} \end{pmatrix} \in K[[X]]^n$$

is a solution of $D\vec{y} = G\vec{y}$ satisfying

1) $y_0, \ldots, y_{n-1}$ *are linearly independent over $K(X)$,*

2) $\sigma(\vec{y}) < \infty$.

Then $D - G$ satisfies the Galočkin condition,

$$\sigma(G) < \infty .$$

The relevance of Chudnovsky's Theorem lies in the following facts. First, from local conditions on the system $D - G$ one obtains a global condition, that is the global nilpotence of $D - G$. Moreover, all solutions at non-singular points are G-series and if $Y X^A$ is a solution matrix at a singular point then each element of Y is a G-series.

2. Preparatory Results.

Consider the differential operator

$$L = A_0 D^n + A_1 D^{n-1} + \cdots + A_n \in K[[X]][D]$$

where $D = d/dX$; then there is a bound (depending only on L) for the order of the zero or pole of z at $X = 0$ for any solution $z \in K((X))$ of $L = 0$. This follows easily from the fact that this order is a zero of the indicial polynomial of L at zero. For this reason the following Proposition is useful.

Proposition 2.1. Let $(\mathcal{F}, D)$ be a differential extension of $(K(X), D)$.

(i) Let $G \in \mathcal{M}_n(K(X))$ where K is an arbitrary field of characteristic zero. Then, for $i = 1, \dots, n$, there exist operators $L_i \in K(X)[D] \setminus \{0\}$ depending only on G, such that $L_i z_i = 0$ for all $z_1, \dots, z_n \in \mathcal{F}$ satisfying

$$D \begin{pmatrix} z_1 \\ \vdots \\ z_n \end{pmatrix} = G \begin{pmatrix} z_1 \\ \vdots \\ z_n \end{pmatrix} .$$

(ii) Given non-zero Q_1, $Q_2 \in K(X)[D]$ there exist non-zero Q_3, $Q_4 \in K(X)[D]$ such that

$$Q_3(z_1 z_2) = 0 , \qquad Q_4(z_1 + z_2) = 0$$

for every z_1, $z_2 \in \mathcal{F}$ such that $Q_1 z_1 = Q_2 z_2 = 0$. Moreover, we can choose Q_3 and Q_4 such that $\operatorname{ord} Q_3 \le (\operatorname{ord} Q_1)(\operatorname{ord} Q_2) - 1$ and $\operatorname{ord} Q_4 \le \operatorname{ord} Q_1 + \operatorname{ord} Q_2$.

(iii) Given non-zero L, $P \in K(X)[D]$ there is a non-zero $\hat{L} \in K(X)[D]$ such that $\hat{L} P z = 0$ for all solutions z of $Lz = 0$.

(iv) Let C be a subfield of K with $[K : C] = n < \infty$. Let $\{\omega_1, \dots, \omega_n\}$ be a C-basis of K. Let $L \in K(X)[D] \setminus \{0\}$. Then there exist operators $L_1, \dots, L_n \in C(X)[D] \setminus \{0\}$ such that if $y \in K[[X]]$, $Ly = 0$ and if

$$y = \sum_{i=1}^n \omega_i y_i$$

where $y_i \in C[[X]]$ $(1 \le i \le n)$, then $L_i y_i = 0$ for each $i = 1, \dots, n$.

Proof. (i) By Theorem III.4.2 there exists $H \in G\ell(n, K(X))$ such that

$$G_{[H]} = \begin{pmatrix} 0 & 1 & 0 & \dots & 0 \\ 0 & 0 & 1 & \dots & 0 \\ \multicolumn{5}{c}{\dotfill} \\ 0 & 0 & & \dots & 1 \\ C_0 & C_1 & & \dots & C_{n-1} \end{pmatrix}$$

with $C_j \in K(X)$. Therefore

$$H \begin{pmatrix} z_1 \\ z_2 \\ \vdots \\ z_n \end{pmatrix} = \begin{pmatrix} z \\ z' \\ \vdots \\ z^{(n-1)} \end{pmatrix}$$

where $Lz = 0$ with $L = D^n - C_{n-1}D^{n-1} - \cdots - C_0$. Then z_i is the element of the i-th row of the matrix

$$H^{-1}\begin{pmatrix} z \\ \vdots \\ z^{(n-1)} \end{pmatrix}$$

so that $z_i \in K(X)z + \cdots + K(X)z^{(n-1)}$ and $z_i, z_i', \ldots, z_i^{(n)}$ are linearly dependent over $K(X)$.

(ii) Let u, v be formal solutions of $Q_1 = 0$, $Q_2 = 0$ respectively. Then

$$(uv)^{(s)} = \sum_{i+j=s} \binom{s}{i} u^{(i)}v^{(j)} \ .$$

If $i \geq n_1 = \operatorname{ord} Q_1$ (resp. $j \geq n_2 = \operatorname{ord} Q_2$) then we reduce $u^{(i)}$ by means of Q_1 (resp. $v^{(j)}$ by Q_2) so that we obtain that $1, uv, \ldots, (uv)^{(n_1n_2-1)}$ are linearly dependent over $K(X)$. Let m be minimal such that $1, uv, \ldots, (uv)^{(m)}$ are linearly dependent over $K(X)$; then there exists $Q_3(D) \in K(X)[D]$ with $\operatorname{ord} Q_3 \leq n_1n_2 - 1$ such that $Q_3(D)(uv) = 0$. Clearly, specializing the preceding identity we find $Q_3(z_1z_2) = 0$ for all z_1, z_2 as in the statement. The proof for $z_1 + z_2$ is similar.

(iii) From (i) it easily follows that for all $s \geq 1$ there exists non-zero $L_s \in K(X)[D]$ such that $L_s D^s z = 0$ for all z such that $Lz = 0$, and therefore, by (ii), for each $A(X) \in K(X)$ there exists $\tilde{L}_s \neq 0$ such that $\tilde{L}_s(A(X)D^s z) = 0$. Again by (ii), we find an $\hat{L} \neq 0$ such that $\hat{L}(P(D)z) = 0$.

(iv) We consider a similar situation involving systems. Let $\mu \geq 1$, $G \in \mathcal{M}_\mu(K(X))$; then

$$G = \sum_{i=1}^{n} G_i\omega_i \ ,$$

with $G_i \in \mathcal{M}_\mu(C(X))$. Let $\vec{z} \in (K[[X]])^\mu$, a μ-column vector. We may write

$$\vec{z} = \sum_{i=1}^{n} \vec{z}^{(i)}\omega_i \ ,$$

with $\vec{z}^{(i)} \in (C[[X]])^\mu$ for each i. Finally let

$$\omega_i\omega_j = \sum_{h=1}^{n} B_h^{(i,j)}\omega_h \ ,$$

$B_h^{(i,j)} \in C$. Then $D\vec{z} = G\vec{z}$ implies that

$$D\vec{z}^{(h)} = \sum_{i,j=1}^{n} B_h^{(i,j)} G^{(i)} \vec{z}^{(j)} \ . \quad h = 1, 2, \ldots, n \ .$$

This shows that the $\mu \times n$ matrix $Z = (\vec{z}^{(1)}, \ldots, \vec{z}^{(n)})$ is the solution vector of a linear system defined over $C(X)$. Let $Z_{\alpha,\beta}$ be a component of Z. By (i) there exists $L_{\alpha,\beta} \in C(X)[D]$, $L_{\alpha,\beta} \neq 0$ such that $L_{\alpha,\beta} Z_{\alpha,\beta} = 0$, i.e. $L_{\alpha,\beta}(\vec{z}^{(\beta)})_\alpha = 0$. To deduce assertion (iv), let

$$\vec{z} = \begin{pmatrix} y \\ y' \\ \vdots \\ y^{(\mu-1)} \end{pmatrix}$$

where μ is the order of L, let G be the obvious matrix associated with L and let $\beta = 1$ in the above discussion.   **Q.E.D.**

The most famous example is given by $C = \mathbf{Q}$, $K = \mathbf{Q}(i)$, $i^2 = -1$. Let $L = D - i$, $L_1 = L_2 = D^2 + 1$. Of course then $y = e^{iX} = \cos X + i \sin X$ satisfies $Ly = 0$ and $L_1 \cos X = 0 = L_2 \sin X$. But the converse is false: if $y_1, y_2 \in C[[X]]$, $L_1 y_1 = 0 = L_2 y_2$ and $y = y_1 + iy_2$ then we need not have $Ly = 0$.

Lemma 2.2 (Shidlovsky). *Let $\mathcal{F}$ be a field extension of $K(X)$, and let $V \subset \mathcal{F}$ be a finite dimensional K-vector space. Then there is a bound on the total degree of the elements of $K(X)$ which may be written in the form ξ/η with $\xi, \eta \in V$.*

Proof. Let $\tilde{V}$ be the $K(X)$-vector space generated by V and let

$$\{\eta_1, \ldots, \eta_l, \eta_{l+1}, \ldots, \eta_t\}$$

be a base of V over K such that $\{\eta_1, \ldots, \eta_l\}$ is a basis of $\tilde{V}$ over $K(X)$. Then we can write

$$\eta_{l+j} = \sum_{i=1}^{l} A_{l+j,i} \eta_i \quad j = 1, \ldots, t - l$$

with $A_{l+j,i} \in K(X)$. Let $h \in K(X)$ be of the form $h = \xi/\eta$ with $\xi, \eta \in V$ and $\eta \neq 0$. If $\eta = \sum_{s=1}^{t} v_s \eta_s$ and $\xi = \sum_{s=1}^{t} w_s \eta_s$ with $v_s, w_s \in K$, then we obtain successively

$$h\eta = h \sum_{s=1}^{t} v_s \eta_s = \sum_{s=1}^{t} w_s \eta_s = \xi$$

$$h \left(\sum_{i=1}^{l} v_i \eta_i + \sum_{j=1}^{t-l} v_{l+j} \sum_{i=1}^{l} A_{j+l,i} \eta_i \right) = \sum_{i=1}^{l} w_i \eta_i + \sum_{j=1}^{t-l} w_{l+j} \sum_{i=1}^{l} A_{j+l,i} \eta_i$$

$$h \cdot \sum_{i=1}^{l} \eta_i \left(v_i + \sum_{j=1}^{t-l} v_{l+j} A_{l+j,i} \right) = \sum_{i=1}^{l} \eta_i \left(w_i + \sum_{j=1}^{t-l} w_{l+j} A_{l+j,i} \right)$$

so that

$$h \left(v_i + \sum_{j=1}^{t-l} v_{l+j} A_{j+l,i} \right) = w_i + \sum_{j=1}^{t-l} w_{j+l} A_{j+l,i} .$$

Now if we had $v_i + \sum_{j=1}^{t-l} v_{l+j} A_{j+l,i} = 0$ for every $i = 1, \ldots, l$, then $\eta = 0$ which gives a contradiction. Hence $v_i + \sum_{j=1}^{t-l} v_{l+j} A_{j+l,i} \neq 0$ for some i. The total degree is then bounded since the rational functions $A_{j+l,i}$ do not depend on the choice of h. **Q.E.D.**

Let

$$\vec{P} = \begin{pmatrix} P_0 \\ \vdots \\ P_{n-1} \end{pmatrix} \in K[X]^n, \quad \vec{P} \neq 0 .$$

Let Y be a matrix solution of $D - G$ in a differential extension field of $K(X)$. Then for $m = 0, 1, \ldots$ we define the column vector

$$\vec{R}_m = Y \frac{1}{m!} D^m \left(Y^{-1} \vec{P} \right) . \tag{2.1}$$

Since

$$D - G = Y \circ D \circ Y^{-1}$$

we have

$$\vec{R}_m = \frac{1}{m!} Y \left(D^m \left(Y^{-1} \vec{P} \right) \right) = \frac{1}{m!} (D - G)^m \vec{P} ,$$

so that

$$\vec{R}_m \in K(X)^n .$$

Now for $s = 0, 1, \ldots$ define the $n \times n$ matrix $R^{(s)}$ with entries in $K(X)$ by

$$R^{(s)} = \left(\binom{s}{s} \vec{R}_s, \binom{s+1}{s} \vec{R}_{s+1}, \ldots, \binom{s+n-1}{s} \vec{R}_{s+n-1} \right) . \qquad (2.2)$$

Let $\nu(\vec{P})$ be the rank of $R^{(0)}$.

Proposition 2.3. *Let*

$$\vec{y} = \begin{pmatrix} y_0 \\ \vdots \\ y_{n-1} \end{pmatrix} \in K[[X]]^n$$

be a solution of $D - G$ with $G \in \mathcal{M}_n(K(X))$ and assume that $y_0, \ldots, y_{n-1}$ are linearly independent over $K(X)$. Then there exists a constant $C(G)$ which depends only on G such that if $\vec{P} \in K[X]^n$ satisfies

$$\mathrm{ord}_X \begin{vmatrix} P_i & P_j \\ y_i & y_j \end{vmatrix} > \deg P + C(G) \qquad \text{for all } i, j ,$$

then $\nu(\vec{P}) = n$.

Proof. Assume $\nu = \nu(\vec{P}) < n$. We have

$$Y^{-1} R^{(0)} =$$

$$\left(Y^{-1} \vec{P}, D(Y^{-1} \vec{P}), \ldots, D^{n-1}(Y^{-1} \vec{P}) \right) \begin{pmatrix} \frac{1}{0!} & & & 0 \\ & \frac{1}{1!} & & \\ & & \ddots & \\ 0 & & & \frac{1}{(n-1)!} \end{pmatrix} .$$

If we define

$$\begin{pmatrix} w_0 \\ \vdots \\ w_{n-1} \end{pmatrix} = Y^{-1} \vec{P}$$

then we can write

$$
Y^{-1}R^{(0)} = \begin{pmatrix} w_0 & w_0' & \cdots & w_0^{(n-1)} \\ \vdots & \vdots & \ddots & \vdots \\ w_{n-1} & w_{n-1}' & \cdots & w_{n-1}^{(n-1)} \end{pmatrix} \begin{pmatrix} \frac{1}{0!} & & & 0 \\ & \frac{1}{1!} & & \\ & & \ddots & \\ 0 & & & \frac{1}{(n-1)!} \end{pmatrix} .
$$

Therefore $\nu = \dim_C \sum_{i=0}^{n-1} C w_i$ where C is the field of constants (of the differential extension containing the elements of Y). Let $\tilde{w}_0, \ldots, \tilde{w}_{\nu-1}$ be a basis of the C-space $\langle w_0, \ldots, w_{n-1} \rangle$. Then there exists $M \in G\ell(n, C)$ such that

$$
M \begin{pmatrix} w_0 \\ w_1 \\ \vdots \\ w_{n-1} \end{pmatrix} = \begin{pmatrix} \tilde{w}_0 \\ \vdots \\ \tilde{w}_{\nu-1} \\ 0 \\ \vdots \\ 0 \end{pmatrix} .
$$

Hence

$$
MY^{-1}R^{(0)} = \begin{pmatrix} A \\ 0 \end{pmatrix}
$$

where A is a $\nu \times n$ matrix (with elements in the above mentioned differential extension of $K(X)$) of rank ν. Since YM^{-1} is still a matrix solution of the system $D - G$ we can replace Y by YM^{-1} so that we may assume from the beginning that

$$
Y^{-1}R^{(0)} = \begin{pmatrix} A \\ 0 \end{pmatrix} .
$$

Since $R^{(0)}$ has rank ν there exist a $\nu \times \nu$ submatrix R_{IJ} of $R^{(0)}$ with $\det R_{IJ} \neq 0$ where $I, J \subset \{0, \ldots, n-1\}$ and $\#I = \#J = \nu$. If we set $I' = \{0, \ldots, n-1\} \backslash I$ and $J' = \{0, \ldots, n-1\} \backslash J$ then we can write

$$
R^{(0)} = \begin{pmatrix} R_{IJ} & R_{IJ'} \\ R_{I'J} & R_{I'J'} \end{pmatrix} .
$$

(In fact we could assume that $J = \{0, \ldots, \nu-1\}$. Indeed, suppose that for some $m \geq 1$

$$
\vec{R}_m = \sum_{j=0}^{m-1} \lambda_j \vec{R}_j
$$

where $\lambda_i \in K(X)$ for $i = 0, \ldots, m - 1$. Then

$$Y^{-1}\vec{R}_m = \sum_{j=0}^{m-1} \lambda_j Y^{-1}\vec{R}_j ,$$

$$D(Y^{-1}\vec{R}_m) = D\left(\sum_{j=0}^{m-1} \lambda_j Y^{-1}\vec{R}_j\right)$$

$$= \sum_{j=0}^{m-1} \left(\lambda_j' Y^{-1}\vec{R}_j + \lambda_j D(Y^{-1}\vec{R}_j)\right)$$

$$= \sum_{j=0}^{m-1} (\lambda_j' Y^{-1}\vec{R}_j + (j+1)\lambda_j Y^{-1}\vec{R}_{j+1}) .$$

On the other hand

$$D(Y^{-1}\vec{R}_m) = (m + 1)Y^{-1}\vec{R}_{m+1}$$

so that

$$(m + 1)\vec{R}_{m+1} = \sum_{j=0}^{m-1} \lambda_j' \vec{R}_j + \sum_{j=0}^{m-1} (j+1)\lambda_j \vec{R}_{j+1} \in \sum_{j=0}^{m-1} K(X)\vec{R}_j .$$

From this by an easy induction argument we can see that $\vec{R}_0, \ldots, \vec{R}_{\nu-1}$ are linearly independent. However, this will play no role in the proof.)

Similarly we write

$$Y^{-1} = \begin{pmatrix} (Y^{-1})_{SI} & (Y^{-1})_{SI'} \\ (Y^{-1})_{S'I} & (Y^{-1})_{S'I'} \end{pmatrix}$$

where $S = \{0, 1, \ldots, \nu - 1\}$, $S' = \{\nu, \ldots, n - 1\}$. Therefore

$$\begin{pmatrix} (Y^{-1})_{SI} & (Y^{-1})_{SI'} \\ (Y^{-1})_{S'I} & (Y^{-1})_{S'I'} \end{pmatrix} \begin{pmatrix} R_{IJ} & R_{IJ'} \\ R_{I'J} & R_{I'J'} \end{pmatrix} = \begin{pmatrix} * & * \\ 0 & 0 \end{pmatrix}$$

and in particular

$$(Y^{-1})_{S'I}R_{IJ} + (Y^{-1})_{S'I'}R_{I'J} = 0$$

so that

$$(Y^{-1})_{S'I} + (Y^{-1})_{S'I'}B = 0$$

where

$$B = R_{I'J} R_{IJ}^{-1} \in \mathcal{M}_{n-\nu,\nu}(K(X)) \ .$$

Therefore

$$((Y^{-1})_{S'I}, (Y^{-1})_{S'I'}) = (Y^{-1})_{S'I'}(-B, I_{n-\nu}) \ .$$

Since the rank of $((Y^{-1})_{S'I}, (Y^{-1})_{S'I'})$ is $n-\nu$ we obtain that $(Y^{-1})_{S'I'}$ has rank $n - \nu$, and so is invertible. Hence

$$-B = (Y^{-1})_{S'I'}^{-1}(Y^{-1})_{S'I} \ .$$

Let V be the vector space of all polynomials in $K[\ldots, (Y^{-1})_{ij}, \ldots]$ of degree $\leq n - \nu$ (this space does not change if we replace Y^{-1} by MY^{-1} with $M \in G\ell(n, C)$). The entries of B are elements of $K(X)$ which are quotients ξ/η with ξ, $\eta \in V$. Hence, by Shidlovsky's Lemma (Lemma 2.2), the elements of B have total degree bounded by a constant which depends only on G.

By permuting the rows of $R^{(0)}$ we can assume the set of indices to be $I = \{0, \ldots, \nu - 1\}$. This amounts to multiplying $R^{(0)}$ on the left by a permutation matrix N, and this can be done by replacing the matrix G by the matrix $N^{-1}GN$. Now define two matrices Q_1 and Q_2 with elements in $K[[X]]$ by setting

$$Q_1 = \begin{pmatrix} y_{n-1} & 0 & 0 & \cdots & 0 \\ y_1 & -y_0 & 0 & \cdots & 0 \\ y_2 & 0 & -y_0 & \cdots & 0 \\ \vdots & \vdots & \vdots & \ddots & \vdots \\ y_{\nu-1} & 0 & 0 & \cdots & -y_0 \end{pmatrix} \in \mathcal{M}_{\nu,\nu}(K[[X]]) \ ,$$

and

$$Q_2 = \begin{pmatrix} 0 & \cdots & 0 & -y_0 \\ 0 & \cdots & 0 & 0 \\ \multicolumn{4}{c}{\cdots\cdots\cdots\cdots\cdots} \\ 0 & \cdots & 0 & 0 \end{pmatrix} \in \mathcal{M}_{\nu,n-\nu}(K[[X]]) \ ,$$

and let

$$T = (Q_1, Q_2)\begin{pmatrix} R_{IJ} \\ R_{I'J} \end{pmatrix}$$

$$= (Q_1, Q_2)\begin{pmatrix} I_\nu \\ B \end{pmatrix} R_{IJ} \ .$$

Then

$$TR_{IJ}^{-1} = (Q_1, Q_2) \begin{pmatrix} I_\nu \\ B \end{pmatrix} = Q_1 + Q_2 B$$

$$= \begin{pmatrix} y_{n-1} & 0 & 0 & \cdots & 0 \\ y_1 & -y_0 & 0 & \cdots & 0 \\ y_2 & 0 & -y_0 & \cdots & 0 \\ \vdots & \vdots & \vdots & \ddots & \vdots \\ y_{\nu-1} & 0 & 0 & \cdots & -y_0 \end{pmatrix}$$

$$+ \begin{pmatrix} -b_0 y_0 & -b_1 y_0 & \cdots & -b_{\nu-1} y_0 \\ 0 & 0 & \cdots & 0 \\ 0 & 0 & \cdots & 0 \\ \vdots & \vdots & \ddots & \vdots \\ 0 & 0 & \cdots & 0 \end{pmatrix}$$

$$= \begin{pmatrix} y_{n-1} - b_0 y_0 & -b_1 y_0 & -b_2 y_0 & \cdots & -b_{\nu-1} y_0 \\ y_1 & -y_0 & 0 & \cdots & 0 \\ y_2 & 0 & -y_0 & \cdots & 0 \\ \vdots & \vdots & \vdots & \ddots & \vdots \\ y_{\nu-1} & 0 & 0 & \cdots & -y_0 \end{pmatrix}$$

where $b_0, \ldots, b_{\nu-1}$ are elements of B. Applying elementary row operations which do not change the determinant we obtain an equivalent matrix of the form

$$\begin{pmatrix} y_{n-1} - b_0 y_0 - \cdots - b_{\nu-1} y_{\nu-1} & 0 & 0 & \cdots & 0 \\ y_1 & -y_0 & 0 & \cdots & 0 \\ y_2 & 0 & -y_0 & \cdots & 0 \\ \vdots & \vdots & \vdots & \ddots & \vdots \\ y_{\nu-1} & 0 & 0 & \cdots & -y_0 \end{pmatrix}.$$

Therefore

$$\det(TR_{IJ}^{-1}) = (-y_0)^{\nu-1}(y_{n-1} - b_0 y_0 - \cdots - b_{\nu-1} y_{\nu-1})$$

which is different from zero since $\nu < n$ and the y_i are linearly independent over $K(X)$. By Proposition 2.1 it follows that $\mathrm{ord}_X y_0$ is bounded by a constant which depends only on G. If we write $b_i = c_i/c_{n-1}$ for $i = 0, \ldots, \nu - 1$ with $c_i, c_{n-1} \in K[X]$ then

$$y_{n-1} - b_0 y_0 - \cdots - b_{\nu-1} y_{\nu-1} = \frac{1}{c_{n-1}}(c_{n-1} y_{n-1} - c_0 y_0 - \cdots - c_{\nu-1} y_{\nu-1}) ;$$

therefore, since the elements of B have bounded total degree by Shidlovsky's Lemma, we have $D^l c_j = 0$ for sufficiently large l, and we find, again by Proposition 2.1, an upper bound for $\mathrm{ord}_X \det(T R_{IJ}^{-1})$, namely there exists a constant $C_1(G)$ depending only on G such that

$$\mathrm{ord}_X \det(T R_{IJ}^{-1}) \leq C_1(G) .$$

We now wish to find a lower bound for $\mathrm{ord}_X \det(T R_{IJ}^{-1}) = \mathrm{ord}_X \det T - \mathrm{ord}_X \det R_{IJ}$. To this end we now find a lower bound for $\mathrm{ord}_X \det T$ and an upper bound for $\mathrm{ord}_X \det R_{IJ}$.

From the definition

$$T = \begin{pmatrix} y_{n-1} & 0 & \cdots & 0 & \cdots & -y_0 \\ y_1 & -y_0 & \cdots & 0 & \cdots & 0 \\ \vdots & \vdots & \ddots & \vdots & \ddots & \vdots \\ y_{\nu-1} & 0 & \cdots & -y_0 & \cdots & 0 \end{pmatrix} \begin{pmatrix} R_{IJ} \\ \\ R_{I'J} \end{pmatrix}$$

we find that the elements in the first row of T are of the form

$$\begin{vmatrix} y_{n-1} & R_{n-1,l} \\ y_0 & R_{0l} \end{vmatrix} , \qquad l \in J$$

while the elements of the i-th row are of the form

$$\begin{vmatrix} y_i & R_{il} \\ y_0 & R_{0l} \end{vmatrix} , \qquad l \in J ;$$

here $R_{ij} = (\vec{R}_j)_i$. We now compute

$$\begin{vmatrix} y_i & R_{il} \\ y_j & R_{jl} \end{vmatrix}$$

in terms of

$$\begin{vmatrix} y_i & R_{i0} \\ y_j & R_{j0} \end{vmatrix} = \begin{vmatrix} y_i & P_i \\ y_j & P_j \end{vmatrix} .$$

We have

$$D\begin{vmatrix} y_i & R_{is} \\ y_j & R_{js} \end{vmatrix} = \begin{vmatrix} (G\vec{y})_i & R_{is} \\ (G\vec{y})_j & R_{js} \end{vmatrix} + \begin{vmatrix} y_i & DR_{is} \\ y_j & DR_{js} \end{vmatrix} .$$

However, since by definition

$$\vec{R}_s = \frac{1}{s!}(D - G)^s \vec{P} ,$$

we have

$$(s+1)\vec{R}_{s+1} = (D - G)\vec{R}_s$$

and so

$$D\vec{R}_s = G\vec{R}_s + (s+1)\vec{R}_{s+1} .$$

Thus

$$D\begin{vmatrix} y_i & R_{is} \\ y_j & R_{js} \end{vmatrix} = \begin{vmatrix} (G\vec{y})_i & R_{is} \\ (G\vec{y})_j & R_{js} \end{vmatrix} + \begin{vmatrix} y_i & (G\vec{R}_s)_i \\ y_j & (G\vec{R}_s)_j \end{vmatrix} + (s+1)\begin{vmatrix} y_i & R_{i,s+1} \\ y_j & R_{j,s+1} \end{vmatrix}$$

and so

$$(s+1)\begin{vmatrix} y_i & R_{i,s+1} \\ y_j & R_{j,s+1} \end{vmatrix} - D\begin{vmatrix} y_i & R_{is} \\ y_j & R_{js} \end{vmatrix}$$

$$= -\begin{vmatrix} \sum_l G_{il}y_l & R_{is} \\ \sum_l G_{jl}y_l & R_{js} \end{vmatrix} - \begin{vmatrix} y_i & \sum_l G_{il}R_{ls} \\ y_j & \sum_l G_{jl}R_{ls} \end{vmatrix}$$

$$= -\sum_l G_{il}\begin{vmatrix} y_l & R_{ls} \\ y_j & R_{js} \end{vmatrix} - \sum_l G_{jl}\begin{vmatrix} y_i & R_{is} \\ y_l & R_{ls} \end{vmatrix} ,$$

as follows from the identity

$$\begin{vmatrix} \sum_l G_{il}y_l & z_i \\ \sum_l G_{jl}y_l & z_j \end{vmatrix} + \begin{vmatrix} y_i & \sum_l G_{il}z_l \\ y_j & \sum_l G_{jl}z_l \end{vmatrix} = \sum_l G_{il}\begin{vmatrix} y_l & z_l \\ y_j & z_j \end{vmatrix} + \sum_l G_{jl}\begin{vmatrix} y_i & z_i \\ y_l & z_l \end{vmatrix} .$$

Therefore

$$(s+1)\begin{vmatrix} y_i & R_{i,s+1} \\ y_j & R_{j,s+1} \end{vmatrix} = D\begin{vmatrix} y_i & R_{is} \\ y_j & R_{js} \end{vmatrix}$$

$$- \sum_l G_{il}\begin{vmatrix} y_l & R_{ls} \\ y_j & R_{js} \end{vmatrix} - \sum_l G_{jl}\begin{vmatrix} y_i & R_{is} \\ y_l & R_{ls} \end{vmatrix} .$$

We can conclude that

$$\inf_{i,j} \mathrm{ord}_X \begin{vmatrix} y_i & R_{i,s+1} \\ y_j & R_{j,s+1} \end{vmatrix} \geq \inf(-1, \mathrm{ord}_X G) + \inf_{i,j} \mathrm{ord}_X \begin{vmatrix} y_i & R_{is} \\ y_j & R_{js} \end{vmatrix}$$

and so

$$\inf_{i,j} \operatorname{ord}_X \begin{vmatrix} y_i & R_{is} \\ y_j & R_{js} \end{vmatrix} \geq s \inf(-1, \operatorname{ord}_X G) + \inf_{i,j} \operatorname{ord}_X \begin{vmatrix} y_i & P_i \\ y_j & P_j \end{vmatrix}.$$

Since $\det T$ is a sum of products of ν elements of the type

$$\begin{vmatrix} y_i & R_{is} \\ y_j & R_{js} \end{vmatrix}$$

and $s \leq n - 1$ we obtain

$$\operatorname{ord}_X \det T \geq \nu \left[\inf_{i,j} \operatorname{ord}_X \begin{vmatrix} y_i & P_i \\ y_j & P_j \end{vmatrix} + (n-1)\inf(-1, \operatorname{ord}_X G) \right].$$

To find an upper bound for $\operatorname{ord}_X \det R_{IJ}$, we find an upper bound for the degree of the numerator of the rational function $\det R_{IJ}$. We write

$$G = \frac{\mathcal{G}}{Q}$$

where $\mathcal{G} \in \mathcal{M}_n(K[X])$ and $Q \in K[X]$. We have $\vec{R}_0 = \vec{P}$ and we prove by induction that

$$\vec{R}_s = \frac{\vec{H}_s}{Q^s} \tag{2.3}$$

where $\vec{H}_s \in K[X]^n$. Indeed, if we assume this last equality to be true we obtain

$$(s+1)\vec{R}_{s+1} = (D - G)\vec{R}_s = \frac{Q\vec{H}_s' - sQ'\vec{H}_s - \mathcal{G}\vec{H}_s}{Q^{s+1}}.$$

Therefore

$$\deg \vec{H}_{s+1} \leq \deg \vec{H}_s + h$$

where $h = \sup(\deg \mathcal{G}, \deg Q - 1)$, and so

$$\deg \vec{H}_s \leq \deg \vec{P} + sh. \tag{2.4}$$

We conclude that

$$\deg(\text{Numerator } \det R_{IJ}) \leq \nu \deg \vec{P}$$
$$+ h((n-1) + (n-2) + \cdots + (n-\nu)).$$

Thus we have proved that there exist two constants $C_1(G)$ and $C_2(G)$ which depend only on G such that

$$C_1(G) \geq \operatorname{ord}_X \det(TR_{IJ}^{-1}) \geq \nu \left(\inf_{i,j} \operatorname{ord}_X \begin{vmatrix} y_i & P_i \\ y_j & P_j \end{vmatrix} - \deg \vec{P} \right) + C_2(G) .$$

From this the result follows immediately.   **Q.E.D.**

Remark 2.4. Let $G \in \mathcal{M}_n(K(X))$. We say that $D - G$ is *reducible* over $K(X)$ if there exists $H \in G\ell(n, K(X))$ such that

$$G_{[H]} = \begin{pmatrix} \tilde{G}_{II} & \tilde{G}_{IJ} \\ 0 & \tilde{G}_{JJ} \end{pmatrix}$$

where $I = \{1, 2, \dots, l\}$, $J = \{l+1, \dots, n\}$ and $0 < l < n$. Since

$$- {}^t(G_{[H]}) = (D \, {}^tH^{-1}) \, {}^tH + {}^tH^{-1}(-{}^tG) \, {}^tH$$
$$= (-{}^tG)_{[{}^tH^{-1}]}$$

clearly we have that $D - G$ is irreducible if and only if $D + {}^tG$ is irreducible.

We now show that if $D - G$ is irreducible then $\det R^{(0)} \neq 0$ for arbitrary choice of $\vec{P} \neq 0$, so that in this case we have no need for Proposition 2.3.

Let $0 \neq \vec{P} \in K(X)^n$. We recall that

$$\vec{R}_m = \frac{1}{m!} Y D^m (Y^{-1} \vec{P}) \in K(X)^n$$

where $Y \in G\ell(n, \mathcal{F})$ is a solution matrix of $D - G$ in a differential field extension $\mathcal{F}$ of $K(X)$.

Lemma 2.5. *With given* $0 \neq \vec{P} \in K(X)^n$, *the matrix* $(\vec{R}_0, \dots, \vec{R}_{n-1})$ *has rank* n *unless there exists* $0 \neq \vec{z} \in \mathcal{F}^n$ *such that*

$$D\vec{z} = -{}^tG\vec{z}$$

and

$${}^t\vec{z} \cdot \vec{P} = 0 .$$

Proof.   If the vectors $\vec{R}_0, \ldots, \vec{R}_{n-1}$ are linearly dependent over $K(X)$ then, *a fortiori*, they are linearly dependent over $\mathcal{F}$, and hence the vectors $Y^{-1}\vec{R}_0, \ldots, Y^{-1}\vec{R}_{n-1}$ are linearly dependent over $\mathcal{F}$. Therefore, the matrix

$$W = \left(Y^{-1}\vec{P}, D(Y^{-1}\vec{P}), \ldots, D^{n-1}(Y^{-1}\vec{P}) \right)$$

has rank $< n$. The matrix W is the Wronskian matrix of

$$\begin{pmatrix} w_1 \\ \vdots \\ w_n \end{pmatrix} = Y^{-1}\vec{P}$$

and has non-zero determinant if and only if $w_1, \ldots, w_n$ are linearly independent over $C_{\mathcal{F}}$, the field of constants of $\mathcal{F}$ (cf. Lemma III.1.1). Now if $w_1, \ldots, w_n$ are linearly dependent, then there exist $\lambda_i \in C_{\mathcal{F}}$, not all zero, such that $\sum_i \lambda_i w_i = 0$. Let $\vec{z}_i$ be the i-th row of Y^{-1} and let $\vec{z} = \sum_i \lambda_i \vec{z}_i$ (note that $\vec{z}_i$ and $\vec{z}$ are row vectors). Then

$$\vec{z} \cdot \vec{P} = 0$$

and

$$D^t\vec{z} = -{}^tG\,{}^t\vec{z} .$$

Q.E.D.

Lemma 2.6.   *The system $D - G$ is reducible if and only if there exists a solution*

$$\vec{y} = \begin{pmatrix} y_1 \\ \vdots \\ y_n \end{pmatrix} \neq 0$$

of $D - G$ with $y_1, \ldots, y_n$ linearly dependent over $K(X)$.

Proof.   Let $l = \dim_{K(X)}\langle y_1, \ldots, y_n \rangle$. Permuting the indices if necessary we may assume that $y_1, \ldots, y_l$ are linearly independent over $K(X)$. Let

$$\vec{y}_I = \begin{pmatrix} y_1 \\ \vdots \\ y_l \end{pmatrix}, \qquad \vec{y}_J = \begin{pmatrix} y_{l+1} \\ \vdots \\ y_n \end{pmatrix},$$

and let $M \in \mathcal{M}_{n-l,l}(K(X))$ be such that

$$\vec{y}_J = M\vec{y}_I .$$

If we set

$$H = \begin{pmatrix} I_l & 0 \\ -M & I_{n-l} \end{pmatrix}$$

then

$$H\begin{pmatrix} \vec{y}_I \\ \vec{y}_J \end{pmatrix} = \begin{pmatrix} \vec{y}_I \\ 0 \end{pmatrix}$$

so that $\begin{pmatrix} \vec{y}_I \\ 0 \end{pmatrix}$ is a solution of $D - G_{[H]}$. If we write

$$G_{[H]} = \begin{pmatrix} \tilde{G}_{II} & \tilde{G}_{IJ} \\ \tilde{G}_{JI} & \tilde{G}_{JJ} \end{pmatrix}$$

then it is easily verified that $\tilde{G}_{JI} = 0$, so that the system is reducible.
   Conversely, assume that there exists $H \in Gl(n, K(X))$ such that

$$G_{[H]} = \begin{pmatrix} \tilde{G}_{II} & \tilde{G}_{IJ} \\ 0 & \tilde{G}_{JJ} \end{pmatrix}$$

with $I = \{1, 2, \dots, l\}$ and $J = \{l + 1, \dots, n\}$. Let $\vec{y}_I \neq 0$ be a solution of the equation $D - G_{II}$. Then $\begin{pmatrix} \vec{y}_I \\ 0 \end{pmatrix}$ is a solution of $D - G_{[H]}$ so that there exists a solution $\vec{y}$ of $D - G$ such that

$$\begin{pmatrix} \vec{y}_I \\ 0 \end{pmatrix} = H\vec{y} .$$

If $\vec{h}$ denotes the last row of H then we have $\vec{h} \cdot \vec{y} = 0$, as desired.   **Q.E.D.**

   The matrices $G_s \in \mathcal{M}_n(K(X))$ are here defined by

$$D^s Y = s! G_s Y$$

where Y is a matrix solution of the system $D - G$. We recall the definition of $R^{(s)}$ given in (2.2). We now show that

$$G_s R^{(0)} = \sum_{i=0}^{s} \frac{(-1)^i}{(s-i)!} D^{s-i} R^{(i)} . \tag{2.5}$$

Proof. We show first that the following identity holds in the ring $\mathcal{M}_n\left(K(X)[D]\right)$

$$G_s = \sum_{j=0}^{s} \binom{s}{j} \frac{(-1)^j}{s!} D^{s-j} \circ (D - G)^j \ . \tag{2.6}$$

From

$$D - G = Y \circ D \circ Y^{-1}$$

we obtain

$$(D - G)^s = Y \circ D^s \circ Y^{-1}$$

and so the second member of (2.6) becomes

$$\sum_{j=0}^{s} \binom{s}{j} \frac{(-1)^j}{s!} D^{s-j} \circ Y \circ D^j \circ Y^{-1}$$

which, since by Leibniz's rule we have

$$D^{s-j} \circ Y = \sum_{i=0}^{s-j} \binom{s-j}{i} Y^{(i)} \circ D^{s-j-i} \ ,$$

equals

$$\sum_{j=0}^{s} \binom{s}{j} \frac{(-1)^j}{s!} \sum_{i=0}^{s-j} \binom{s-j}{i} Y^{(i)} \circ D^{s-i} \circ Y^{-1}$$

$$= \sum_{i=0}^{s} \left(\sum_{j=0}^{s-i} \frac{(-1)^j}{s!} \binom{s}{j} \binom{s-j}{i} \right) Y^{(i)} \circ D^{s-i} \circ Y^{-1} \ .$$

Now

$$\frac{(-1)^j}{s!} \binom{s}{j} \binom{s-j}{i} = \frac{(-1)^j}{j!(s-j)!} \frac{(s-j)!}{i!(s-j-i)!}$$

$$= \frac{(-1)^j}{i!j!(s-j-i)!}$$

$$= \frac{(-1)^j}{i!(s-i)!} \binom{s-i}{j} \ .$$

But

$$\sum_{j=0}^{s-i} \binom{s-i}{j}(-1)^j = \begin{cases} 0 & \text{if } s-i \geq 1 \\ 1 & \text{if } s-i = 0 \end{cases}$$

so that the right hand side of (2.6) equals $Y^{(s)} \circ Y^{-1} = G_s$, as desired.
Our second step is to show that

$$R^{(s)} = \frac{1}{s!}(D-G)^s R^{(0)} , \qquad (2.7)$$

that is

$$\binom{s+j}{j} \vec{R}_{s+j} = \frac{1}{s!}(D-G)^s \vec{R}_j$$

for $j = 0, 1, \ldots, n-1$. But this is a trivial consequence of $\vec{R}_m = \frac{1}{m!}(D-G)^m \vec{P}$. Putting (2.6) and (2.7) together gives

$$G_s R^{(0)} = \sum_{j=0}^{s} \binom{s}{j} \frac{(-1)^j}{s!} D^{s-j} \circ (D-G)^j R^{(0)}$$

$$= \sum_{j=0}^{s} \binom{s}{j} \frac{(-1)^j}{s!} D^{s-j} j! R^{(j)}$$

$$= \sum_{j=0}^{s} \frac{(-1)^j}{(s-j)!} D^{s-j} R^{(j)} ,$$

as desired. **Q.E.D.**

3. Siegel's Lemma.

We define the (*multiplicative*) *height* ht $\vec{x}$ of a vector $\vec{x} = (x_1, \ldots, x_s) \in K^s$ by

$$\text{ht } \vec{x} = \prod_{v \in \mathcal{P}} \sup(1, |x_1|_v, \ldots, |x_s|_v) . \qquad (3.1)$$

Similarly, if $0 \neq q = \sum_{j=0}^{N} q_j X^j \in K[X]$ with $\deg q \leq N$, we define

$$\text{ht } q = \text{ht } (q_0, \ldots, q_N) = \prod_{v \in \mathcal{P}} \sup(1, |q_0|_v, \ldots, |q_N|_v) \qquad (3.2)$$

and

$$h(q) = h(q_0, \ldots, q_N) = \log \text{ht } q = \sum_{v \in \mathcal{P}} \sup_{0 \leq j \leq N} \log^+ |q_j|_v . \qquad (3.3)$$

Lemma 3.1 (Siegel's Lemma). *Let K be a number field. There exist positive constants c and γ depending only on K having the property* *that for any positive integers M and N with $M < N$, and any $A \in \mathcal{M}_{M,N}(K)$, there exists a vector $\vec{x} \in K^N$, $\vec{x} \neq 0$ such that*

$$A\vec{x} = 0$$

and such that

$$\operatorname{ht} \vec{x} \leq c\gamma^{M/(N-M)} \left(N \operatorname{ht} A\right)^{M/(N-M)} .$$

Proof. We recall the conventional normalizations for the valuations of K: for $z \in K$ we take

$$\|z\|_v = |z|_{\mathbf{R}} \quad \text{if } v \text{ is real,}$$
$$\|z\|_v = |z|_{\mathbf{C}}^2 \quad \text{if } v \text{ is complex,}$$
$$\|z\|_v = |\mathrm{N}_{K_v/\mathbf{Q}_p} z|_p \quad \text{if } v \text{ is finite, } v|p, \ |p|_p = 1/p.$$

Note that $\|x\|_v = |x|_v^d$.

We then have for $z_1, z_2 \ldots, z_N \in K$

$$\|z_1 + z_2 + \cdots + z_N\|_v \leq \begin{cases} N \sup_i(\|z_i\|_v) & \text{if } v \text{ is real} \\ N^2 \sup_i(\|z_i\|_v) & \text{if } v \text{ is complex} \\ \sup_i(\|z_i\|_v) & \text{if } v \text{ is finite.} \end{cases}$$

Let J be an idèle of K, that is $J = (J_v)_v$ where for each $v \in \mathcal{P}$, $J_v \in K_v^\times$ and $\|J_v\|_v = 1$ for almost all v. We define

$$V(J) = \prod_{v \in \mathcal{P}} \|J_v\|_v$$
$$\mathcal{S}(J) = \left\{ z \in K \,\middle|\, \|z\|_v \leq \|J_v\|_v \text{ for all } v \right\} .$$

It is well known (cf. Lang [1], pg. 102) that there exist positive constants c_1 and c_2 independent of J such that

$$c_1 V(J) < \#\mathcal{S}(J) \leq \sup\left(1, c_2 V(J)\right) . \qquad (3.4)$$

Clearly we may (and do) assume that $c_1 < 1$ and $c_2 > 1$.

Let I be an idèle of K with $V(I) \geq 1$. Consider the map

$$(\mathcal{S}(I))^N \longrightarrow K^M$$
$$\vec{x} \longmapsto A\vec{x} .$$

If $l \in \mathbf{N}$, $l \neq 0$, let l^∞ be the idèle defined by

$$(l^\infty)_v = \begin{cases} l & \text{if } v \text{ is real,} \\ l^2 & \text{if } v \text{ is complex,} \\ 1 & \text{if } v \text{ is finite.} \end{cases}$$

Let us assume that J is such that all elements A_{ij} of A belong to $\mathcal{S}(J)$; if $A = 0$ we take $J = 1$. Then in any case $V(J) \geq 1$. Moreover

$$\left\| \sum_{j=1}^{N} A_{ij} x_j \right\|_v \leq \begin{cases} N\|J_v\|_v\|I_v\|_v & \text{if } v \text{ is real,} \\ N^2\|J_v\|_v\|I_v\|_v & \text{if } v \text{ is complex,} \\ \|J_v\|_v\|I_v\|_v & \text{if } v \text{ is finite,} \end{cases}$$

and therefore $\sum_j A_{ij} x_j \in \mathcal{S}(IJN^\infty)$ so that the image of our map lies in $(\mathcal{S}(IJN^\infty))^M$. From (3.4) it follows that

$$\#\left(\mathcal{S}(I)\right)^N > (c_1 V(I))^N$$

and

$$\#\left(\mathcal{S}(IJN^\infty)\right)^M \leq \sup\left(1, \left(c_2 V(IJN^\infty)\right)^M\right) .$$

We know that each of the four numbers $V(J)$, $V(N^\infty)$, $V(I)$ and c_2 is ≥ 1, and hence

$$\#\left(\mathcal{S}(IJN^\infty)\right)^M \leq (c_2 V(IJN^\infty))^M .$$

Hence if

$$(c_1 V(I))^N > (c_2 V(IJN^\infty))^M$$

then by the Pigeon Hole Principle there exist two distinct elements $\vec{y}$, $\vec{z} \in \mathcal{S}(I)^N$ such that $A\vec{y} = A\vec{z}$. Thus if $\vec{x} = \vec{y} - \vec{z}$ then clearly $A\vec{x} = 0$ and $\vec{x} \in \mathcal{S}(I2^\infty)$. Let $L = I2^\infty$. Then the conditions on L which assure the existence of an $\vec{x} \in \mathcal{S}(L)^N \backslash \{0\}$ such that $A\vec{x} = 0$ are the following

$$V(L(2^\infty)^{-1}) \geq 1$$

and

$$(c_1 V(L(2^\infty)^{-1}))^{N-M} > \left(\frac{c_2 V(LJN^\infty(2^\infty)^{-1})}{c_1 V(L(2^\infty)^{-1})} \right)^M$$
$$= \left(\theta V(J)N^d\right)^M ,$$

where $\theta = c_2/c_1 > 1$ and $d = [K : \mathbf{Q}]$. This last condition is equivalent to

$$V(L(2^\infty)^{-1}) > \frac{1}{c_1}\theta^{M/(N-M)} \left(V(J)N^d\right)^{M/(N-M)} ,$$

which also includes the first condition.

Let J be such that $\|J_v\|_v = \sup_{i,j}(1, \|A_{ij}\|_v)$. Then

$$V(J) = (\operatorname{ht} A)^d .$$

Thus if

$$V(L(2^\infty)^{-1}) > \frac{1}{c_1}\theta^{M/(N-M)}(N\operatorname{ht} A)^{dM/(N-M)}$$

there exists a non-trivial solution $\vec{x} \in \mathcal{S}(L)^N$ such that $A\vec{x} = 0$.

Let us now fix a $v_0 \in \mathcal{P}_\infty$, and a $\tilde{c} > 2c_1^{-1/d}$. If L is such that

$$L_v = 1 \quad \text{for all } v \neq v_0$$
$$L_{v_0} = \tilde{c}^d\theta^{M/(N-M)}(N\operatorname{ht} A)^{dM/(N-M)} ,$$

then there exists a non-trivial solution $\vec{x} \in \mathcal{S}(L)^N$, so that $\|\vec{x}\|_v \leq 1$ for all $v \neq v_0$ and

$$(\operatorname{ht} \vec{x})^d = \prod_v \sup(1, \|x_1\|_v, \ldots, \|x_N\|_v)$$
$$= \sup(1, \|x_1\|_v, \ldots, \|x_N\|_v)$$
$$\leq L_{v_0} .$$

Therefore

$$\operatorname{ht} \vec{x} \leq c\gamma^{M/(N-M)}(N\operatorname{ht} A)^{M/(N-M)}$$

where $c = 2c_1^{-1/d}$ and $\gamma = \theta^{1/d}$.   **Q.E.D.**

If

$$\vec{y} = \sum_{s=0}^\infty \vec{y}_s X^s \in K[[X]]^n$$

with $\vec{y}_s \in K^n$ we define

$$\tilde{h}(s, v) = \sup_{l \leq s} \log^+ |\vec{y}_l|_v .$$

For $N \in \mathbf{N}$ we define the *truncation operator* T_N as the K-linear endomorphism of $K[[X]]$ such that

$$T_N X^i = \begin{cases} X^i & \text{if } i \leq N \\ 0 & \text{if } i > N . \end{cases}$$

If I denotes the identity map of $K[[X]]$ we set

$$\mathcal{R}_N = I - T_N .$$

In the application we will take $\vec{y}$ to be the column vector in the statement of Theorem 1.5.

Lemma 3.2. *Let $\tau \in (0,1)$ and let*

$$\vec{y} = \begin{pmatrix} y_0 \\ \vdots \\ y_{n-1} \end{pmatrix} = \sum_{s=0}^{\infty} \vec{y}_s X^s \in K[[X]]^n ,$$

with $\vec{y}_s \in K^n$. For each $N \in \mathbf{N}$ there exists a $q \in K[X] \backslash \{0\}$, depending on N, such that

$$\deg q \leq N ,$$

$$\operatorname{ord}_X \mathcal{R}_N (q\vec{y}) \geq 1 + N + N \frac{1 - \tau}{n} ,$$

$$h(q) \leq \text{const} + \frac{1 - \tau}{\tau} \left(\log N + \sum_{v \in P} \tilde{h} \left(N + \left[N \frac{1 - \tau}{n} \right], v \right) \right) .$$

Proof. Let $q = \sum_{j=0}^{N} q_j X^j$ with $q_0, \ldots, q_N$ indeterminates in K. Then the second condition in the statement of the lemma is equivalent to requiring that the coefficients of $X^{N+1}, X^{N+2}, \ldots, X^{N+[N(1-\tau)/n]}$ in $q\vec{y}$ be zero, that is

$$\sum_{j=0}^{N} \vec{y}_{s-j} q_j = 0 \tag{3.5}$$

for $s = N + 1, \ldots, N + \left[N \frac{1-\tau}{n} \right]$. Consider the matrix of blocks

$$A = \begin{pmatrix} A_{N+1} \\ \vdots \\ A_{N+[N(1-\tau)/n]} \end{pmatrix}$$

where the block A_s is given by

$$A_s = (\vec{y}_s \quad \cdots \quad \vec{y}_{s-N}) \in \mathcal{M}_{n,N+1}(K) .$$

Then A is an $M \times (N+1)$ matrix where $M = n\left[N\frac{1-\tau}{n}\right]$ and the linear system (3.5) can be written as

$$A \begin{pmatrix} q_0 \\ \vdots \\ q_N \end{pmatrix} = 0 .$$

Since $N + 1 > M$ for all positive integers N and

$$\frac{M}{1+N-M} = \frac{n\left[N\dfrac{1-\tau}{n}\right]}{N+1-n\left[N\dfrac{1-\tau}{n}\right]} \leq \frac{N(1-\tau)}{1+N\tau} \leq \frac{1-\tau}{\tau}$$

we may apply Siegel's Lemma with N replaced by $N + 1$ to obtain

$$h(q) \leq \text{const} + \frac{1-\tau}{\tau}\left(\log \gamma + \log N + \log \text{ht } A\right) .$$

Since

$$\text{ht } A \leq \prod_{v \in \mathcal{P}} \sup\left(1, |\vec{y}_0|_v, \ldots, |\vec{y}_{N+[N\frac{1-\tau}{n}]}|_v\right)$$

we have

$$\log \text{ht } A \leq \sum_{v \in \mathcal{P}} \tilde{h}\left(N + \left[N\frac{1-\tau}{n}\right], v\right) ,$$

thus obtaining the desired inequality. **Q.E.D.**

4. Conclusion of the Proof of Chudnovsky's Theorem.

Let $\vec{y}$ satisfy the hypotheses of Theorem 1.5. If we take

$$\vec{P} = T_N(q\vec{y}) = q\vec{y} - \mathcal{R}_N(q\vec{y}) ,$$

then from Lemma 3.2 it follows that

$$\begin{vmatrix} P_i & y_i \\ P_j & y_j \end{vmatrix} = \begin{vmatrix} \mathcal{R}_N(y_iq) & y_i \\ \mathcal{R}_N(y_jq) & y_j \end{vmatrix} \in X^{1+N+[N(1-\tau)/n]}K[[X]]$$

and so

$$\operatorname{ord}_X \begin{vmatrix} P_i & y_i \\ P_j & y_j \end{vmatrix} - \deg \vec{P} > \left[\frac{N(1-\tau)}{n} \right] \to \infty$$

as $N \to \infty$. By Proposition 2.3 this implies that if $N > N_0$ then $\det R^{(0)} \neq 0$ provided that $\vec{P} \neq 0$. This is indeed the case since otherwise we would have

$$\operatorname{ord}_X (q\vec{y}) = \operatorname{ord}_X (\mathcal{R}_N(q\vec{y})) \geq 1 + N + \left[\frac{N(1-\tau)}{n} \right]$$

so that since $\deg q \leq N$ we would obtain

$$\operatorname{ord}_X \vec{y} > \left[N \frac{1-\tau}{n} \right]$$

and this is impossible for N sufficiently large ($\operatorname{ord}_X \vec{y}$ is a constant independent of N).

Let $Q \in \mathcal{O}_K[X]$ be such that $QG \in \mathcal{M}_n(K[X])$ and let

$$t = 1 + \sup (\deg QG, \deg Q - 1) .$$

Our goal is now to calculate $\vec{R}_s$ in terms of $\vec{y}$ and q.

Lemma 4.1. *Let τ, n, N and q be as in Lemma 3.2, and let $\vec{P} = T_N(q\vec{y})$. Then for any integer s such that*

$$s \leq \frac{N}{t} \frac{1-\tau}{n}$$

we have

$$T_{N+s(t-1)} \left(\frac{Q^s}{s!} (D^s q)\vec{y} \right) = \vec{R}_s Q^s \in K[X]^n .$$

Proof. We have already proved (see (2.3)) that $\vec{R}_s Q^s \in K[X]^n$. Let

$$\vec{L}_s = \frac{Q^s}{s!} (D^s q) \vec{y} - \vec{R}_s Q^s \in K[[X]]^n .$$

We claim that

$$(s+1)\vec{L}_{s+1} = (QD - s(DQ) - QG) \vec{L}_s .$$

Indeed, we have

$$QD\vec{L}_s = QD\left(\frac{Q^s}{s!}(D^s q)\vec{y} - \vec{R}_s Q^s\right)$$
$$= Q\left(\frac{Q^s}{s!}(D^{s+1}q)\vec{y} + \frac{Q^s}{s!}(D^s q)D\vec{y} + \frac{Q^{s-1}}{(s-1)!}DQ(D^s q)\vec{y}\right)$$
$$- Q\left(Q^s D\vec{R}_s + sQ^{s-1}\vec{R}_s DQ\right).$$

Since

$$D\vec{R}_s = (s+1)\vec{R}_{s+1} + G\vec{R}_s$$

we obtain

$$QD\vec{L}_s = Q\left(\frac{Q^s}{s!}(D^{s+1}q)\vec{y} + \frac{Q^s}{s!}(D^s q)G\vec{y} + \frac{Q^{s-1}}{(s-1)!}DQ(D^s q)\vec{y}\right.$$
$$\left. - (s+1)Q^s \vec{R}_{s+1} - Q^s G\vec{R}_s - sQ^{s-1}\vec{R}_s DQ\right)$$

so that

$$\frac{QD\vec{L}_s}{s+1} = \left(\frac{Q^{s+1}}{(s+1)!}(D^{s+1}q)\vec{y} - Q^{s+1}\vec{R}_{s+1}\right)$$
$$+ DQ\left(\frac{Q^s}{(s-1)!(s+1)}(D^s q)\vec{y} - Q^s\frac{s}{s+1}\vec{R}_s\right)$$
$$+ \frac{QG}{s+1}\left(\frac{Q^s}{s!}(D^s q)\vec{y} - Q^s\vec{R}_s\right)$$
$$= \vec{L}_{s+1} + \frac{s}{s+1}DQ\vec{L}_s + \frac{QG}{s+1}\vec{L}_s,$$

as desired. However, QG and Q are polynomials so that

$$\text{ord}_X\, \vec{L}_{s+1} \geq \text{ord}_X\, \vec{L}_s - 1.$$

But

$$\vec{L}_0 = q\vec{y} - \vec{P} = \mathcal{R}_N(q\vec{y})$$

so that by Lemma 3.2

$$\text{ord}_X\, \vec{L}_0 \geq 1 + N + \left[N\frac{1-\tau}{n}\right]$$

and therefore

$$\operatorname{ord}_X \vec{L}_s \geq \operatorname{ord}_X \vec{L}_0 - s \geq 1 + N + \left[N \frac{1 - \tau}{n} \right] - s \ .$$

Since by hypothesis

$$\frac{N}{t} \frac{1 - \tau}{n} \geq s$$

we have

$$\left[N \frac{1 - \tau}{n} \right] \geq st$$

so that

$$\operatorname{ord}_X \vec{L}_s \geq 1 + N + s(t - 1) \ .$$

Therefore, if s satisfies the hypotheses of the present Lemma we find

$$\frac{Q^s}{s!} (D^s q) \vec{y} \equiv \vec{R}_s Q^s \quad \operatorname{mod} X^{1+N+s(t-1)} K[[X]] \ .$$

From (2.4) (where h is to be replaced by $t - 1$) it follows that

$$\deg \vec{R}_s Q^s \leq \deg \vec{P} + s(t - 1) \leq N + s(t - 1) \ .$$

This gives the desired result.   **Q.E.D.**

We recall that for $v \in \mathcal{P}$

$$h(s, v) = \sup_{l \leq s} \log^+ |G_l|_{v, \text{gauss}} = \sup_{l \leq s} \log |G_l|_{v, \text{gauss}}$$

$$\tilde{h}(s, v) = \sup_{l \leq s} \log^+ |\vec{y}_l|_v \ ,$$

where the matrices G_s satisfy

$$\frac{D^s}{s!} \vec{y} = G_s \vec{y}$$

(since $G_0 = I_n$ we can use either log or $\log^+$ in $h(s, v)$) and

$$\vec{y} = \sum_{s=0}^{\infty} \vec{y}_s X^s \ .$$

We are now ready to begin the final steps in the proof of Chudnovsky's Theorem, that is to prove that

$$\sigma(G) = \limsup_{s \to \infty} \sum_{v \in \mathcal{P}} \frac{1}{s} h(s, v) < \infty .$$

We recall that by hypothesis $\sigma(\vec{y}) < \infty$ so that we have

$$\sigma_f(\vec{y}) = \limsup_{s \to \infty} \frac{1}{s} \sum_{v \in \mathcal{P}_f} \tilde{h}(s, v) < \infty$$

$$\sigma_\infty(\vec{y}) = \limsup_{s \to \infty} \frac{1}{s} \sum_{v \in \mathcal{P}_\infty} \tilde{h}(s, v) < \infty .$$

Let s be a large positive integer, and let N be sufficiently large so that $R^{(0)}$ is invertible and also that

$$N > \frac{nt(s + n - 1)}{1 - \tau} .$$

It then follows from Lemma 4.1 that

$$T_{N+l(t-1)} \left(\frac{Q^l}{l!} (D^l q) \vec{y} \right) = \vec{R}_l Q^l$$

for $l \le s + n - 1$. From equation (2.5) we then obtain

$$G_s = \left(\sum_{j=0}^{s} \frac{(-1)^j}{(s - j)!} D^{s-j} R^{(j)} \right) \left(R^{(0)} \right)^{-1} .$$

Therefore for a finite v we have (on suppressing the suffix "gauss" on the right side of our inequalities)

$$|G_s|_{v,\text{gauss}} \le \sup_{j \le s} \left| \frac{D^{s-j}}{(s - j)!} R^{(j)} \right|_v \left| \text{adj } R^{(0)} \right|_v | \det R^{(0)} |_v^{-1}$$

$$\le \sup_{j \le s} |R^{(j)}|_v |\text{adj } R^{(0)}|_v | \det R^{(0)} |_v^{-1} .$$

But

$$|R^{(j)}|_{v,\text{gauss}} \le \sup \left(|\vec{R}_j|_v, |\vec{R}_{j+1}|_v, \ldots, |\vec{R}_{j+n-1}|_v \right)$$

so that

$$|G_s|_{v,\text{gauss}} \leq \sup_{j \leq s+n-1} |\vec{R}_j|_v |R^{(0)}|_v^{n-1} |\det R^{(0)}|_v^{-1}$$

$$\leq \sup_{j \leq s+n-1} |\vec{R}_j|_v \left(\sup_{j \leq n-1} |\vec{R}_j|_v \right)^{n-1} \frac{1}{|\Delta|_v}$$

where $\Delta = \det R^{(0)}$. Now from

$$\vec{R}_l = Q^{-l} T_{N+l(t-1)} \left(\frac{Q^l}{l!} (D^l q) \vec{y} \right)$$

it follows that

$$|\vec{R}_l|_{v,\text{gauss}} \leq |Q|_v^{-l} |Q|_v^l |q|_v |T_{N+l(t-1)} \vec{y}|_v \leq |q|_v |T_{N+l(t-1)} \vec{y}|_v .$$

Thus if we set

$$h(q, v) = \sup_{0 \leq j \leq N} \log^+ |q_j|_v ,$$

we obtain

$$\log^+ |\vec{R}_l|_{v,\text{gauss}} \leq \tilde{h}(N + l(t-1), v) + h(q, v)$$

so that

$$\log |G_s|_{v,\text{gauss}} \leq \tilde{h} \left(N + (s+n-1)(t-1), v \right)$$
$$+ (n-1) \tilde{h} \left(N + (n-1)(t-1), v \right)$$
$$+ n h(q, v) + \log \frac{1}{|\Delta|_v} .$$

Recall that (see equation (2.3))

$$\vec{R}_s Q^s = \vec{H}_s$$

and let $\overline{\Delta} \in K[X]$ be defined by

$$\overline{\Delta} = \vec{H}_0 \wedge \vec{H}_1 \wedge \cdots \wedge \vec{H}_{n-1}$$
$$= \vec{R}_0 \wedge Q\vec{R}_1 \wedge \cdots \wedge Q^{n-1}\vec{R}_{n-1}$$
$$= Q^{\binom{n}{2}} \det R^{(0)}$$
$$= Q^{\binom{n}{2}} \Delta .$$

Since we chose Q in $\mathcal{O}_K[X]$ we have

$$|\overline{\Delta}|_{v,\text{gauss}} = |\Delta|_v |Q|_v^{\binom{n}{2}} \leq |\Delta|_v .$$

Therefore we obtain

$$\log |G_s|_{v,\text{gauss}} \leq \tilde{h}\left(N + (s+n-1)(t-1), v\right)$$
$$+ (n-1)\tilde{h}\left(N + (n-1)(t-1), v\right) + nh(q,v) + \log \frac{1}{|\overline{\Delta}|_v}$$

whence we find that

$$\frac{1}{s} \sum_{v \in \mathcal{P}_f} h(s,v) \leq$$

$$\frac{N + (s+n-1)(t-1)}{s\left(N + (s+n-1)(t-1)\right)} \sum_{v \in \mathcal{P}_f} \tilde{h}\left(N + (s+n-1)(t-1), v\right)$$

$$+ (n-1)\frac{N + (n-1)(t-1)}{s\left(N + (n-1)(t-1)\right)} \sum_{v \in \mathcal{P}_f} \tilde{h}\left(N + (n-1)(t-1), v\right)$$

$$+ \frac{n}{s} \sum_{v \in \mathcal{P}_f} h(q,v) + \frac{1}{s} \sum_{v \in \mathcal{P}_f} \log \frac{1}{|\overline{\Delta}|_v} .$$

Now let $1 - \varepsilon_s$ be the fractional part of $s \cdot n \cdot t/(1-\tau)$, so $\varepsilon_s \in (0,1]$, and let

$$\frac{N}{s} = \frac{nt}{1-\tau} + \frac{k + \varepsilon_s}{s} \tag{4.1}$$

where $k > n^2 t/(1-\tau)$ is a fixed positive integer. Then, letting $s \longrightarrow \infty$ we find

$$\sigma(G) \leq \sigma_f(\vec{y})\left(\frac{nt}{1-\tau} + t - 1 + (n-1)\frac{nt}{1-\tau}\right)$$

$$+ \limsup_{s \to \infty} \left(\frac{n}{s} \sum_{v \in \mathcal{P}_f} h(q,v) + \frac{1}{s} \sum_{v \in \mathcal{P}_f} \log \frac{1}{|\overline{\Delta}|_v}\right) \tag{4.2}$$

$$\leq \left(t - 1 + \frac{n^2 t}{1-\tau}\right)\sigma_f(\vec{y}) + \Omega$$

where

$$\Omega = \limsup_{s \to \infty} \frac{1}{s}\left(\sum_{v \in \mathcal{P}_f} \log \frac{1}{|\overline{\Delta}|_v} + n \sum_{v \in \mathcal{P}_f} h(q,v)\right) . \tag{4.3}$$

To conclude the proof of Chudnovsky's Theorem we must find an upper bound for Ω.

Let ζ be a root of unity such that $\overline{\Delta}(\zeta) \neq 0$ and $Q(\zeta) \neq 0$. We may assume that $\zeta \in K$ since the height is independent of K. Since

$$|\overline{\Delta}(\zeta)|_v \leq |\overline{\Delta}|_{v,\text{gauss}} \tag{4.4}$$

for all roots of unity ζ, from the Product Formula we obtain

$$\sum_{v \in P_f} \log \frac{1}{|\overline{\Delta}|_v} \leq \sum_{v \in P_f} \log \frac{1}{|\overline{\Delta}(\zeta)|_v}$$
$$= \sum_{v \in P_\infty} \log |\overline{\Delta}(\zeta)|_v \ .$$

We recall that

$$\overline{\Delta} = \det \begin{pmatrix} \vec{R}_0 & \vec{R}_1 Q & \cdots & \vec{R}_{n-1} Q^{n-1} \end{pmatrix}$$

and that for $j \leq n-1$ we have

$$Q^j \vec{R}_j = T_{N+j(t-1)} \left(Q^j \frac{D^j q}{j!} \vec{y} \right) \ .$$

Since

$$\frac{1}{j!} D^j q = \sum_{s=j}^{N} \binom{s}{j} q_s X^{s-j} \ ,$$

we have

$$Q^j \frac{D^j q}{j!} \vec{y} = \sum_{m=0}^{\infty} X^m \sum_{h+k+l=m} (Q^j)_h \left(\frac{D^j q}{j!} \right)_k \vec{y}_l$$
$$= \sum_{m=0}^{\infty} X^m \sum_{h+k+l=m} (Q^j)_h \binom{j+k}{j} q_{j+k} \vec{y}_l \ .$$

Therefore $(Q^j \vec{R}_j)(\zeta)$ is a sum of terms of the form

$$(Q^j)_h \binom{j+k}{j} q_{j+k} \vec{y}_l \zeta^{h+k+l}$$

with h, k, l and j satisfying the following inequalities:

$$0 \leq j \leq n - 1, \quad 0 \leq h \leq a, \quad 0 \leq k \leq N, \quad j + k \leq N,$$
$$0 \leq l \leq N + (n - 1)(t - 1)$$

where $a = \deg Q^{n-1}$. The v-value of each of these terms is bounded by

$$\binom{N}{j} \sup_{l \leq N+(n-1)(t-1)} |\vec{y}_l|_v \sup_{l \leq N} |q_l|_v \sup(1, c_1)^{n-1}$$

where $c_1 = \sup$ of the v-values of the coefficients of Q. Therefore we obtain the following estimate for $|(Q^j \vec{R}_j)(\zeta)|_v$, for $j \leq n - 1$:

$$|(Q^j \vec{R}_j)(\zeta)|_v \leq (1 + a)(1 + N)\,(1 + N + (n - 1)(t - 1))\,\binom{N}{j} \cdot$$
$$\cdot \left(\sup_{l \leq N+(n-1)(t-1)} |\vec{y}_l|_v \sup_{l \leq N} |q_l|_v \sup(1, c_1)^{n-1} \right).$$

In the preceding estimate we tacitly used the triangle inequality for the valuation v; for a proof see the Appendix at the end of this Section.

We now observe that for N sufficiently large we have

$$\binom{N}{j} \leq \binom{N}{n-1}$$

for $j \leq n - 1$ and, again for N sufficiently large,

$$\binom{N}{n-1} \leq c_2 N^{n-1}$$

for a suitable constant c_2. Moreover, once again for N sufficiently large, we also have

$$(1 + a)(1 + N)\,(1 + N + (n - 1)(t - 1)) \leq c_3 N^2$$

for a suitable c_3. Taking $c_v = c_2 c_3 \sup(1, c_1)^{n-1}$ we find that

$$|(Q^j \vec{R}_j)(\zeta)|_v \leq c_v N^{n+1} \sup_{l \leq N} |q_l|_v \sup_{l \leq N+(n-1)(t-1)} |\vec{y}_l|_v \,,$$

and so

$$|\overline{\Delta}(\zeta)|_v \leq (n!)c_v^n N^{n(n+1)} \left(\sup_{l \leq N} |q_l|_v \right)^n \left(\sup_{l \leq N+(n-1)(t-1)} |\vec{y}_l|_v \right)^n .$$

Therefore, on passing to logarithms we have

$$\log |\overline{\Delta}(\zeta)|_v \leq \log n! + c'_v + n(n+1)\log N$$
$$+ nh(q,v) + n\tilde{h}\left(N + (n-1)(t-1), v\right)$$

and on summing over all infinite primes we find

$$\sum_{v \in \mathcal{P}_\infty} \log |\overline{\Delta}(\zeta)|_v \leq \text{const} + n(n+1)(\#\mathcal{P}_\infty)\log N + n \sum_{v \in \mathcal{P}_\infty} h(q,v)$$
$$+ n \sum_{v \in \mathcal{P}_\infty} \tilde{h}\left(N + (n-1)(t-1), v\right) .$$

From this last inequality and from (4.3) and (4.4) we have

$$\Omega \leq \limsup_{s \to \infty} \frac{1}{s} \left(\sum_{v \in \mathcal{P}_\infty} \log |\overline{\Delta}(\zeta)|_v + n \sum_{v \in \mathcal{P}_f} h(q,v) \right)$$
$$\leq \limsup_{s \to \infty} \frac{1}{s} \left(n(n+1)(\#\mathcal{P}_\infty)\log N + n \sum_{v \in \mathcal{P}} h(q,v) \right.$$
$$\left. + n \sum_{v \in \mathcal{P}_\infty} \tilde{h}\left(N + (n-1)(t-1), v\right) \right)$$
$$= \limsup_{s \to \infty} \frac{1}{s} \left(n(n+1)(\#\mathcal{P}_\infty)\log N + nh(q) \right.$$
$$\left. + n \sum_{v \in \mathcal{P}_\infty} \tilde{h}\left(N + (n-1)(t-1), v\right) \right) .$$

Now from (4.1) it follows that

$$\frac{\log N}{s} \longrightarrow 0$$

as $s \longrightarrow \infty$. From Lemma 3.2 and (4.1) it follows that

$$\limsup_{s \to \infty} \frac{1}{s} h(q) \leq \frac{1-\tau}{\tau} \limsup_{s \to \infty} \frac{1}{s} \sum_{v \in \mathcal{P}} \tilde{h}\left(N + \left[N\frac{1-\tau}{n}\right], v\right)$$
$$= \frac{1-\tau}{\tau}\left(1 + \frac{1-\tau}{n}\right)\frac{nt}{1-\tau}\sigma(\vec{y}) .$$

Similarly,

$$\limsup_{s \to \infty} \frac{1}{s} \sum_{v \in \mathcal{P}_\infty} \tilde{h}\left(N + (n-1)(t-1), v\right) = \frac{nt}{1-\tau}\sigma_\infty(\vec{y}) \ .$$

Therefore from (4.2) it follows that

$$\sigma(G) \leq \left(t - 1 + \frac{n^2 t}{1-\tau}\right)\sigma_f(\vec{y}) + \frac{n^2 t}{1-\tau}\sigma_\infty(\vec{y})$$

$$+ \frac{1-\tau}{\tau}\left(1 + \frac{1-\tau}{n}\right)\frac{n^2 t}{1-\tau}\sigma(\vec{y})$$

$$\leq \left[t - 1 + \frac{n^2 t}{1-\tau} + n^2 t\left(\frac{n+1}{n}\frac{1}{\tau} - \frac{1}{n}\right)\right]\sigma(\vec{y})$$

$$= \left[t - 1 - nt + n^2 t\left(\frac{1}{1-\tau} + \frac{1}{\tau}\frac{n+1}{n}\right)\right]\sigma(\vec{y}) \ .$$

The time has now come to choose a suitable $\tau \in (0,1)$. Let

$$\theta = \sqrt{\frac{n}{n+1}} \ , \qquad \tau = \frac{1}{1+\theta} \ .$$

Then, for $n \geq 2$,

$$\frac{1}{1-\tau} + \frac{1}{\tau}\frac{n+1}{n} = (1 + \frac{1}{\theta})^2 \leq 2 + \frac{1}{n} + 2\sqrt{1 + \frac{1}{n}}$$

$$\leq 2 + \frac{1}{2} + 2\sqrt{1.5} \leq 4.95 \ .$$

Therefore,

$$\sigma(G) \leq (4.95\, n^2 t - 1 - (n-1)t)\sigma(\vec{y}) \ .$$

For $n = 1$ we obtain $\sigma(G) \leq (5.9\,t - 1)\sigma(\vec{y})$. This completes the proof of Chudnovsky's Theorem.

We briefly indicate the converse of Theorem VII.4.2.

Corollary 4.2. *Let $G \in \mathcal{M}_n(K(X))$ and let YX^A be a solution matrix with $Y \in G\ell(n, K((X)))$, $A \in \mathcal{M}_n(K)$. If $\sigma(Y) < \infty$, $\rho(A/X) < \infty$ then $\sigma(G) < \infty$.*

Proof. Let α be an eigenvalue of A, so $\alpha \in \mathbf{Q}$. Then there exists a linear combination $\vec{y}$ of columns of Y such that $\vec{y}X^\alpha$ is a solution of

$D - G$ and so $\vec{y}$ is a solution of $D - (G - \alpha I_n/X)$ with $\sigma(\vec{y}) < \infty$. If this last equation is irreducible then by Lemma 2.6 and Chudnovsky's Theorem we have $\rho(G - \alpha I_n/X) < \infty$ and since $\alpha \in \mathbf{Q}$ we conclude that $\rho(G) < \infty$. Thus we may assume that $D - (G - \alpha I_n/X)$ is reducible and hence that $D - G$ is reducible. Then we use induction on n.   **Q.E.D.**

Appendix to Chapter VIII

Let $v \in \mathcal{P}_\infty$, and assume that K is imbedded in $\mathbf{C}$. For $x \in K$ we put

$$|x|_v = |x|_{\mathrm{cl}}^{d_v/d}$$

where $|x|_{\mathrm{cl}}$ means the classical archimedean valuation in both the real and complex cases. We wish to show that the triangle inequality holds for this valuation.

First we observe that for $\alpha \in (0, 1)$, and $x \geq 0$, $x \in \mathbf{R}$ we have

$$(1 + x)^\alpha \leq 1 + x^\alpha .$$

Indeed, the inequality surely holds for $x = 0$; checking the derivatives it suffices to show that

$$\alpha(1 + x)^{\alpha-1} \leq \alpha x^{\alpha-1}$$

which holds since $\alpha - 1 < 0$ and $1 + x > x$.

Now for $x, y \in K$ and $x \neq 0$,

$$
\begin{aligned}
|x + y|_v &= |x|_v |1 + y/x|_v \\
&= |x|_v |1 + y/x|_{\mathrm{cl}}^{d_v/d} \\
&\leq |x|_v \left(1 + |y/x|_{\mathrm{cl}}\right)^{d_v/d} \\
&\leq |x|_v \left(1 + |y/x|_{\mathrm{cl}}^{d_v/d}\right) \\
&= |x|_v + |y|_v ,
\end{aligned}
$$

as asserted.

APPENDIX I

CONVERGENCE POLYGON FOR DIFFERENTIAL EQUATIONS

Let $K \subset \Omega$ be complete fields satisfying the conditions of Sections IV.4, IV.5. Let $R_0 \in (0,1)$ and let C be the annulus $C = \{x \mid |x| \in (R_0, 1)\}$. Let $\mathcal{F}_0$ be the ring of elements of $K(X)$ without poles in C and let $\mathcal{F}$ be the completion of $\mathcal{F}_0$ under the sup norm on C. For $r \in [R_0, 1]$ let t_r denote the generic point over K such that $|t_r| = r$. We write t for t_1. If $\xi \in \mathcal{F}$ then we have

$$|\xi|_0(r) = |\xi(t_r)|$$

for all $r \in (R_0, 1)$. However the sequences of elements of $\mathcal{F}_0$ converging uniformly on C to an element of $\mathcal{F}$ also converge uniformly on $D(t, 1^-)$ and on $D(t_{R_0}, R_0^-)$ and so we may view the elements of $\mathcal{F}$ as extending naturally to analytic elements on $C \cup D(t, 1^-) \cup D(t_{R_0}, R_0^-)$. Furthermore

$$|\xi(t)| = \lim_{r \to 1} |\xi|_0(r)$$

and we define $|\xi|_0(1)$ as this limit. Furthermore each $\xi \in \mathcal{F}$ may be represented by a Laurent series with coefficients in K, convergent everywhere on C

$$\xi = \sum_{j \in \mathbf{Z}} \xi_j X^j \ , \ \xi_j \in K$$

and for $r \in [R_0, 1]$

$$|\xi|_0(r) = \sup_j |\xi_j| r^j \ ,$$

the assertion for $r \in (R_0, 1)$ coming from Proposition IV.1.1 while for $r = 1$ the assertion follows by taking limits as $r \to 1$.

301

Let $G \in \mathcal{M}_n(\mathcal{F})$. We construct a polygon for $D - G$, where $D = d/dX$. We start with the usual sequence $\{G_s\}$ defined recursively by $G_0 = I_n$, $G_{s+1} = G_sG + DG_s$. We write $A_s = G_s/s!$ and use the Laurent series representation on C

$$A_s(X) = \sum_{j \in \mathbf{Z}} A_{s,j} X^j \ , \ A_{s,j} \in \mathcal{M}_n(K) \ .$$

For $M \geq 1$ we define a function f_M on $\mathbf{Q}$ by setting for each $z \in \mathbf{Q}$

$$f_M(z) = \sup_{\substack{s \in \mathbf{N}^\times \\ sz \in \mathbf{Z}}} \frac{1}{s} \log \frac{|A_{s,sz}|}{M\{s, n-1\}_p} \ .$$

We know that

$$\|G_s\|_C \leq M_0^s$$

where

$$M_0 = \sup\left(\frac{1}{R_0}, \|G\|_C\right)$$

and hence by Proposition IV.1.1

$$f_M(z) \leq \log\left(\frac{M_0}{|\pi|}\right) + \begin{cases} z \log \dfrac{1}{R_0} & \text{if } z \leq 0 \\ 0 & \text{if } z \geq 0 \end{cases}$$

It follows that all points $(z, f_M(z))$ lie on or below the polygon indicated

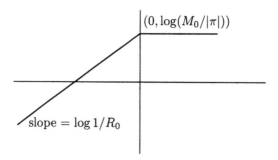

Such a polygon will be referred to as *bounded*. For each M we obtain a "polygon" by taking the convex closure of the set consisting of $(0, -\infty)$ and all $(z, f_M(z))$ such that $z \in \mathbf{Q}$.

We shall take M sufficiently large, specifically

$$M \geq M_1 = M_0^{n-1} \left(\frac{1}{R_0}\right)^{\mu}$$

where $\mu = 0$ unless the trace of G is an element of E_0' with positive integral residues at each pole in $D(0, 1^-)$ in which case μ is the sum of these residues.

For each $m \in [0, \log \frac{1}{R_0}]$ we define the *line of contact of slope m* to be the line

$$Y = mZ + b_m$$

where $b_m = \sup_{z \in Q}(f_M(z) - mz)$. In particular then b_m is the Y-intercept of the line of contact of slope m. We do not know if the periphery of the "polygon" consists of lines.

For $r \in [R_0, 1]$, let $\mathcal{U}_{G,t_r}$ be the solution matrix at t_r of $D - G$. Let $\rho(r)$ denote the radius of convergence of $\mathcal{U}_{G,t_r}$.

We treat the case $\rho(1) > 1$ separately. In that case we associate with G the polygon

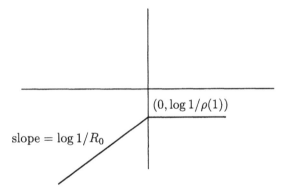

If $\rho(1) = \infty$ we interpret this polygon as being the line $Y = -\infty$.

Theorem. *There exists a bounded convex polygon containing $(0, -\infty)$ such that for each $r \in [R_0, 1]$ the line of contact of slope $\log \frac{1}{r}$ has Y-intercept precisely equal to $\log \frac{1}{\rho(r)}$.*

Proof. For $\rho(1) > 1$ the assertion has been shown. Hence we may assume $\rho(1) \leq 1$. In this case we choose $M \geq M_1$ and define the polygon to be the convex closure of all points $(z, f_M(z))_{z \in \mathbf{Q}}$. Let $\log(1/\sigma(r))$ be the Y-intercept of the line of contact of slope $\log(1/r)$. We first show that $\rho(r) \geq \sigma(r)$. By hypothesis

$$\log \frac{1}{\sigma(r)} \geq \sup_{z \in \mathbf{Q}} f_M(z) - z \log \frac{1}{r}$$

and hence

$$\{s, n-1\}_p M \geq \sup_{\substack{sz \in \mathbf{Z} \\ s \in \mathbf{N}^\times}} |A_{s,sz}| r^{sz} \sigma(r)^s$$

i.e., since the case $s = 0$ holds trivially for $M > 1$,

$$\{s, n-1\}_p M \geq \sup_{s \in \mathbf{N}} \sup_{j \in \mathbf{Z}} |A_{s,j}| r^j \sigma(r)^s$$

$$\geq \sup_{s \in \mathbf{N}} |A_s(t_r)| \sigma(r)^s .$$

But

$$\mathcal{U}_{t_r, G}(X) = \sum_{s=0}^{\infty} A_s(t_r)(X - t_r)^s$$

which shows that $\rho(r) \geq \sigma(r)$.

To obtain an inequality in the reverse direction we use $\|-\|_{t_r, b}$, for $b > 0$, to denote the boundary seminorm for ratios of functions analytic on $D(t_r, b^-)$ (see Section IV.1). If $b \leq 1$ then $\|G_s\|_{t_r, b}$ is well defined; indeed

$$\|G_s\|_{t_r, b} = \begin{cases} |G_s|_0(r) & \text{if} \quad r \geq b , \\ |G_s|_0(b) & \text{if} \quad b \geq r . \end{cases}$$

Thus

$$\|G_s\|_{t_r, b} = |G_s|_0 (\sup(r, b)) .$$

By Remark IV.3.2, we obtain

$$\|A_s\|_{t_r, \rho(r)} \leq \rho(r)^{-s} \{s, n-1\}_p M_2$$

where

$$M_2 = \sup_{i \leq n-1} \left(\rho(r)^i \|G_i\|_{t_r, \rho(r)} \right)$$

$$\leq \sup_{i \leq n-1} \left(|G_i|_0 (\sup(r, \rho(r))) \right)$$

$$\leq \sup_{i \leq n-1} \|G_i\|_C \leq M_0^{n-1}$$

the last inequality following from the fact that $\| d/dX \|_C$, the operator norm of d/dX on C, is bounded by $1/R_0$.

Our object is to obtain an upper bound for $|A_s|_0(r)$ and we have obtained an estimate for $|A_s|_0 \,(\sup(r, \rho(r)))$. If $\rho(r) \leq r$ this is adequate, while if $\rho(r) > r$ then we have, by (VII.3.2),

$$|A_s|_0(r) \leq |A_s|_0 \,(\rho(r)) \left(\frac{\rho(r)}{r} \right)^\mu \leq \left(\frac{1}{R_0} \right)^\mu \|A_s\|_{t_r, \rho(r)}$$

using the bound $1 \geq \rho(r)$ and the uniform bound μ for the number of poles of A_s in $D(0, 1^-)$ (if $\rho(r) > r$; recall that if $\rho(r) > r$, then $D - G$ has only trivial singularities in $D(0, 1^-)$). Thus in all cases we have

$$|A_s|_0(r) \leq \rho(r)^{-s} \{s, n-1\}_p M_1 .$$

Thus, for $s \in \mathbf{N}$, $j \in \mathbf{Z}$,

$$|A_{s,j}| r^j \leq |A_s|_0(r) \leq \rho(r)^{-s} \{s, n-1\}_p M_1 .$$

Hence, for $s \in \mathbf{N}^\times$, $z = j/s$, we have

$$\log \frac{|A_{s,sz}|}{M_1 \{s, n-1\}_p} \leq s \log \frac{1}{\rho(r)} + sz \log \frac{1}{r}$$

i.e. for $M \geq M_1$,

$$f_M(z) \leq \log \frac{1}{\rho(r)} + z \log \frac{1}{r} .$$

Hence

$$\log \frac{1}{\rho(r)} \geq \sup_{z \in \mathbf{Q}} \left(f_M(z) - z \log \frac{1}{r} \right) = \log \frac{1}{\sigma(r)}$$

which gives $\sigma(r) \geq \rho(r)$. **Q.E.D.**

Corollary. Let $R_0 \leq r_1 < r_2 \leq 1$. If $\rho(r_i) = r_i$ $(i = 1, 2)$, then $\rho(r) = r$ for all $r \in [r_1, r_2]$.

Proof. Let ℓ_i be the line of contact of slope $\log(1/r_i)$ $(i = 1, 2)$. Let P be the point of intersection of ℓ_1 and ℓ_2, and let ℓ_r be the line through P of slope $\log(1/r)$. Then ℓ_r lies above the polygon and hence lies above the line of contact of the same slope. Thus if q is the Y-intercept of ℓ_r then

$$q \geq \log \frac{1}{\sigma(r)} .$$

In the present situation both ℓ_1 sand ℓ_2 pass through $(-1,0)$. Hence $q = \log(1/r)$ and so $\sigma(r) \geq r$. But $\sigma(r) > r$ implies $\sigma(r_1) = \sigma(r) > r > r_1$ a contradiction. Hence $r = \sigma(r)$ as asserted.   **Q.E.D.**

Remark. By the same argument one can deduce that $\log(1/\rho(r))$ is a convex function of $\log(1/r)$ for $r \in [R_0, 1]$ (see Christol [2]) and indeed the Corollary is a consequence of this convexity (to avoid any ambiguity regarding what we mean by a convex function, we cite $f(x) = x^2$ as an example of such a function). It seems reasonable to believe that Christol's convexity theorem implies the Theorem of this Appendix. It would be enough to define

$$g(z) = \inf_{m \in [m_1, m_2]} (b_m + mz)$$

for all $z \in \mathbf{R}$. Then show that if $m \to b_m$ is a convex function on $[m_1, m_2]$ then $g(z)$ is a convex function of z and then finally show that, for $m \in [m_1, m_2]$, we have $b_m = \sup_z(g(z) - mz)$.

The variation of $\rho(r)$ with r has been studied by P. Robba (see Robba [2]) with a view towards calculation of index. A posthumous work of Philippe Robba, *Cohomologies p-adiques et sommes exponentielles*, should appear shortly in the series "Travaux en cours" of Hermann and Co. We refer the reader to that work for further discussion of index calculations.

APPENDIX II

ARCHIMEDEAN ESTIMATES

In this Appendix $|-|$ will denote the classical complex absolute value.

If $A = (A_{ij}) \in \mathcal{M}_n(\mathbf{C})$ we shall denote by $|A|$ the sup of the $|A_{ij}|$. Likewise if $v \in \mathbf{C}^n$, we put $|v| = \sup_i |v_i|$. But when viewing A as a transformation of $\mathbf{C}^n$, we must recognize that

$$|Av| = \|A\||v|$$

with $\|A\| \leq n|A|$. Likewise, if A, B lie in $\mathcal{M}_n(\mathbf{C})$ then

$$|AB| \leq n|A||B| .$$

Let $G \in \mathcal{M}_n(\mathbf{C}[[X]])$ and suppose that G converges in the closed disk $D(0, R^+)$ in $\mathbf{C}$. We assert that the matrix $W \in \mathcal{M}_n(\mathbf{C}[[X]])$ constructed in the proof of Proposition III.8.5 converges in the open disk $D(0, R^-)$. We use the notation of that proof. By hypothesis there exists $M > 0$ such that $|G_j| \leq M/R^j$ for all $j \geq 0$. For $m \geq 0$ let

$$\alpha_m \geq \|(mI - \psi_{G_0})^{-1}\| ,$$

the choice of α_m to be made later. (Recall that $mI - \psi_{G_0}$ is a transformation of $\mathcal{M}_n(\mathbf{C})$, that is of $\mathbf{C}^{n^2}$.) From the relation

$$W_m = (mI - \psi_{G_0})^{-1}(G_1 W_{m-1} + \cdots + G_{m-1} W_1 + G_m)$$

we deduce

$$
\begin{aligned}
|W_m| &\leq \alpha_m \left(|G_1 W_{m-1}| + \cdots + |G_{m-1} W_1| + |G_m| \right) \\
&\leq \alpha_m \left(n|G_1||W_{m-1}| + \cdots + n|G_{m-1}||W_1| + |G_m| \right) \\
&\leq \alpha_m \left(\frac{nM}{R}|W_{m-1}| + \cdots + \frac{nM}{R^{m-1}}|W_1| + \frac{M}{R^m} \right)
\end{aligned}
$$

and so

$$R^m |W_m| \leq \alpha_m n M \left(R^{m-1} |W_{m-1}| + \cdots + R|W_1| + 1 \right) .$$

Thus $R^m |W_m| \leq z_m$ where z_m is defined inductively by

$$z_0 = 1 , \quad z_m = n M \alpha_m (1 + z_1 + \cdots + z_{m-1}) .$$

We assert that α_m may be chosen such that the series

$$\sum_{m=0}^{\infty} z_m X^m$$

converges for $|x| < 1$. Indeed, since

$$z_{m-1} = n M \alpha_{m-1}(1 + z + \cdots + z_{m-2})$$

we have

$$\frac{z_m}{n M \alpha_m} = z_{m-1} + \frac{z_{m-1}}{n M \alpha_{m-1}} ,$$

that is

$$z_m = z_{m-1} \left(\frac{\alpha_m}{\alpha_{m-1}} + n M \alpha_m \right) .$$

Thus by the ratio test it is enough to show that we may choose the sequence $\{\alpha_m\}$ such that

$$\lim_{m \to \infty} \left(\frac{\alpha_m}{\alpha_{m-1}} + n M \alpha_m \right) = 1 .$$

For this we do not need estimates for α_m with m small. Now observe that $\mathrm{adj}\,(mI - \psi_{G_0})$ is an element of $\mathcal{M}_{n^2}\,(\mathbf{C}[m])$ which as a polynomial in m is of degree bounded by $n^2 - 1$. It follows that

$$|\mathrm{adj}\,(mI - \psi_{G_0})| \leq k m^{n^2 - 1}$$

for all $m > m_0$ and suitable $k > 0$. Likewise $\det (mI - \psi_{G_0})$ is a monic polynomial of degree n^2 and hence

$$|\det (mI - \psi_{G_0})| \geq \frac{1}{2} m^{n^2}$$

for all $m > 0$. Thus for $m > m_0$ we have

$$\|(mI - \psi_{G_0})^{-1}\| \leq n^2 |(mI - \psi_{G_0})^{-1}|$$

$$= n^2 \left| \frac{1}{\det(mI - \psi_{G_0})} \operatorname{adj}(mI - \psi_{G_0}) \right|$$

$$\leq \frac{n^2 k m^{n^2 - 1}}{\frac{1}{2} m^{n^2}} = \frac{2kn^2}{m} \ .$$

We put

$$\alpha_m = k'/m$$

for $m > m_0$, where $k' = 2kn^2$. Thus

$$\lim_{m \to \infty} \alpha_m = 0$$

and

$$\lim_{m \to \infty} \alpha_m / \alpha_{m-1} = 1$$

as desired.

APPENDIX III

CAUCHY'S THEOREM

Let K be a field of characteristic zero and consider the system of ordinary differential equations

$$\frac{dY_i}{dX} = f_i(X, Y_1, \ldots, Y_n)$$

with $f_i(X, Y_1, \ldots, Y_n) \in K[[X, Y_1, \ldots, Y_n]]$, for $i = 1, \ldots, n$. We seek solutions $y_1(X), \ldots, y_n(X) \in K[[X]]$ of the system with initial conditions $y_i(0) = 0$ for $i = 1, \ldots, n$. We will write the system in vector form

$$\frac{d\vec{Y}}{dX} = \vec{f}(X, \vec{Y}) \tag{1}$$

where

$$\vec{Y} = \begin{pmatrix} Y_1 \\ \vdots \\ Y_n \end{pmatrix}, \qquad \vec{f}(X, \vec{Y}) = \begin{pmatrix} f_1(X, \vec{Y}) \\ \vdots \\ f_n(X, \vec{Y}) \end{pmatrix}.$$

Then we seek

$$\vec{y} = \begin{pmatrix} y_1 \\ \vdots \\ y_n \end{pmatrix} \in (K[[X]])^n$$

such that

$$\frac{d\vec{y}}{dX} = \vec{f}(X, \vec{y}), \qquad \vec{y}(0) = 0.$$

We assert that there exists one and only one such formal solution. Clearly, taking $\vec{y} = \lim_s \vec{y}_s$ in the X-adic topology, this is equivalent to the existence and uniqueness of a sequence $\{\vec{y}_s\}_{s \geq 0}$, where $\vec{y}_0 = 0$ and $\vec{y}_s \in (K[X])^n$ is a polynomial of degree s, such that

$$\vec{y}_s \equiv \vec{y}_{s-1} \mod X^s$$

and that

$$\frac{d\vec{y}_s}{dX} \equiv \vec{f}(X, \vec{y}_s) \mod X^s \tag{2}$$

for all $s \geq 1$. To prove the existence and uniqueness of such a sequence we proceed by induction. Let $\vec{y}_s = \vec{y}_{s-1} + \vec{B}_s X^s$ with $\vec{B}_s \in K^n$. Then

$$\frac{d\vec{y}_s}{dX} = \frac{d\vec{y}_{s-1}}{dX} + s\vec{B}_s X^{s-1}$$

so that, since clearly $\vec{f}(X, \vec{y}_s) \equiv \vec{f}(X, \vec{y}_{s-1}) \mod X^s$, we have

$$s\vec{B}_s X^{s-1} \equiv \vec{f}(X, \vec{y}_{s-1}) - \frac{d\vec{y}_{s-1}}{d X} \mod X^s .$$

From (2) it follows that

$$\vec{B}_s = \frac{1}{s}\left(X^{-s+1}\left(\vec{f}(X, \vec{y}_{s-1}) - \frac{d\vec{y}_{s-1}}{dX}\right)\right)\Big|_{X=0} .$$

We now write

$$\vec{y} = \vec{A}_1 X + \frac{\vec{A}_2}{2!}X^2 + \frac{\vec{A}_3}{3!}X^3 + \cdots \tag{3}$$

and determine the $\vec{A}_j$'s using Taylor's formula. Clearly we have

$$\vec{A}_1 = \vec{f}(0, 0, \ldots, 0)$$
$$\vec{A}_2 = \frac{d^2\vec{y}}{dX^2}\Big|_{X=0} = \frac{d}{dX}\left(\vec{f}(X, \vec{y})\right)\Big|_{X=0}$$
$$= \left(\frac{\partial\vec{f}}{\partial X}(X, \vec{y}) + \sum_{i=1}^{n}\frac{\partial\vec{f}}{\partial Y_i}(X, \vec{y})\frac{dy_i}{dX}\right)\Big|_{X=0}$$
$$= \left(\frac{\partial\vec{f}}{\partial X}(X, \vec{Y}) + \sum_{i=1}^{n}f_i(X, \vec{Y})\frac{\partial\vec{f}}{\partial Y_i}(X, \vec{Y})\right)\Big|_{(X, \vec{Y})=0}$$
$$= \left(\frac{\partial}{\partial X} + \sum_{i=1}^{n}f_i\frac{\partial}{\partial Y_i}\right)\vec{f}\,\Big|_{(X, \vec{Y})=0} ,$$

and, in general, by induction, we find

$$\vec{A}_s = \frac{d^s\vec{y}}{dX^s}\Big|_{X=0} = \left(\frac{\partial}{\partial X} + \sum_{i=1}^{n}f_i\frac{\partial}{\partial Y_i}\right)^{s-1}\vec{f}\,\Big|_{(X, \vec{Y})=0} \tag{4}$$

From (4) it easily follows that if we write

$$\vec{f}(X, \vec{Y}) = \sum_{\mu \in \mathbf{N}^{n+1}} \vec{C}_\mu X^{\mu_0} Y_1^{\mu_1} \cdots Y_n^{\mu_n} , \tag{5}$$

then, for each $s \geq 1$,

$$\vec{A}_s = \vec{P}_s \left(\ldots , (\vec{C}_\mu)_h , \ldots \right)$$

where $\vec{P}_s \left(\ldots , (\vec{C}_\mu)_h , \ldots \right)$ is a universal polynomial in $\mathbf{Z}[\ldots , (\vec{C}_\mu)_h , \ldots]$ with positive coefficients.

The Classical Case.

We take $K = \mathbf{C}$, the field of complex numbers. For $i = 1, \ldots , n$ we consider functions $f_i(X, Y_1, \ldots , Y_n)$, which are assumed to be analytic in the polydisk $D(0, a^+) \times (D(0, b^+))^n \subset \mathbf{C}^{n+1}$ where $a, b \in \mathbf{R}_+$, the positive real numbers. Cauchy's problem consists in proving the convergence in a non–trivial disk of the formal solution $\vec{y}$ obtained above. After scaling the variables we may assume that $a = b = 1$. Let $\vec{f}(X, \vec{Y})$ be as in (5) and let

$$\vec{g}(X, \vec{Y}) = \sum_{\mu \in \mathbf{N}^{n+1}} \vec{B}_\mu X^{\mu_0} Y_1^{\mu_1} \cdots Y_n^{\mu_n} \in \mathbf{R}_+[[X, Y_1, \ldots , Y_n]] ,$$

to be explicit, $(\vec{B}_\mu)_j$ is a positive real number for all μ and j.

We say that $\vec{g}$ *dominates* $\vec{f}$ and write $\vec{f} \prec \vec{g}$ if for each μ we have

$$|(\vec{C}_\mu)_j| \leq (\vec{B}_\mu)_j$$

for $j = 1, 2, \ldots , n$, that is each component of every coefficient in $\vec{g}$ is greater than or equal to the absolute value of the corresponding coefficient component in $\vec{f}$.

Let us suppose that $\vec{f} \prec \vec{g}$ and let

$$\vec{z} = \sum_{s=1}^{\infty} \vec{A}_s \frac{X^s}{s!}$$

be the formal solution of the system

$$\frac{d\vec{Y}}{dX} = \vec{g}(X, \vec{Y}) \tag{6}$$

with initial conditions $\vec{z}(0) = 0$. Therefore

$$\vec{A}_s = \left(\frac{\partial}{\partial X} + \sum_{i=1}^{n} g_i \frac{\partial}{\partial Y_i} \right)^{s-1} \vec{g} \Big|_{(X,\vec{Y})=0}$$

and clearly

$$\vec{A}_s = \vec{P}_s \left(\ldots, (\vec{B}_\mu)_h, \ldots \right)$$

whence

$$|(\vec{A}_s)_j| \le (\vec{A}_s)_j ,$$

and so if the solution $\vec{z}$ of system (6) is convergent on some disk centered at zero, then the same is true of $\vec{y}$.

By hypothesis $\vec{f}$ is convergent for $|x| \le 1, |y_1| \le 1, \cdots, |y_n| \le 1$, so that there exists a positive M such that

$$|(\vec{C}_\mu)_h| \le M$$

for all $\mu \in \mathbf{N}^{n+1}$ and $h = 1, \ldots, n$. If we set

$$\vec{g} = \begin{pmatrix} M \\ \vdots \\ M \end{pmatrix} \frac{1}{(1-X)(1-Y_1)\cdots(1-Y_n)}$$

then clearly $\vec{g}$ dominates $\vec{f}$. The system (6) then becomes

$$\frac{dY_i}{dX} = \frac{M}{(1-X)(1-Y_1)\cdots(1-Y_n)} \tag{7}$$

for $i = 1, 2, \ldots, n$,

Since the right hand side of (7) is independent of i, if $z_1, \ldots, z_n$ are the formal solutions to (7) with initial conditions $z_i(0) = 0$, then we have

$$z_1 = \cdots = z_n = z$$

where $z \in \mathbf{C}[[X]]$ satisfies the equation

$$\frac{dZ}{dX} = \frac{M}{(1-X)(1-Z)^n}$$

with initial condition $z(0) = 0$. Therefore we obtain

$$(1-z)^{n+1} = 1 + M(n+1)\log(1-X)$$

so that

$$z = 1 - \{1 + M(n+1)\log(1-X)\}^{1/(n+1)}$$

which clearly converges on a suitably small disk centered at zero.

The p-adic Case.

We now assume K to be a complete p-adic field. By a function analytic on the disk $D(0, a^+)$, $a > 0$, we mean a power series

$$f(X) = \sum_{j=0}^{\infty} q_j X^j \in K[[X]]$$

such that

$$|q_j| a^j \to 0 \quad \text{as } j \to \infty .$$

The key point in what follows is that the set of all functions analytic on $D(0, a^+)$ forms a Banach space $\mathcal{B}_a$ under the norm

$$\left\| \sum_{j=0}^{\infty} q_j X^j \right\| = \sup_j |q_j| a^j .$$

The operator

$$\frac{d}{dX} : \mathcal{B}_a \longrightarrow \mathcal{B}_a$$

is a linear map and

$$\left\| \frac{d}{dX} \sum_j q_j X^j \right\| = \sup_j |j q_j| a^{j-1} \le \frac{1}{a} \left\| \sum_j q_j X^j \right\|$$

so that

$$\left\| \frac{d}{dX} \right\|_{\mathcal{B}_a} \le \frac{1}{a} .$$

A function analytic on the polydisk $D(0, a^+) \times D(0, b^+)^n$, $a, b > 0$, is a power series

$$\sum_{u \in \mathbf{N}^{n+1}} q_u X^{u_0} Y_1^{u_1} \cdots Y_n^{u_n}$$

satisfying

$$|q_u| a^{u_0} b^{u_1 + \cdots + u_n} \to 0$$

as $u_0 + u_1 + \cdots + u_n \to \infty$. The set $\mathcal{B}_{a,b}$ of functions analytic in the above polydisk is a Banach space under the norm

$$\left\| \sum_{u \in \mathbf{N}^{n+1}} q_u X^{u_0} Y_1^{u_1} \cdots Y_n^{u_n} \right\| = \sup_u |q_u| a^{u_0} b^{u_1 + \cdots + u_n}$$

and

$$\left\|\frac{\partial}{\partial X}\right\|_{\mathcal{B}_{a,b}} \le \frac{1}{a}, \qquad \left\|\frac{\partial}{\partial Y_i}\right\|_{\mathcal{B}_{a,b}} \le \frac{1}{b}.$$

Moreover the evaluation map

$$\mathcal{B}_{a,b} \longrightarrow \Omega$$
$$f(X, \vec{Y}) \longmapsto f(0,0)$$

is a linear map with norm ≤ 1, and the multiplication map

$$f \longmapsto fg$$

where $g \in \mathcal{B}_{a,b}$ is a linear map with norm equal to $\|g\|$.

We now assume the $f_i(X, \vec{Y})$ in (1) to be elements of $\mathcal{B}_{a,b}$. Therefore there exists $M > 0$ such that $\|\vec{f}\| \le M$, that is $\|f_i\| \le M$ for $i = 1, 2, \ldots, n$. Then the linear map $\dfrac{\partial}{\partial X} + \displaystyle\sum_{i=1}^{n} f_i \dfrac{\partial}{\partial Y_i}$ satisfies

$$\left\|\frac{\partial}{\partial X} + \sum_{i=1}^{n} f_i \frac{\partial}{\partial Y_i}\right\|_{\mathcal{B}_{a,b}} \le \sup\left(\frac{1}{a}, M\frac{1}{b}\right)$$

so that, by (4) we obtain

$$|\vec{A}_j| \le M \left(\sup\left(\frac{1}{a}, \frac{M}{b}\right)\right)^{j-1}.$$

If, as usual, π is defined by the condition $\pi^{p-1} = -p$, then from (II.4.6) it follows that

$$|j!| \ge |\pi|^j$$

so that $\vec{y}(X)$ converges for

$$|x| < \frac{|\pi|}{\sup\left(\dfrac{1}{a}, \dfrac{M}{b}\right)}.$$

Remark 1.   In the case of a linear system

$$\frac{d\vec{Y}}{dX} = G(X)\vec{Y}$$

where $G(X)$ is an $n \times n$ matrix analytic in $D(0, r)$ one may find a solution

$$\vec{y} = \vec{A}_0 + \vec{A}_1 X + \cdots$$

with an arbitrary choice of the vector $\vec{A}_0$ as one sees by setting $\vec{z} = \vec{y} - \vec{A}_0$ and applying the result just proved (classical or non-archimedean) to the system

$$\frac{d\vec{Z}}{dX} = G(X)\vec{Z} + G(X)\vec{A}_0 \ .$$

Remark 2. The first treatment of the non–archimedean Cauchy Theorem was given by Lutz [1]. Later the topic was considered by Igusa [1], Clark [1] and Serre [1].

BIBLIOGRAPHY

Adolphson, A., Dwork, B. and Sperber, S.
[1] Growth of Solutions of Linear Differential Equations at a Logarithmic Singularity, *Trans. Amer. Math. Soc. 271* (1982), 245–252.

Amice, Y.
[1] *Les nombres p-adiques*, Presses Universitaire de France, collection SUP 14, Paris, 1975.

André, Y.
[1] *G-functions and Geometry*, Aspects of Mathematics E3, Viehweg, Braunschweig/Wiesbaden, 1989.

Artin, E.
[1] *Algebraic Numbers and Algebraic Functions*, Gordon and Breach, New York, 1967.

Baldassarri, F. and Chiarellotto, B.
[1] On André's transfer theorem, *Contemporary Math. 133* (1992), 25–37.

Berthelot, P. and Messing, W.
[1] Théorie de Dieudonné Cristalline, I, in *Journées de Géométrie Algébrique de Rennes (Juillet 1978)*, Astérisque 63, Soc. Math. de France, 1979, 17–38.

Bombieri, E.
[1] On *G*-functions, in *Recent progress in Analytic Number Theory*, vol. 2, Academic Press, New York, 1981, 1–67.

Christol, G.
[1] Un théorème de transfert pour les disques singuliers réguliers, in *Cohomologie p-adique*, Astérisque 119–120, Soc. Math. de France, 1984, 151–168.
[2] Modules différentielles sur des couronnes, *Ann. Inst. Fourier (Grenoble)*, to appear.

Christol, G. and Dwork, B.
[1] Effective p-adic bounds at regular singular points, *Duke Math. J. 62*
    (1991), 689–720.

Chudnovsky D. and Chudnovsky G.
[1] Applications of Padé Approximations to diophantine inequalities in
    values of G-functions, in *Number Theory. A Seminar held at the
    Graduate School and University Center of the City of New York
    1982*, Lect. Notes in Math. 1052, Springer-Verlag, Berlin, 1984,
    1–51.
[2] Applications of Padé Approximations to the Grothendieck conjecture
    on linear differential equations, in *Number Theory. A Seminar held
    at the Graduate School and University Center of the City of New
    York 1983–84*, Lect. Notes in Math. 1135, Springer-Verlag, Berlin,
    1985, 52–100.

Clark, D.N.
[1] A note on the p-adic convergence of solutions of linear differential
    equations, *Proc. Amer. Math. Soc. 17* (1966), 262–269.

Davenport, H.
[1] *Multiplicative number theory*, Grad. Texts in Math. 74, Springer-
    Verlag, Berlin, 1980.

Deligne, P.
[1] *Equations différentielles à points singuliers réguliers*, Lect. Notes in
    Math. 163, Springer-Verlag, Berlin, 1970.

Dwork, B.
[1] On the rationality of the zeta function of an algebraic variety, *Amer.
    J. Math. 82* (1960), 631–648.
[2] On the zeta function of a hypersurface, *Inst. Hautes Études Sci. Publ.
    Math. 12* (1962), 5–68.
[3] Differential operators with nilpotent p-curvature, *Amer. J. Math.
    112* (1990), 749–786.

Dwork, B. and van der Poorten, A.
[1] The Eisenstein constant, *Duke Math. J. 65* (1992), 23–43.

Dwork, B. and Robba, P.
[1] Effective p-adic bounds for solutions of homogeneous linear differen-
    tial equations, *Trans. Amer. Math. Soc. 259* (1980), 559–577.

Hardy, G.H. and Wright,E.M.
[1] *An introduction to the theory of numbers*, Clarendon Press, Oxford,
    1960.

Honda, T.
[1] Algebraic Differential Equations, *Symposia Mathematica 24* (1981), 169–204.

Igusa, J.
[1] Analytic groups over complete fields, *Proc. Nat. Acad. Sci. U.S.A. 42* (1956), 540–541.

Katz, N.
[1] Nilpotent connections and the Monodromy Theorem: application of a result of Turritin, *Inst. Hautes Études Sci. Publ. Math. 35* (1970), 175–232.
[2] Algebraic solutions of Differential Equations (p–curvature and the Hodge filtration), *Invent. Math. 18* (1972), 1–118.

Lang, S.
[1] *Algebraic Number Theory*, Addison-Wesley, Reading, 1970.

Leopoldt, H.–W.
[1] Zur Approximation des p-adischen logarithmus, *Abh. Math. Sem. Hamburg 25* (1962), 77–81.

Lutz, E.
[1] Sur l'equation $y^2 = x^3 - Ax - B$ dans les corps $\mathfrak{p}$–adiques, *J. Reine Angew. Math. 177* (1937), 238–247.

Poole, E.G.C.
[1] *Introduction to the Theory of Linear Differential Equations*, Oxford University Press, Oxford, 1936.

Robba, P.
[1] Index of p-adic differential operators, III: Application to twisted exponential sums, in *Cohomologie p –adique*, Astérisque 119-120, Soc. Math. de France, 1984, 191–266.
[2] Indice d'un opérateur différentiel p–adique, IV: Cas des systèmes. Mesure de l'irrégularité dans un disque, *Ann. Inst. Fourier (Grenoble) 35* (1985), 13–55.

Schmitt, H.
[1] Operators with Nilpotent p-curvature, *Proc. Amer. Math. Soc.*, to appear.

Serre, J.P.
[1] *Lie Algebras and Lie Groups*, Benjamin, Reading, 1965.

INDEX

accessory parameter, 158
additive character, 56
   trivial, 56
additive radius of convergence, 42
Amice ring, 129
analytic element, 128, 129
   analytic in an annulus, 131
analytic function, 117
   bounded, 117
annulus of convergence, 47
apparent singularity, 171
Artin-Hasse exponential, 55
associated scalar operator, 87

binomial series, 49, 97, 140
Bombieri condition, 226
   local, 234
Bombieri estimate, 261
boundary seminorm, 117
bounded polygon, 303

Cauchy estimates, 114
Cauchy sequence, 6
characteristic of a field, 4
Chudnovsky's Theorem, 267
complete field, 6
completion, 7
cyclic differential module, 87
cyclic vector
   Theorem of, 89

denominator, 224
dense, 7

derivation, 77
   non-trivial, 77
Dieudonné's Theorem, 54
differential extension, 79
differential field, 77
differential module, 86
   cyclic, 87
   nilpotent, 88
differential operator, 78
   globally nilpotent, 98
   nilpotent, 81
   trivial, 79
differential system, 86
   nilpotent, 88
   reducible, 280
   trivial, 87
dilogarithm, 125
discrete valuation, 21
Dwork-Frobenius Theorem, 156

effective growth
   Theorem of, 120
entire function, 68
equivalent
   norms, 9
   valuations, 4
exponential series, 50
exponents
   of an operator, 84, 98
   of a system, 113
extension of a valuation, 17

field, 3

GPSR Authorized Representative: Easy Access System Europe - Mustamäe tee
50, 10621 Tallinn, Estonia, gpsr.requests@easproject.com

www.ingramcontent.com/pod-product-compliance
Ingram Content Group UK Ltd.
Pitfield, Milton Keynes, MK11 3LW, UK
UKHW041909310325
456954UK00002B/28